At Olduvai Gorge in northern Tanzania natural erosion exposed a deep series of superimposed geological beds containing rich artefact and fossil assemblages spanning the last 1.8 million years. The site is famous as a result of excavations conducted there since 1951 under the direction of Mary Leakey and her husband, the late Louis Leakey.

The first definitive report on these excavations was published in 1965, followed by three further volumes over the next twenty-four years. Volume 5, written largely by Mary Leakey herself, is the last of these reports, and records the archaeological finds in the upper part of the Olduvai sequence from excavations carried out from the end of 1968 until 1971. The period covered here is from about 1.2 to 0.4 million years ago, and the finds include artefacts and faunal remains excavated from sites in Beds III, IV and the Masek Beds. The volume follows on from the archaeological record in Beds I and II published in 1971 in Volume 3 of the series.

In addition to the chapters by Mary Leakey, Richard Hay has written a brief summary of the geology as a background to the archaeology, Derek Roe provides a metrical analysis of the handaxes and cleavers, Paul Callow describes the technology and raw materials, and Peter Jones details experimental work on the manufacture and use of tools, in particular those associated with butchering and skinning. Celia Nyamweru's appendix describes the mapping out of the JK Pits archaeological site at Olduvai. An overview by Derek Roe sums up the entire volume and draws the contributions together, interpreting and expanding upon their conclusions.

OLDUVAI GORGE
VOLUME 5

Frontispiece: Olduvai Gorge, view across the Main Gorge

OLDUVAI GORGE

VOLUME 5

EXCAVATIONS IN BEDS III, IV AND THE MASEK BEDS, 1968–1971

M. D. LEAKEY WITH D. A. ROE

WITH CONTRIBUTIONS BY
P. CALLOW, R. L. HAY, P. R. JONES,
CELIA K. NYAMWERU AND D. A. ROE

CAMBRIDGE UNIVERSITY PRESS
Cambridge, New York, Melbourne, Madrid, Cape Town, Singapore, São Paulo, Delhi

Cambridge University Press
The Edinburgh Building, Cambridge CB2 8RU, UK

Published in the United States of America by Cambridge University Press, New York

www.cambridge.org
Information on this title: www.cambridge.org/9780521105200

© Cambridge University Press 1994

This publication is in copyright. Subject to statutory exception
and to the provisions of relevant collective licensing agreements,
no reproduction of any part may take place without the written
permission of Cambridge University Press.

First published 1994
This digitally printed version (with additions) 2009

A catalogue record for this publication is available from the British Library

Library of Congress Cataloguing in Publication data
(REVISED FOR VOLUME 5)
Olduvai Gorge.
Includes bibliographical references and index.
v. 5. Excavations in Beds III, IV and the Masek Beds,
1968–1971 / M. D. Leakey and D. A. Roe.
1. Fossil man – Tanzania – Olduvai Gorge.
2. Olduvai Gorge (Tanzania) – Antiquities.
3. Tanzania – Antiquities.
GN282.043 1991   569'.9   88-20398

ISBN 978-0-521-33403-7 hardback
ISBN 978-0-521-10520-0 paperback

Additional resources for this publication at www.cambridge.org/9780521105200

Figures 2.3, 4.8, 5.3 and 5.18 are available for download

CONTENTS

List of figures	page viii
List of tables	x
List of plates	xii
Introductory Note	xiii
Acknowledgements	xv
List of abbreviations	xvi

Introduction M. D. LEAKEY	1
1 Geology and dating of Beds III, IV and the Masek Beds R. L. HAY *University of Illinois at Urbana-Champaign*	8
2 Bed III. Site JK (Juma's Korongo) M. D. LEAKEY	15
3 The base of Bed IV. WK Hippo Cliff, PDK Trench IV, WK Lower Channel M. D. LEAKEY	36
4 Lower Bed IV. HEB East, HEB and HEB West, WK Intermediate Channel M. D. LEAKEY	45
5 Upper Bed IV. WK Upper Channel, WK East A and C, PDK Trenches I–III, HEB West Level I M. D. LEAKEY	75
6 The Masek Beds and sites in uncertain stratigraphic positions M. D. LEAKEY	116
7 The fauna M. D. LEAKEY	130
8 A metrical analysis of selected sets of handaxes and cleavers from Olduvai Gorge D. A. ROE *Donald Baden-Powell Quaternary Research Centre, University of Oxford*	146
9 The Olduvai bifaces: technology and raw materials P. CALLOW *University of Cambridge*	235
10 Results of experimental work in relation to the stone industries of Olduvai Gorge P. R. JONES	254
11 Summary and overview D. A. ROE	299
Appendix A. Modified bones from Beds III and IV M. D. LEAKEY	311
Appendix B. Mapping of an archaeological site at Olduvai Gorge CELIA K. NYAMWERU *St Lawrence University, New York*	315
References	321
Index	323

FIGURES

Figures 2.3, 4.8, 5.3 and 5.18 are available for download from www.cambridge.org/9780521105200

Int. 1	Sketch map showing the sites excavated in Beds III, IV and the Masek Beds	page 2
Int. 2	Diagrammatic section of Beds III, IV, and the Masek, Ndutu and Naisiusiu Beds to show the stratigraphic positions of the excavated sites and hominid remains	3
1.1	Map showing major geologic and topographic features in the area surrounding Olduvai Gorge	9
1.2	Palaeogeography of Bed III with inferred drainage pattern	11
1.3	Palaeogeography of Bed IV with inferred drainage pattern	11
1.4	Palaeogeography of the Masek Beds with inferred drainage pattern	13
2.1	Section from west to east along the main drainage channel in Beds III and IV from TK to JK	16
2.2	JK: the surface with pits and the surrounding area in the JK Gully	26
2.3	JK: contoured plan of pits and furrows	
3.1	WK Hippo Cliff: plan showing positions of the hippo bones and associated finds	37
4.1	HEB East: section along the south face of the trench	46
4.2	HEB East: plan of finds above the channel (Spit 1)	47
4.3	HEB East: plan of finds in the upper part of the channel (Spit 2)	47
4.4	HEB East: plan of finds in the middle part of the channel (Spit 3)	48
4.5	HEB East: plan of finds in the lower part of the channel (Spit 4)	49
4.6	HEB East: graph to show proportionate occurrences of artefacts, faunal remains and cobbles at different levels in the channel	50
4.7	HEB and HEB West: plan of excavations and channels in Level 4	53
4.8	HEB and HEB West: section along the south face of the excavations	
4.9	HEB West: plan of finds associated with the sand lens	65
5.1	Bed IV: section along the south side of the gorge showing sites PDK Trenches I–III, PDK Trench IV, WK East, the hippo butchery site and WK	76
5.2	WK: stratigraphic section showing positions of the three channels with artefacts	77
5.3	WK: plan of finds in the channels and on the eroded surface associated with the channel	
5.4	WK: pitted anvils and hammerstones	83
5.5	WK East A: sections along north and east faces of Trench I and the Trial Trench	88
5.6	WK East A: plan of finds in the upper part of the channel filling (Spits 1 to 3)	88
5.7	WK East A: plan of finds in Spits 4 and 5	89
5.8	WK East A: plan of finds in the upper part of Spit 6	90
5.9	WK East A: plan of finds in Spit 6b (the middle part of Spit 6)	91
5.10	WK East A: plan of finds in the lower part of Spit 6 (Spit 6c)	92
5.11	WK East A: plan of finds in basal part of channel fill (Spits 6d and 7)	93
5.12	WK East A: graph to show proportionate occurrences of artefacts, faunal remains and cobbles in the channel	94
5.13	WK East A: sundry small tools	95
5.14	WK East A: punches	97
5.15	WK East A: pitted anvils and hammerstones	98
5.16	WK East C: section along the east face of the excavation	103
5.17	WK East C: graph to show proportionate occurrences of artefacts, faunal remains and cobbles in the channel	104
5.18	PDK Trenches I–III: plan of finds in the channel	
6.1	FLK Masek Beds: section along the west face of the excavations	117
6.2	FLK Masek Beds: plan of finds in the lower part of the channel	117
6.3	FLK Masek Beds: superimposed outlines of five handaxes showing similarity in size and form	118
6.4	Scrapers from FLK Masek Beds	119
6.5	HK: section along north face of trench showing stratigraphic position of artefacts	124
6.6	TK Fish Gully: section to show the relationships of artefacts *in situ* to those in disturbed context	127
8.1	Measurements taken from the bifaces	152
8.2	(a) Framework for the handaxe-shape diagrams. (b) Array of plan-forms on the handaxe-shape diagrams	155
8.3	Framework for the cleaver-shape diagrams	156
8.4	Array of plan-forms on the cleaver-shape diagrams	157
8.5	Cleaver butt-shape symbols	157
8.6	Handaxe-shape diagrams: HK	205
8.7	Handaxe-shape diagram: TK FG	206
8.8	Handaxe-shape diagram: FLK Masek	207
8.9	Handaxe-shape diagram: PDK Trenches I–III	208

LIST OF FIGURES

8.10	Handaxe-shape diagram: WK East A	209
8.11	Handaxe-shape diagram: WK East C	210
8.12	Handaxe-shape diagram: WK	211
8.13	Handaxe-shape diagram: HEB West Level 2a	212
8.14	Handaxe-shape diagram: HEB West Level 2b	213
8.15	Handaxe-shape diagram: HEB West Level 3	214
8.16	Handaxe-shape diagram: HEB East	215
8.17	Handaxe-shape diagram: PDK Trench IV	216
8.18	Handaxe-shape diagram: BK	217
8.19	Handaxe-shape diagram: TK Upper Level	218
8.20	Handaxe-shape diagram: TK Lower Level	219
8.21	Handaxe-shape diagram: SHK	220
8.22	Handaxe-shape diagram: EF–HR	221
8.23	Handaxe-shape diagram: MLK	222
8.24	Cleaver-shape diagram: HK	223
8.25	Cleaver-shape diagram: FLK Masek	224
8.26	Cleaver-shape diagram: WK	225
8.27	Cleaver-shape diagram: HEB West Level 2a	226
8.28	Cleaver-shape diagram: HEB West Level 2b	227
8.29	Cleaver-shape diagram: HEB West Level 3	228
8.30	Cleaver-shape diagram: HEB East	229
8.31	Cleaver-shape diagram: PDK Trench IV	230
8.32	Cleaver-shape diagram: BK	231
8.33	Cleaver-shape diagram: SHK	232
8.34	Cleaver-shape diagram: EF–HR	233
8.35	Cleaver-shape diagram: MLK	234
9.1	Bifaces: areas of cortex, primary flake scar and secondary flaking	246
9.2	Bifaces: as Fig. 9.1B, but for the two most common raw materials	247
10.1	Sources and distribution of raw materials used for stone tools found at sites in Beds I–IV and the Masek Beds	255
10.2	Lava bifaces: charts showing tool frequency within weight classes and the range of edge length preserved on them	264
10.3	Lava bifaces: charts showing tool frequency in various weight categories and the range of edge length preserved on them	264
10.4	Phonolite bifaces: charts showing tool frequency in various weight categories and the range of edge length preserved on them	265
10.5	Quartzite bifaces: charts showing the tool frequency in various weight categories and the range of edge lengths preserved on them	266
10.6	The different average weight/edge-length relationships of basalt bifaces from Bed IV made on cores and on large flakes	268
10.7	Cross sections of quartzite slab bifaces as compared to a quartzite biface made on a large flake	269
10.8	The fourfold increase of shape and area as perimeter length is doubled and also the greater area of a circle than that of a slim triangle of the same perimeter length	270
10.9	Edge length available for use on a triangular and a disc-shaped tool	270
10.10	Percentages of blank types on which bifaces are made in Bed IV	272
10.11	The changing ratio of weight to edge length of handaxes as they are re-sharpened several times	274
10.12 and 10.13	The maximum and minimum size ranges from sites in Bed I, Lower and Lower Middle Bed II for polyhedrons, subspheroids and spheroids	276
10.14	The maximum and minimum size ranges for polyhedrons, subspheroids and spheroids from sites in Upper Middle and Upper Bed II	276
10.15 and 10.16	The maximum and minimum size ranges for polyhedrons and the subspheroid group from sites in Bed III and the base of Bed IV, and from Lower Bed IV	277
10.17	The maximum and minimum size ranges for polyhedrons and the subspheroid group for Upper Bed IV, Masek and post Masek sites	277
10.18	Weight frequency charts for subspheroids from Beds III, IV, Masek and post Masek sites	278
10.19	Weight frequency charts for subspheroids from Bed IV sites	279
10.20	Numbers of quartzite pieces with spherical index of 2 and less	280
10.21	Weight frequencies for Beds III, IV and the Masek Beds samples of subspheroids and hammerstones compared with hammerstones from PRJ flaking floor	281
10.22 and 10.23	Scatter diagrams showing the width/length and thickness/length ratios for the *outils écaillés* and punches from JK, PDK Trench IV, HEB East, HEB West Level 1 and HEB Level 3	284
10.24 and 10.25	Scatter diagrams to show the width/length and thickness/length ratios of *outils écaillés* and punches from WK Upper and Intermediate Channels, WK East C and PDK Trenches I–III	285
10.26	Width/length and thickness/length ratios for *outils écaillés* and punches from WK East A.	286
10.27	Comparable ratios for *outils écaillés* and punches from experimental flaking	286
10.28	How the concavo-convex edge is formed through battering the end of a flake	287
10.29 and 10.30	The three main ways in which flakes split while being battered	288
10.31	The two main ways in which an oval cobble can be held for bipolar flake battering	289
10.32	The angled pits produced on cobbles used for bipolar battering	290
10.33	The time required to penetrate elephant skin 1.5 cm thick with un-retouched flakes of quartzite, phonolite, basalt and chert	292
10.34	Rates of skin cutting by different flake types	293
A.1	Detail of map of the main Pits surface	314
A.2	Arrangement of surveyed strips on main Pits surface	316
A.3	Control points used for measuring a single strip	316
A.4	Modification to levelling staff in order to make measurements in narrow grooves	319

TABLES

2.1 Soil samples from Olduvai: the JK Pits and vicinity ... 33
7.1 List of fauna from sites excavated in Beds III, IV and the Masek Beds ... 131–2
7.2 List of all known fauna from Beds III, IV and the Masek Beds ... 133
7.3 Minimum numbers of individual mammals from sites excavated in Beds III, IV and the Masek Beds ... 134
7.4 Minimum numbers and percentages of mammals from sites excavated in Beds III, IV and the Masek Beds ... 134
7.5 Fauna from JK ... 135–6
7.6 Fauna from WK Lower Channel ... 137
7.7 Fauna from HEB East ... 138
7.8 Fauna from HEB and HEB West ... 139
7.9 Fauna from WK Intermediate Channel ... 140
7.10 Fauna from WK Upper Channel ... 141
7.11 Fauna from WK East A ... 142
7.12 Fauna from WK East C ... 143
7.13 Fauna from PDK Trenches I–III ... 143
7.14 Fauna from FLK Masek ... 144
8.1 The biface samples studied ... 150
8.2 Metrical analysis of bifaces: handaxes and cleavers: length ... 158–9
8.3 Metrical analysis of bifaces: handaxes and cleavers: weight ... 160–1
8.4 Metrical analysis of bifaces: handaxes and cleavers: ratio Th/B ... 160–1
8.5 Metrical analysis of bifaces: handaxes and cleavers: ratio T_1/L ... 162–3
8.6 Metrical analysis of bifaces: handaxes and cleavers: ratio B/L ... 162
8.7 Metrical analysis of bifaces: handaxes and cleavers: ratio B_1/B_2 ... 164–5
8.8 Metrical analysis of bifaces: handaxes and cleavers: ratio L_1/L ... 164–5
8.9 Metrical analysis of bifaces: handaxes and cleavers: t-values and estimates of significance: length ... 166
8.10 Metrical analysis of bifaces: handaxes and cleavers: t-values and estimates of significance: weight ... 167
8.11 Metrical analysis of bifaces: handaxes and cleavers: t-values and estimates of significance: ratio Th/B ... 168
8.12 Metrical analysis of bifaces: handaxes and cleavers: t-values and estimates of significance: ratio T_1/L ... 169
8.13 Metrical analysis of bifaces: handaxes and cleavers: t-values and estimates of significance: ratio B/L ... 170
8.14 Metrical analysis of bifaces: handaxes and cleavers: t-values and estimates of significance: ratio B_1/B_2 ... 171
8.15 Metrical analysis of bifaces: handaxes and cleavers: t-values and estimates of significance: ratio L_1/L ... 172
8.16 Metrical analysis of bifaces: summary of all the statistical comparisons of the handaxe and cleaver samples ... 173
8.17 Metrical analysis of bifaces: handaxes: length ... 174–5
8.18 Metrical analysis of bifaces: handaxes: weight ... 174–5
8.19 Metrical analysis of bifaces: handaxes: ratio Th/B ... 176–7
8.20 Metrical analysis of bifaces: handaxes: ratio T_1/L ...
8.21 Metrical analysis of bifaces: handaxes: ratio B/L ... 176–7
8.22 Metrical analysis of bifaces: handaxes: ratio B_1/B_2 ... 178–9
8.23 Metrical analysis of bifaces: handaxes: ratio L_1/L ... 180–1
8.24 Metrical analysis of bifaces: handaxes: t-values and estimates of significance: length ... 180
8.25 Metrical analysis of bifaces: handaxes: t-values and estimates of significance: weight ... 181
8.26 Metrical analysis of bifaces: handaxes: t-values and estimates of significance: ratio Th/B ... 182
8.27 Metrical analysis of bifaces: handaxes: t-values and estimates of significance: ratio T_1/L ... 183
8.28 Metrical analysis of bifaces: handaxes: t-values and estimates of significance: ratio B/L ... 184
8.29 Metrical analysis of bifaces: handaxes: t-values and estimates of significance: ratio B_1/B_2 ... 185
8.30 Metrical analysis of bifaces: handaxes: t-values and estimates of significance: ratio L_1/L ... 186
8.31 Metrical analysis of bifaces: summary of all the statistical comparisons of the handaxe samples ... 187
8.32 Metrical analysis of bifaces: cleavers: length ... 188–9
8.33 Metrical analysis of bifaces: cleavers: weight ... 188–9
8.34 Metrical analysis of bifaces: cleavers: ratio Th/B ... 190–1
8.35 Metrical analysis of bifaces: cleavers: ratio T_1/L ... 190–1
8.36 Metrical analysis of bifaces: cleavers: ratio B/L ... 192
8.37 Metrical analysis of bifaces: cleavers: ratio B_1/B_2 ... 192–3
8.38 Metrical analysis of bifaces: cleavers: ratio L_1/L ... 194–5
8.39 Metrical analysis of bifaces: cleavers: cleaver edge angle ... 194
8.40 Metrical analysis of bifaces: cleavers: ratio CEL/B ... 195

LIST OF TABLES

8.41 Metrical analysis of bifaces: cleavers: t-values and estimates of significance: length — 196

8.42 Metrical analysis of bifaces: cleavers: t-values and estimates of significance: weight — 196

8.43 Metrical analysis of bifaces: cleavers: t-values and estimates of significance: ratio Th/B — 197

8.44 Metrical analysis of bifaces: cleavers: t-values and estimates of significance: ratio T_1/L — 197

8.45 Metrical analysis of bifaces: cleavers: t-values and estimates of significance: ratio B/L — 198

8.46 Metrical analysis of bifaces: cleavers: t-values and estimates of significance: ratio B_1/B_2 — 198

8.47 Metrical analysis of bifaces: cleavers: t-values and estimates of significance: ratio L_1/L — 199

8.48 Metrical analysis of bifaces: cleavers: t-values and estimates of significance: cleaver edge angle — 199

8.49 Metrical analysis of bifaces: cleavers: t-values and estimates of significance: ratio CEL/B — 200

8.50 Metrical analysis of bifaces: summary table for all the statistical comparisons of the cleaver samples — 201

8.51 Metrical analysis of bifaces: summary table for all the statistical comparisons made — 201

9.1 Handaxes: frequency of occurrence of the various raw materials, by site — 236

9.2 Cleavers: frequency of occurrence of the various raw materials, by site — 236

9.3 Handaxes and cleavers: frequency of cortex and primary scars, by biface type and raw material — 238

9.4 Handaxes and cleavers: occurrence of different combinations of cortex and primary scars on the two faces — 239

9.5 Handaxes and cleavers: occurrence of different combinations of cortex and primary scars on the two faces, by bed and industry type — 239

9.6 Handaxes and cleavers: occurrence of different combinations of cortex and primary scars on the two faces, by raw material and industry type — 240

9.7 Handaxes: means and standard deviations for quantitative attributes, by site — 242

9.8 Cleavers: mean and standard deviations for quantitative attributes, by site — 243

9.9 Handaxes from Bed IV only: means and standard deviations for quantitative attributes, by raw material and industry type — 244

9.10 Cleavers from Bed IV only: means and standard deviations for quantitative attributes by raw material and industry type — 245

9.11 Handaxes: medians and interquartile ranges for types of surface and for secondary scar ratios, by site — 248

9.12 Cleavers: medians and interquartile ranges for types of surface and for secondary scar ratios, by site — 249

9.13 Handaxes from Bed IV only: medians and interquartile ranges for types of surface and for secondary scar ratios, by raw material and industry type — 250

9.14 Cleavers from Bed IV only: medians and interquartile ranges for types of surface and for secondary scar ratios, by raw material and industry type — 251

9.15 Handaxes and cleavers from Bed IV only: typological frequencies for each raw material — 252

10.1 The archaeological distribution of polyhedrons, spheroids and subspheroids in the Olduvai sequence — 275

10.2 Polyhedrons, spheroids and subspheroids in Beds III and IV, the Masek Beds and post Masek occurrences — 278

10.3 The archaeological distribution of *outils écaillés*, punches and pitted anvils in the Olduvai sequence — 283

10.4 *Outils écaillés*, punches and pitted anvils in Bed III, Upper and Lower Bed IV — 283

10.5 Results of experimental battering of flakes using bipolar techniques — 291

10.6 Experimental butchery of large carcasses with stone tools showing the amount of meat removed and time involved — 294

A.1 Booking of the control points for strip C of the main Pits surface. — 318

A.2 Booking of the levelling — 318

PLATES

Frontispiece Olduvai Gorge

Between pages 322 and 323

1. The north side of the gorge showing Beds I, II and the red Bed III overlain by Bed IV
2. Bed III, JK: photographic mosaic of the pits and furrows
3. JK, Pit 2
4. JK: two pairs of convergent furrows
5. Aerial photograph of Magado Crater
6. Magado Crater: salt evaporation pits
7. Magado Crater: irrigation channels
8. Bed IV, HEB Level 3: cleaver and handaxes
9. Bed IV, HEB West, Level 2b: cleavers and handaxes
10. Looking west down the gorge from WK East
11. Bed IV, site WK at an early stage in the excavation
12. Bed IV, WK Upper Channel: pitted anvils, handaxes and other artefacts
13. Bed IV, WK Upper Channel: handaxes, cleavers and other artefacts, with faunal remains
14. Handaxes from WK Upper Channel
15. Cleavers from WK Upper Channel
16. WK Upper Channel: handaxes and bifaces
17. Three pitted hammerstones or anvils with single pits from WK Upper Channel
18. Pitted hammerstones or anvils from WK Upper Channel
19. A pitted anvil from WK Upper Channel
20. Quartzite handaxes from FLK Masek Beds
21. Two large quartzite handaxes from FLK Masek Beds
22. Elephant acetabulum from JK
23. Elephant acetabulum from HEB Level 3
24. Fragments of elephant limb bone shafts flaked to pointed ends, from HEB and LLK
25. Three distal ends of humeri, probably of hippopotamus
26. Three proximal condyles of hippopotamus femora

Introductory Note to the 50ᵗʰ Anniversary of the Discovery of 'Zinjanthropus'

The Olduvai Gorge in the Republic of Tanzania came to the attention of the world shortly after my mother Mary discovered the 'Zinjanthropus boisei' skull on July 17ᵗʰ 1959. The field of African prehistory, and in particular the study of human evolution, has changed and developed dramatically over the past 50 years. I am particularly pleased that Cambridge University Press have decided to republish the 5 monographs that comprehensively cover the many scientific studies that have been undertaken on the Olduvai material collected by my parents, Louis and Mary, working with a number of colleagues. As the Golden Anniversary of the discovery approaches, it is timely to reflect on the importance of that find.

I was lucky to arrive at Olduvai two days after the discovery and I well recall the excitement of the occasion. My parents were operating on a very tight budget and the field season was short. Fortunately, on hand was world-renowned photographer Des Bartlett who, aided by his wife Jen, fully recorded on film the first few days of excavations and reassembly of bone fragments back in camp. As pieces were glued back together, and the shape of the skull and its morphology became clear, my parents showed uncharacteristic and unrestrained emotion! At the time, ages for fossils were wild guesses and radiometric dating had not been done anywhere in Africa. The best, guessed age for Zinj was a little more than 500,000 years. Some months later, a real Potassium/Argon date was obtained by Jack Evenden and Garniss Curtis, and the 1,750,000 age was announced. This ignited huge excitement worldwide and for the first time my father was able to raise financial support for extended field work at Olduvai. Everything changed. The unqualified enthusiasm and support of the National Geographic Society from 1960 onwards had a major impact on the later work at Olduvai, and indeed on the growing international interest of Africa as the cradle of humanity.

Since those first exciting years at Olduvai, the investigation of human origins has gone forward and extended to many other sites in Africa. The age of hominins has been taken back to beyond five million years and the collected fossils and lithic records are now numerous. International multi-disciplinary teams are working in many parts of the world and, with the exception of a few fundamentalist 'flat earth' types, the acceptance of the fossil record of our past is widely accepted. Much of this has come about because of the initial Olduvai finds.

The pioneering work at Olduvai was the launch of this fantastic 50-year period when we as a species have come to realize and appreciate our common evolutionary past. Olduvai, conserved and protected by the Republic of Tanzania, remains as a landmark in the epic story of humanity, and these monographs are a wonderful testimony to that landmark.

Richard Leakey, FRS

ACKNOWLEDGEMENTS

Once more I wish to express my gratitude to the United Republic of Tanzania for permission to continue working at Olduvai Gorge, as well as to Mr A. A. Mturi, Director of Antiquities, and Mr A. J. N. Mgina, former Conservator of the Ngorongoro Conservation Authority, for their help and cooperation.

The National Geographic Society, Washington, DC has been largely responsible for funding the work at Olduvai over many years. I am deeply indebted to the Committee for Research and Exploration for their generosity and to the late Dr Melvin M. Payne, then Chairman, for his interest and encouragement. The L. S. B. Leakey Foundation has also made generous grants, particularly for the purchase of vehicles. Other persons, who wish to remain anonymous, have made most welcome annual gifts. To all those whose financial aid has enabled me to work at Olduvai I tender my most grateful thanks.

Mr Peter Jones worked for several years at Olduvai as my assistant, I am greatly indebted to him for his skilful photography and help in camp logistics. His chapter in this volume describing his experimental work in the manufacture and uses of stone tools is a most valuable contribution which throws new light on some of the features that have long puzzled those of us studying stone industries.

I am once more deeply indebted to Dr Richard Hay for his help and cooperation in solving the stratigraphic problems of the sites excavated in Beds III, IV and the Masek Beds. Drs Andrew Brock, the late Alan Cox and Frank Brown have all contributed greatly to elucidating the geomagnetic sequence at Olduvai; their work has been invaluable. My particular thanks are due to Dr Raymonde Bonnefille for her study of the Olduvai fossil pollen spectra. When she began her work at the gorge it was widely considered to be a waste of time and money since the consensus of opinion held that pollen grains were almost certainly unobtainable from the highly alkaline Olduvai sediments. Dr Bonnefille's identification of many hundreds of specimens and comparison with the extant flora has been an invaluable contribution to our knowledge of the past environment. Dr Derek Roe and Dr Paul Callow merit my special thanks for voluntarily undertaking to analyse the bifacial tools; this has been of very great help in studying the industries. Mr Gordon Hanes has made valuable contributions to Olduvai by financing the building of two site museums and two windmills to generate electricity. The late Mr George Dove, former owner of the Ndutu Safari Lodge, most kindly devoted a great deal of time to building the camp and also supplied furniture from his own house before leaving Tanzania for Australia. Mr R. I. M. Campbell and Mr John Reader, both professional photographers, have made available their skill to photograph sites and specimens; they have my particular thanks.

Many others have helped the work at Olduvai, directly and indirectly. My thanks are especially due to Mrs John Brindeis, Mrs Janet Leakey, Dr R. J. Clarke, Mrs M-A. Harms, Dr John Harris, Miss Mary Jackes and Dr Celia Nyamweru, for their active assistance at Olduvai.

By 1968 the late Mr Heslon Mukiri, who had been my excavation foreman since 1937 during my first dig in Kenya at the Neolithic site of Hyrax Hill, sadly found himself unable to continue active field work. He was sorely missed but I am greatly indebted to my Wakamba staff for their skill and patience in excavation.

Since I left Olduvai and returned to live in Kenya the Governors of the National Museums of Kenya and my son Richard have made available to me study space to prepare this volume for publication. I am most grateful for their courtesy.

M. D. LEAKEY

ABREVIATIONS FOR ARTEFACTS SHOWN IN SITE PLANS

AWL	Awl
CH	Chopper
D	Debitage
DC	Discoid
HM	Hammerstone
HX	Handaxe
LTF	Laterally trimmed flake
OE	*Outil écaillé*
PAV	Pitted anvil/hammerstone
PU	Punch
SC	Scraper
SPH or SP	Spheroid
SSP	Subspheroid
ST	Sundry tool
UT	Utilised
UTH	Utilised heavy-duty
UTL	Utilised light-duty

INTRODUCTION

This volume describes the results of excavation in the upper part of the Olduvai sequence, in Beds III, IV and the Masek Beds, carried out from the end of 1968 until 1971 (see Pl. 1). The delay in publication has been due to a number of factors. Chief among these were the Laetoli field seasons which took place from 1975 until 1981 and the delay in receiving chapters from other contributors. Analysis of the cultural material and identification of the fauna as well as the preparation of figures had been virtually completed by 1973 but it was clearly inadvisable to publish the volume without including these contributions.

During the excavations sites were selected for exploration as far as possible in successive stratigraphic levels following the method used for Beds I and II, published in Volume 3 of the Olduvai series (Leakey 1971). A total of twelve different sites was excavated comprising twenty-two archaeological occurrences. Twenty were within the Olduvai stratigraphic sequence, ranging from JK in Bed III to FLK in the Masek Beds. Two additional sites, that yielded the most elegant and highly finished of the bifacial tools, were in disturbed contexts and could not be fitted into the sequence although it is evident that they postdated the Masek Beds and perhaps included material derived from the Lower Ndutu Beds. Figs. Int.1 and Int.2 show the locations of sites excavated and provide a diagrammatic summary of the overall sequence of the upper beds at Olduvai. The twenty occurrences are as follows:

Post Masek Beds	TK Fish Gully
	HK
Masek Beds	FLK
Upper Bed IV	HEB West Level 1
	PDK Trenches I–III
	WK East C
	WK East A
	WK Upper Channel
Lower Bed IV	WK Intermediate Channel
	HEB West Level 2b
	HEB West Level 2a
	HEB
	HEB Level 3
	HEB Level 4
	HEB East
Base of Bed IV	WK Lower Channel
	PDK Trench IV
	WK Hippo Cliff
Bed III	JK clay above siltstone
	JK pink siltstone
	JK ferruginous sand
	JK grey sand

Preliminary results of the excavations in the upper part of the Olduvai sequence have been given in various papers at symposia and conferences (Leakey 1975a, 1978). These results were based on samples only and are superseded by the data published in the present volume in which the entire collections have been reviewed.

In describing the excavations the earliest site of JK in Bed III is discussed first and the remainder in ascending order. When a locality had more than one level or site the description is given when it is first mentioned in the text or preceding the most important archaeological occurrence. This follows the same system used for describing Beds I and II and permits the introduction of new elements in the stone industries to be assessed in their correct sequence. The bifacial tools from those sites in which there is an adequate sample are described and analysed in chapters 8 and 9 by Dr D. A. Roe and Dr Paul Callow, with the exception of the JK bifaces from Bed III since these are clearly a mixed assemblage with many derived specimens. For completeness Drs Roe and Callow have also included bifaces from Bed II and those from the two most recent sites, HK and TK Fish Gully. In my descriptions only the

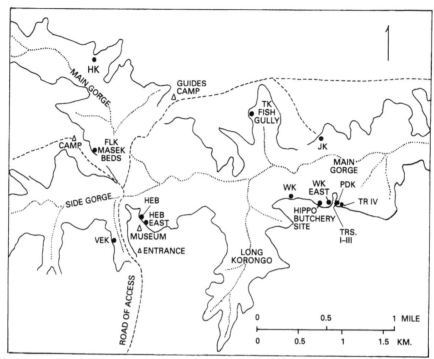

Fig. Int.1 Sketch map showing the sites excavated in Beds III, IV and the Masek Beds

numbers of bifaces in relation to the industries as a whole are noted, except in those cases where they were not described by Drs Roe and Callow.

For convenience Bed IV has been divided into three parts, the base, at the contact with Bed III, and Upper and Lower units demarcated by siltstones which occur in certain areas. Two marker tuffs, IVa and IVb, are recognised but are of little practical use for stratigraphic correlation of different sites owing to their localised distribution and absence in certain key areas. However, the combined evidence of the siltstones, marker tuffs and Dr Hay's geologic and mineralogic study has enabled the relationships of the more important sites to be established satisfactorily.

Dating of Beds III, IV and the Masek Beds is based on palaeomagnetic readings and rates of sedimentation. None of the three beds has yielded uncontaminated tuffs suitable for K-Ar dating. It is estimated that Bed III probably spans the period between 1.15 to 0.8 m.y., Bed IV from 0.8 to 0.6 m.y. and the Masek Beds from 0.6 to 0.4 m.y.

When excavation in the higher levels of the Olduvai sequence was begun in 1978 it was believed that only the Acheulean industrial complex would be found in Bed IV although the possibility that the Developed Oldowan technology of biface manufacture persisted into Bed III had been suggested by unpublished material recovered from site JK by Dr M. R. Kleindienst. In fact the first site with bifacial tools to be excavated in Bed IV (HEB East) was described as Acheulean but detailed analysis has since shown that it should more correctly be assigned to the Developed Oldowan (chapters 8–10). Sites later excavated on the south side of the Gorge in the WK area have proved of importance in demonstrating the contemporaneity of the Acheulean and Developed Oldowan technologies.

Dissimilarities noted between the bifacial tools from Acheulean sites not separated by any appreciable time interval were puzzling at first and considered possibly to be due to individual styles of biface manufacture or traditionally preferred methods (Leakey 1971). Experimental biface manufacture by Mr P. R. Jones (described in chapter 10) using the same rocks has shown that the different techniques were more likely

INTRODUCTION

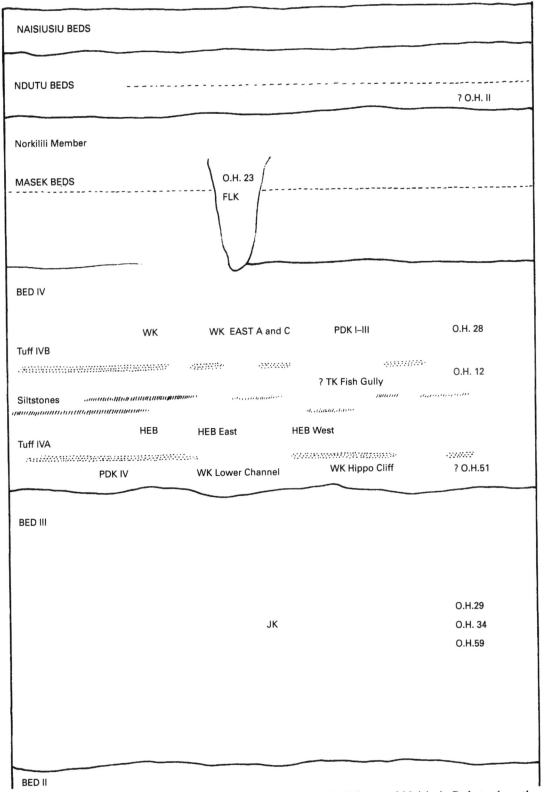

Fig. Int.2 Diagrammatic section of Beds III, IV, and the Masek, Ndutu and Naisiusiu Beds to show the stratigraphic positions of the excavated sites and hominid remains

practised in order to obtain the optimum results from rocks with different flaking properties (Jones 1979). If flaking techniques were specially adapted to particular rocks, it is not surprising that the bifacial tools from sites separated by a substantial time interval should sometimes closely resemble each other, such as those from EF-HR in Middle Bed II and WK in Upper Bed IV. In the descriptions of artefacts given in this volume an attempt has been made to identify the raw material of each specimen. This has been on a visual basis only and some erroneous identifications may have been made, especially when the rocks are heavily weathered.

In Bed III the only site excavated was JK, also partially excavated by M. Kleindienst in 1962. It consists of a relatively deep channel within the main Bed III drainageway (see chapter 2). Artefacts and faunal remains were abundant only in the lower part of the channel, where they were widely dispersed vertically and horizontally. The variable physical condition of the specimens indicates that they are probably not a homogeneous assemblage. A siltstone in the upper part of the channel, above the principal artefact-bearing levels, contained a complex of pits and furrows that have been the subject of speculation and discussion since their discovery in 1971. Even now their interpretation must be regarded with some reserve but the discovery of analogous pits and runnels within a small crater in northern Kenya, used today for salt extraction, suggests that the JK pits may have served a similar purpose (see chapter 2).

The three sites at the base of Bed IV, just above the III–IV interface – WK Lower Channel, WK Hippo Cliff and PDK IV – have yielded only very scanty material. The artefacts from both the WK sites are in conglomerates but those at PDK IV appear to be in undisturbed context. Sites HEB and HEB West in Lower Bed IV yielded a series of Acheulean industries in which the bifacial tools differed markedly in technology and were made from a variety of raw materials, while as noted above the industry from HEB East, also in Lower Bed IV, appears to be of Developed Oldowan facies. The series of artefacts obtained from the Intermediate Channel at WK, within the same stratigraphic unit, is too limited to be of value. Excavations in Upper Bed IV were carried out mainly in the WK and WK East area where both Acheulean and Developed Oldowan sites occur within one or more former river channels which have cut into Tuff IVB. The upper or main channel at WK yielded an abundant Acheulean industry associated with postcranial remains of *Homo erectus*. The WK East sites and PDK I–III that lie less than 1 km east of WK have yielded a Developed Oldowan industry in which the technique of making bifaces on large flakes, so characteristic of the Acheulean from the WK upper channel, is totally lacking.

The prevalence of pitted anvils and hammerstones in all sites of Upper Bed IV indicates the widespread use of the bipolar technique. It has resulted in large quantities of shattered quartzite debris as distinct from the whole or broken flakes found at sites where lava was the principal raw material. Only one site is known in the Masek Beds. It is situated at FLK and has yielded an Acheulean industry with large, elaborately trimmed handaxes. Site HK which was partially excavated during the 1931–2 expedition to Olduvai (Leakey 1951) yielded many finely retouched quartzite handaxes and cleavers. Test trenches dug in 1969 revealed that the artefacts were in a disturbed deposit postdating the Masek and the Lower Ndutu Beds although the artefacts may perhaps have been derived from the latter. Dr Roe has analysed material from the 1931–2 expedition as well as that recovered in 1969 but in the description of the industry as a whole only the more recent collection has been reviewed. Similar disturbed conditions prevailed at one level in TK Fish Gully excavated in 1962 by the late Dr J. Waechter, where elaborately retouched bifaces also occurred in a relatively recent deposit. These tools have been analysed by Dr Roe and, although there is no question that most were contained in a recent hill wash, it is possible that some specimens from other excavated levels may also be included in the series since the field notes are now lost. In view of the possibility that this may be a mixed assemblage a description has not been included.

INTRODUCTION

With the exception of PDK IV the sites excavated in Beds III, IV and the Masek Beds were in river or stream channels where the artefacts and faunal remains had probably been transported and displaced by water action. Sites with virtually undisturbed remains, such as were uncovered in Bed I, were not found in these beds with the possible exception of PDK IV. In this volume plans are published only of sites where a relatively large area was exposed or where a reasonably complete section across a channel was preserved, for example WK Upper Channel and WK East, Area A.

Definition of terms

The terminology for the artefacts defined in Volume 3 of the Olduvai series is retained in the present volume with some additions and modifications. For example, the subdivision between polyhedrons and spheroids/subspheroids is less distinct in the Beds III–IV industries than in those from Beds I and II. Polyhedrons are also much less common and there are relatively few that can confidently be separated from the more angular of the subspheroids. The question of revising these tool categories to approximate more closely to the Beds III–IV material was considered but it was deemed preferable to retain the same terms as for Beds I and II in order to facilitate comparison between the industries. The grouping of the artefacts into tools, utilised material and debitage is retained and both tools and utilised material have again been subdivided into heavy and light-duty categories. The term manuport applied to the unmodified cobbles found on the Oldowan lake margin sites has been omitted, since it was not applicable to cobbles in the stream and river channels of Beds III and IV, where it was impossible to determine whether such cobbles had been introduced by man or were a natural element of the conglomerates. Terms for artefacts connected with the bipolar technique of flaking, that is the pitted anvils/hammerstones and punches, are defined for the first time in this volume since none was noted in Beds I and II with the exception of two anvils. Other tool types remain virtually the same, although there is a reduction in both the numbers and varieties of choppers, only side and end choppers occurring in Beds III–IV.

Tools

1. *Awls* These tools are characterised by short, rather thick, pointed projections, generally at the distal ends of flakes, but sometimes on a lateral edge. In the majority the points are formed by a trimmed notch, on either one or both sides, but occasionally by straight convergent trimmed edges. The points are often blunted by use and have sometimes been snapped off at the base.

2. *Bifaces* Handaxes and cleavers are dealt with comprehensively by Drs Roe and Callow in chapters 8 and 9. Readers are referred to these sections of the book.

3. *Burins* Although rare, burins occur at a proportion of the sites. Angle burins are the usual form and are made on transverse broken edges or on trimmed edges, which are usually slightly concave and flaked from the primary surface. Some specimens are double-ended and there are a few with a working edge on either side.

4. *Choppers* These are usually made on cobbles with rounded cortex surfaces forming the butt ends. When they are made from blocks of quartzite the butts are often formed by a flat vertical surface, trimmed and blunted along the upper and lower edges. In the majority the trimming is bifacial, with multidirectional flaking of the working edges. These are essentially jagged and lack secondary trimming, although utilisation has often resulted in the edges having been chipped and blunted.

Side The maximum dimension is transverse, exceeding the length from the working edge to the butt; they are often made on oblong cobbles with the working edge along one lateral edge. Bifacial examples with alternative flaking predominate but there are also a few unifacial specimens and some in which there is multiple flaking on one face of the working edge and a single scar on the obverse.

End The maximum length is from the working edge to the butt; they are usually made on oblong cobbles with the working edge at one extremity.

5. *Discoids* These are often irregular, but a bifacially flaked working edge is present on the whole or the greater part of the circumference. Specimens made from cobbles are usually planoconvex in cross section with an area of cortical surface retained in the central part of the convex face.

6. *Laterally trimmed flakes* The flakes are generally elongate and end-struck with one or both lateral edges trimmed for the whole or part of their lengths. The retouch is usually somewhat uneven and the flakes are not symmetrical.

7. *Outils écaillés* Both single and double-ended specimens occur. They exhibit the scaled utilisation characteristic of these tools. The edges are blunted and one face is usually slightly concave, whilst the opposite is straight or slightly convex.

8. *Picks* These are massive tools with thick, heavy butts tapering rapidly to relatively narrow, sharply pointed tips.

9. *Polyhedrons* Angular tools with three or more working edges, usually intersecting. The edges project considerably when fresh but when extensively used sometimes become so reduced that the tools resemble subspheroids.

10. *Punches* These are small and rod-shaped, never more than a few centimetres long, invariably of quartzite and battered at both ends. They may be either tools or the final stages of bipolar cores before being discarded. In view of their conformity in size and shape they are classed here as tools, perhaps used for punching through tough hides.

11. *Scrapers* Most of the light-duty scrapers are made from flakes and other small fragments of quartz and quartzite. Many of the heavy-duty scrapers are impossible to assign to any particular type and consist merely of amorphous pieces of lava or quartzite with at least one flat surface from which steep trimming has been carried out along one edge. The light-duty group may be subdivided into the following categories but it is doubtful whether the various forms are of any great significance and it might be more realistic to lump together end, side and discoidal, retaining nosed and hollow scrapers as separate groups.

End. These are almost exclusively within the light-duty group. They are made on flakes or oblong fragments with a working edge at one extremity. The edges are generally curved, but are sometimes nearly straight and often exhibit small projections at the intersection of the trimming scars, or else a slight spur at one side.

Side. This is one of the most common forms of scrapers in both the heavy and light-duty groups. The working edges vary considerably with either shallow or steep trimming. They are usually curved but some are nearly straight and there is sometimes a slight medial projection, as in nosed scrapers.

Discoidal. These occur in both the heavy and light-duty groups. The general form is discoidal although the tools are seldom entirely symmetrical and they are usually trimmed on only part of the circumference.

Nosed. These are mainly confined to light-duty scrapers. There is a medial projection on the working edge, either bluntly pointed, rounded or occasionally spatulate, flanked on either side by a trimmed notch or, more rarely, by straight, convergent trimmed edges.

Hollow. Specimens in which the notch is unquestionably prepared are relatively scarce in both the heavy and light-duty groups, although light-duty flakes and other fragments with notches apparently caused by utilisation are common. In the few specimens which have been deliberately shaped the notches tend to be wide and shallow rather than deeply indented. They are variable in size.

INTRODUCTION

12. *Spheroids/subspheroids* These include some stone balls, smoothly rounded over the whole exterior, but faceted specimens in which the projecting ridges remain or have been only partly removed are more numerous. Subspheroids are similar to the spheroids but less symmetrical and more angular. They grade into worn polyhedrons and it is often difficult to distinguish the two categories.

Utilised material

1. *Pitted anvils/hammerstones* These occur at both Acheulean and Developed Oldowan sites and are clearly an essential element of the bipolar flaking technique. They range in size from small boulders to fist-sized cobbles bearing pecked depressions which are either oblong or circular, occurring singly or in pairs and sometimes with as many as eight on the same stone. The diameters are variable but the majority are between 1 and 1.5 cm in diameter and a few millimetres deep. Pits usually occur on water-worn cobbles but are also present on some heavy-duty tools and quartzite blocks. It is assumed that the boulders with pits served as anvils resting on the ground and that the fist-sized cobbles were held in the hand and used as hammerstones.

2. *Anvils* These consist of cuboid blocks or broken cobblestones with edges of approximately 90° on which there is battered utilisation, usually including plunging scars. Rarely found in Beds III and IV but common in Beds I and II.

3. *Hammerstones* The hammerstones consist of water-worn cobblestones (generally lava) with bruising and slight shattering at the extremities or on other projecting parts. Common in Beds I and II, they are relatively scarce in Beds III and IV and at FLK Masek.

4. *Cobbles and blocks* These are water-worn cobbles, weathered nodules and angular fragments that have some evidence of utilisation, either chipping and blunting of the edges or smashing and battering, but no evidence of artificial shaping.

5. *Light-duty flakes and other fragments* Flakes and other small fragments with chipping and blunting on the edges are quite abundant. They fall into three groups; with straight edges, with concave or notched edges and with convex edges. At a few sites there are some in which the utilisation tends to be scaled, recalling *outils écaillés*.

Debitage

The term debitage has been employed in preference to 'waste' for the unmodified flakes and other fragments, since there are indications at certain sites that some, at least, are not merely discarded by-products of tool manufacture but were made expressly, presumably to serve as sharp cutting tools.

1. *Flakes* The flakes are almost exclusively irregular. They may be subdivided into three groups, as follows: (a) divergent, splayed outwards from the striking platform (the most common type); (b) convergent with the maximum width at the striking platform; (c) approximately parallel-sided (rare). Broken flakes are far more numerous than whole flakes.

2. *Core fragments* Angular quartzite fragments are the most abundant debitage at sites where the bipolar technique was practised.

3. *Cores* There are few clearly recognisable cores with well-defined striking platforms from which flakes have been detached. These are mostly from Acheulean sites but are rare owing to the fact that the bifacial tools appear to have been blocked out at the quarry sites. In bipolar flaking many of the jettisoned blocks are cores by definition but it has been impossible to distinguish the parent blocks from other cuboid fragments and the majority of all angular waste has been placed in the heterogeneous category of core fragments.

GEOLOGY AND DATING OF BEDS III, IV AND THE MASEK BEDS

R. L. HAY

The geology of Beds III, IV and the Masek Beds (see Fig. 1.1) has been described in detail by Dr R. L. Hay in his volume on the geology of the gorge (Hay 1976). For convenience, an abbreviated version of his text is printed here by courtesy of the University of California Press.

Beds III and IV

Stratigraphy and distribution

Beds III and IV are distinguishable stratigraphic units only in the eastern parts of the Main and Side Gorges. Here they are the same units described by Reck (1951). The cliffs at JK can serve as a type section for both Beds III and IV. The contact between Bed II and Bed III is in most places disconformable and easy to locate. Beds II and III can, however, be very difficult to separate to the east, where both are chiefly reddish-brown in colour. Beds III and IV are generally easy to distinguish as far west as FLK and JK where Bed III interfingers over a broad zone with sediments similar to those of Bed IV. The interfingering was first demonstrated at JK through a series of excavations (Kleindienst 1964). Farther west, where Bed III is lithologically indistinguishable from Bed IV, the two units are combined into Beds III–IV undivided.

Bed III. Bed III is dominantly a reddish-brown deposit, chiefly of volcanic detritus which is about 85 per cent claystone, and most of the remainder is sandstone and conglomerate. Its thickness ranges from 4.5 to 11 m and varies systematically where deposited on different fault blocks, pointing to contemporaneous fault movements. Within the same fault blocks, it is thicker on the south side of the Main Gorge than it is on the north. Four different tuffs were found at more than one locality in the eastern part of the Main Gorge. They are numbered from 1, the oldest, up to 4, the youngest. Only the lowermost has aided appreciably in correlating, and the rest have thus far been recognised in only a few places.

Bed IV. Bed IV is chiefly claystone (68 per cent), most of which is soft-weathering and grey to brown. The remainder comprises sandstone (19 per cent), siltstone (7 per cent), conglomerate (4 per cent) and tuff (2 per cent). The contact with Bed III is generally an erosional surface, and the basal bed of Bed IV is in most places a sandstone or conglomerate. The contact of Bed IV with the Masek Beds is sharp, and in a few places the Masek Beds fill channels eroded into Bed IV or through Bed IV into Bed III. Bed IV generally ranges from 2.4 to 7.3 m in thickness in the Main Gorge, and it is as much as 10 m thick in the Side Gorge. The variation in thickness is a result of fault movements during the deposition of Bed IV as well as post-depositional erosion.

Within Bed IV are two marker tuffs, termed Tuffs IVA and IVB. Tuff IVA is the lower of the two and is present at widely separated localities in the Main and Side Gorges. Tuff IVB is found only in the vicinity of WK and JK.

Tuff IVA is a fine-grained vitric trachyte tuff 15 to 30 cm thick. It is typically laminated and yellowish-grey, but is massive and reddish-brown at one locality, TK. The principal primary minerals are biotite and feldspar, and analcime is the

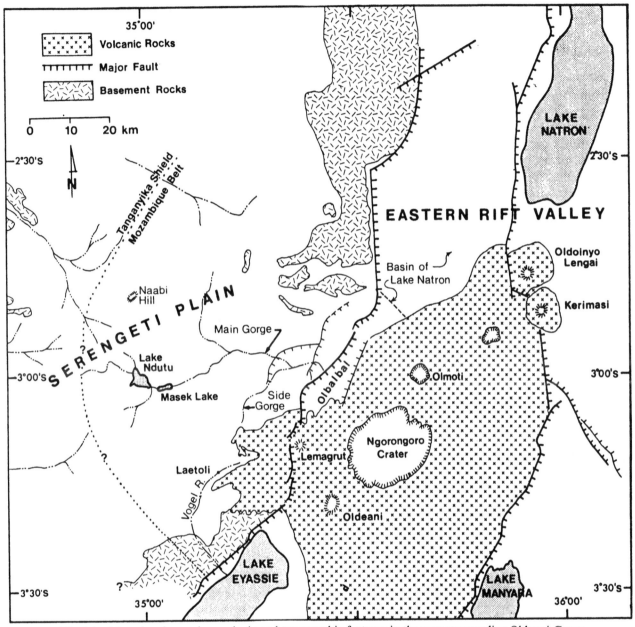

Fig. 1.1 Map showing major geologic and topographic features in the area surrounding Olduvai Gorge
(By courtesy of R. L. Hay and the University of California Press)

alteration product of the glass. The tuff lies in the lower part of Bed IV at two localities, whereas it is in the middle of Bed IV at FLK and GTC. This difference suggests that Bed IV accumulated more rapidly below the tuff towards the east than it did towards the west.

Tuff IVB is a fine to medium-grained crystal-lithic tuff of probably trachyandesite composition which is 15 cm to 2 m thick. It is hard and reddish-brown, and where thickest it is laminated or thin-bedded and contains thin layers of claystone. Its crystals are chiefly angular fragments of plagioclase and augite originating in the explosive fragmentation of crystalline lava. The tuff contains a high proportion of zeolites (analcime and chabazite), which may be reaction products of fine-grained vitric ash whose texture has been lost in the alteration process.

Siltstone and silty sandstone are widespread in the Main Gorge and have proved useful in correlating the various sites and exposures. Most of these silt-rich deposits lie in the lower half of Bed IV, and a thick bed underlies Tuff IVB at both JK and WK and very likely extends eastwards at the same level. A thin siltstone locally overlies Tuff IVB at JK.

Beds III–IV (*undivided*). Beds III and IV are combined in a single unit not only to the west of FLK and JK, but to the south, near Kelogi, and to the north along the east–west extension of the Fifth Fault. Beds III–IV undivided range in thickness from about 4 to 29 m, with the thickest sections in the graben between the FLK fault and the Fifth Fault. The variation in thickness is controlled largely by structural position, but it is also determined to some extent by the depth of erosion into Bed II, and the depth to which Beds III–IV undivided have been eroded beneath the Masek Beds.

Age of Beds III and IV

Magnetic stratigraphy is presently the only geophysical method to provide information about the age of Beds III and IV. Polarity studies were first undertaken by A. Brock, who made laboratory measurements on eight samples from Bed III, eight from Bed IV, and four from Beds III–IV (undivided). Drs F. H. Brown and M. D. Leakey were largely responsible for the field sampling. Most of the samples are of hard, reddish-brown zeolitic claystone, and the remainder are of limestone. Three of the eight Bed III samples have reversed polarity, four are normal, and one is magnetically unstable. The reversely polarised samples were collected both high and low in Bed III, demonstrating that Bed III was deposited during the Matuyama epoch, more than 700,000 years ago. The reversely polarised rocks probably acquired their polarity penecontemporaneous with deposition, as they did their reddish-brown colour. The normal polarity of the other samples is probably a result of the continued growth of haematite crystals in a later period of normal polarity. Four of the Bed IV samples have normal polarity, and the other four were magnetically unstable or of low magnetic intensity. These preliminary results led to the tentative conclusion that Beds III–IV contact coincided approximately with the Matuyama-Brunhes boundary at 700,000 years before the present (Brock, Hay and Brown 1972).

Dr A. Cox in 1972 initiated a more detailed sampling programme in Bed IV and the upper part of Bed III with the aim of locating the Brunhes-Matuyama boundary more precisely. The most significant result is the discovery of reversed polarity in three samples from the siltstone beneath Tuff IVB at the hominid excavation of WK. Tuff and calcrete samples from various levels in the Masek Beds, measured by both Brock and Cox, have normal polarity, and presumably they were deposited during the Brunhes normal epoch. Thus, the Brunhes-Matuyama boundary probably lies within Bed IV, and no lower than Tuff IVB.

The contact between Beds II and III is estimated as about 1.15 m.y.a. on the basis of stratal thicknesses in the gorge and dated fault movements to the east. If Tuff IVB is taken as 0.7 m.y. old, then relative stratal thicknesses in Beds III and IV can be used to obtain an age of about 0.8 m.y.a. for the top of Bed III. The top of Bed IV is about 0.6 m.y. old if one assumes that Bed IV sediments above Tuff IVB accumulated at the same rate as the underlying Bed IV sediments. If the Brunhes-Matuyama boundary lies above Tuff IVB, then the age limits of Bed IV should be slightly older than 0.6 to 0.8 m.y.a.

Environmental synthesis and geologic history

A widespread episode of faulting affected the Olduvai basin about 1.15 m.y.a., causing erosion of Bed II and drastically changing the palaeogeography. Renewed faulting about 0.8 m.y.a. shifted the drainage axis of the basin southwards

GEOLOGY AND DATING

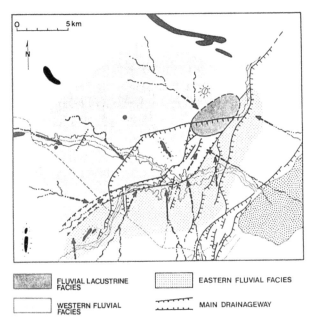

Fig. 1.2 Palaeogeography of Bed III with inferred drainage pattern (By courtesy of R. L. Hay and the University of California Press)

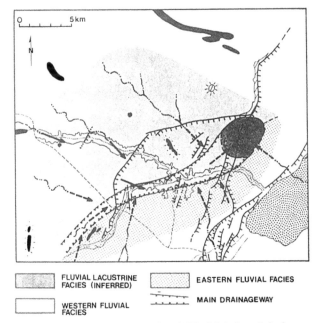

Fig. 1.3 Palaeogeography of Bed IV with inferred drainage pattern (By courtesy of R. L. Hay and the University of California)

about 0.5 km and caused widespread, generally shallow erosion of Bed III in the main drainageway. The alluvial-plain environments of Bed III age continue relatively unchanged through the deposition of Bed IV, and the deposition of Bed IV was terminated about 0.6 m.y.a. by fault movements (see Figs 1.2 and 1.3).

The eastern fluvial facies of Beds III and IV accumulated on an alluvial plain sloping gently to the north and west from the foot of the volcanic highlands. Sediment was transported mostly by braided streams that flowed only intermittently throughout the year. The western fluvial facies accumulated on a broad alluvial plain that sloped southeastwards. Most of the sediment was deposited by meandering streams as much as 2.5 m deep. Water flowed in at least the largest streams throughout the year. The palaeogeography of Beds III and IV differs primarily in the location of the main drainageway. The drainageway for Bed III and equivalent sediments of Beds III–IV undivided flowed from west to east, then changed its course to the northeast. The Bed IV drainageway lay about 0.5 km south of the Bed III drainageway, and it flowed in an easterly direction to the eastern limit of exposures. The eastern margin of the axial sump lay to the east of a locality 4.5 km north of the Third Fault, and the northern limits of the Bed III drainage sump can be only conjectured.

Vegetation in the Olduvai basin was chiefly bush, scrub and open grassland. Trees and perhaps strips of forest fringed the meandering rivers of the western alluvial plain and the main drainageway. Vegetation was relatively sparse on the alluvial plain to the south of the drainageway.

The climate was semi-arid while Beds III and IV were deposited. Mineralogic alterations in both fluvial and lacustrine sediments are evidence of a relatively dry climate, and a fluctuating and generally low water table is suggested by the reddish-brown colour and caliche limestones of the eastern fluvial deposits.

Masek Beds

Stratigraphy and distribution

The Masek Beds are the latest deposits prior to erosion of the gorge. Where exposed in the gorge, these deposits comprise roughly equal amounts of aeolian tuff and detrital sediments. They have a maximum exposed thickness of about 25 m and occur over a slightly larger area than Bed I or Bed II. They crop out over the length of the Main and Side gorges and are exposed in fault scarps along the southwestern part of Olbalbal.

Main and Side Gorges. The Masek Beds are found along the rim of the gorge, and they underlie the plain at shallow depth. The thickness of the Masek Beds varies strikingly in relation to faults, with 12 to 15 m common on the downthrown and 1.5 to 4.5 m on the upthrown side of faults. The Masek Beds disconformably overlie Beds III–IV in most places, although the contact can be difficult to locate where claystones of the Masek Beds overlie claystones of Beds III–IV. In a few places the Masek Beds fill channels 5 to 6 m deep cut into Bed IV, or else through Bed IV into Bed III.

The Masek Beds are subdivided into two units, the lower of which is the thicker. The upper unit is named the Norkilili Member. The type locality for both units is Kestrel K. Aeolian tuff and claystone in roughly equal amounts constitute 85 per cent of the lower unit, and the remainder comprises sandstone and conglomerate. Tuffs are dominantly pale yellowish-brown and may be stained with red. Despite reworking and contamination, two tuff units can be recognised widely and are used as stratigraphic markers within the lower unit. The lower widespread tuff unit is brownish-yellow, rich in pumice, and found in many sections of the Masek Beds in the Main Gorge to the west of the Second Fault. It was noted in the Side Gorge only at one locality.

The lower marker tuff lies at or near the base of the Masek Beds in most places. However, a 5 to 11 m thickness of detrital sediments, principally claystone, underlies the lower unit in the western part of the gorge. These detrital sediments may be correlative with channel fillings of the lower unit in the eastern part of the Main Gorge, two of which are overlain by pumice-rich aeolian tuff which probably represents the lower unit. The upper marker tuff varies greatly in stratigraphic position. It is most commonly at or near the top (e.g. sites PLK, HK and FK) but it may also lie near the middle of the lower unit of the Masek Beds.

The Norkilili Member is generally between 1 and 4.5 m thick and extends for the length of the Main and Side Gorges. It consists chiefly of reworked tuff, both aeolian and fluvial, and includes a substantial proportion of sandstone and conglomerate. The uppermost 60 to 90 cm are in most places brown and highly cemented by calcite. Red spotting characterises most of the tuffs, which range from yellowish-brown to pinkish-brown. Tuffs of the Norkilili Member differ from those of the lower unit of the Masek Beds in a wide variety of mineralogic respects. The red spotting and white, cottony dawsonite generally seem to identify tuffs of the Norkilili Member in the field.

The Norkilili Member conformably overlies the lower unit in most places. However, in the excavations at FLK, tuffaceous sandstone of the Norkilili Member fills a channel 4 m deep cut into the lower unit, which fills an older channel cut into Bed IV. An angular unconformity locally separates the two units on the anticlinal crest to the east of the Second Fault.

Age of the Masek Beds

The age of the Masek Beds can be estimated from several lines of evidence. One line is the dating of Kerimasi, the source of the tuffs in the Masek Beds. That Kerimasi was the source is shown by the mineralogic similarity of its tephra deposits to the Masek tuffs. Moreover, Kerimasi is situated to the northeast, upwind from Olduvai, in a location ideal for supplying Masek tephra.

The age of Kerimasi has been bracketed within limits of 1.1 to 0.4 m.y.a. by K-Ar dating of younger and older volcanic rocks in the vicinity of Kerimasi (Macintyre, Mitchell and Dawson 1974). The most crucial dating in this regard is 0.37 m.y.a. for tephra deposits of the eruptions that produced Swallow Crater and a nearby, related tuff cone. These deposits directly overlie the Kerimasi tephra.

Magnetic polarity suggests that the Masek Beds lie entirely with the Brunhes Normal Epoch, which extends from the present to 0.7 m.y.a. All six samples measured in the laboratory by A. Brock have normal polarity. These samples are from various levels in the Masek Beds and include both calcrete and zeolitic aeolian tuff. The same rock types in the Lemuta Member gave reliable results. Thus the Masek Beds appear to fall within dates of 0.7 and 0.4 m.y.a.

The amount of time represented by the Masek Beds can be estimated by comparing the volume of its detrital sediment with that in Beds III–IV, which span a period of roughly 0.55 m.y. Where exposed in the gorge, the Masek Beds contain about one-third the volume of detrital sediment found in Beds III–IV. This comparison probably errs in including the eastern fluvial facies of Beds III–IV, in as much as the comparable facies of the Masek Beds is not adequately represented in the gorge. If the comparison is restricted to the western fluvial facies of both units, then the ratio of Masek sediments to sediments of Beds III–IV is between 0.35 and 0.4, suggesting a time span on the order of 0.2 m.y.

In summary, the Masek Beds fall within limits of 0.7 and 0.4 m.y.a. on magnetic polarity and K-Ar dating. If the estimate of 0.6 m.y.a. for the top of Bed IV is correct, then the limits are further narrowed to 0.6 to 0.4 m.y.a. The time span estimated from the volume of sediment neatly fills the entire interval of 0.6 to 0.4 m.y.a.

Environmental synthesis and geologic history

The Masek Beds were deposited during a lengthy eruptive episode of Kerimasi and during a period of faulting in the Olduvai region. Displacements are documented for all faults along which the thickness of the Masek Beds is known (see Fig. 1.4).

The Olduvai basin was largely an alluvial plain of low relief while the Masek Beds were deposited. The palaeogeography was similar in many respects to that of Beds III–IV. The northwestern part of the basin sloped gently to the east and

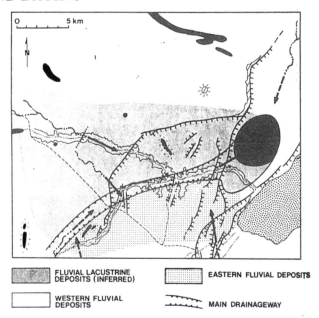

Fig. 1.4 Regional palaeogeography of the Masek Beds showing inferred drainage pattern (By courtesy of R. L. Hay and the University of California Press)

southeast joining the main drainageway which flowed eastward into the drainage sump.

About half of the lower unit was deposited by streams, and about half by wind. Except for the fillings of deep channels in the main drainageway, fluvial sedimentation was generally similar to that of Beds III–IV. As evidence, the proportion of claystone, sandstone and conglomerate is about the same in equivalent facies of both units. Aeolian tuffs of the lower unit were deposited partly in barchan dunes and partly as widespread, relatively thin layers over the ground surface.

The overall drainage pattern of the Norkilili Member seems to have been essentially the same as for the lower unit, but the lack of floodplain clays shows that the nature of the fluvial environment has changed. Evidently the detrital clay was carried to the drainage sump, which points to an increase in stream gradient and valleys lacking floodplains. Earth movements presumably caused this change in fluvial regime.

The climate must have been semi-arid, as judged from the large amount of aeolian sediment, the calcrete and mineralogic alterations, and the paucity and nature of faunal remains. Very likely the climate was drier than that of Beds

III–IV and rather like that of the present. Aeolian tuffs do not by themselves show that the Masek climate was drier than that of Beds III–IV, for their presence or absence might simply record whether or not large volumes of tephra were available. However, the high content of wind-worked *detrital* sediment in floodplain claystones, etc., shows that aeolian processes were relatively more active in Masek time. The oxygen-isotopic composition of pedogenic calcite concretions also points to a drier climate for the Masek Beds (Cerling, Hay and O'Neil 1977).

2
BED III
SITE JK (JUMA'S KORONGO)

M. D. LEAKEY

This site was among those recorded during the 1931–2 expedition to Olduvai. It lies on the north side of the gorge, approximately 2.5 km downstream from the confluence of the Main and Side Gorges. Some preliminary excavations were undertaken at JK in 1932. They yielded faunal material and an Acheulean industry ascribed to Stage 9 of the Chelles-Acheul sequence in the 1951 volume on Olduvai Gorge. The tools were reported to have come from Bed IV, at the base of a deposit of red sand. It was noted that the assemblage was not homogeneous and it was suggested that material of different ages might be present. The locality from which these tools were obtained has since been pointed out by L. S. B. Leakey and, on the basis of the present interpretation of the stratigraphy, they appear to have been in a level of red sand that lies 8.5 m below Tuff IVB and stratigraphically within Bed III.

During 1961 and 1962 Dr M. R. Kleindienst undertook further exploration of the JK site at my invitation. She carried out excavations over a period of four months and obtained artefacts and faunal remains at two localities, JK West and JK East, both of which are within Bed III. JK East yielded an industry with choppers and small tools but lacking bifaces, while JK West yielded bifacial tools, faunal material and some hominid remains. Neither of these artefact assemblages has yet been described by Dr Kleindienst although she published a paper setting out her interpretation of the geological background (Kleindienst 1964).

Excavations at JK were next undertaken by the writer in 1969 and 1970 when exploration of Beds III and IV was begun. A number of trenches at JK West yielded artefacts and faunal remains. For the most part, they were recovered from the sandy filling of a former river channel which cuts through the JK area, orientated approximately southwest to northeast. This material varied greatly in physical condition, containing both fresh and heavily rolled specimens which were clearly not in primary context. The horizon where artefacts had been obtained in 1932 is believed to have been at a rather higher level.

Work was once more resumed at the site in 1971 when a trench was dug specifically to search for further parts of the hominid O.H.34, represented by a femur and part of an associated tibia shaft, found by Dr Kleindienst in 1962 (Day 1971). The excavation led to the inadvertent discovery of a series of pits and associated furrows or runnels dug into a pink siltstone which overlies the coarser fluvial sediments containing artefacts. In this trench the upper part of the pink siltstone was removed as overburden, since it was barren of remains. When a fragment of bone and a chopper were noted the surface was swept clean and some circular patches of grey clay were observed, filling hollows in the siltstone. After cleaning, four shallow, almost exactly circular pits measuring some 30 to 60 cm in diameter and two smaller pits about 20 cm in diameter were revealed. Subsequent excavation of the area to the northwest of this trench revealed a complicated pattern of pits and runnels extending over an area of approximately 8 by 12 m.

The interpretation of the geological sequence at JK and its relationship to the Beds III–IV stratigraphy, as seen in the eastern part of the gorge, was the subject of a great deal of discussion and conjecture, ever since the site was first recorded. It is only in recent years, following a detailed re-examination of the site by Dr R. L.

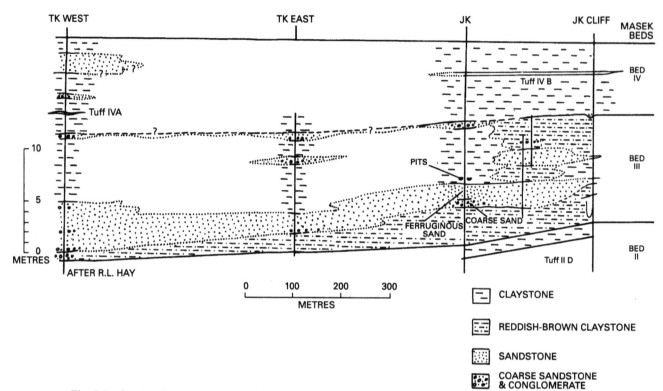

Fig. 2.1 Section from west to east along the main drainage channel in Beds III and IV from TK to JK

Hay and a series of geomagnetic readings by Dr Allan Cox, that the stratigraphic position of the site has been established satisfactorily (Hay 1976). It is, in fact, the only site of appreciable size to be excavated in Bed III; had the archaeological material been in primary context it would have provided a valuable link between the industries of Upper Bed II and Lower Bed IV.

To the east of the JK sites the northern wall of the gorge has eroded into an almost vertical cliff in which all the deposits from Bed I to the Masek Beds are exposed, providing one of the most complete sections in the gorge, to which the shorter sections exposed at JK can be related. At JK East, where the chopper/small-tool industry was recovered by Dr Kleindienst, there is a series of fluvial deposits, in part of which the artefacts occurred. These deposits are locally overlain by a red bed typical of the eastern facies of Bed III. The marker Tuff IVB can be seen at a higher level in the same gully and can also be traced to the east in the main JK cliff, as well as to the west in the JK West gully.

The section exposed in the JK West gully and in the trenches is as follows:

(a) recent hillwash
(b) Tuff IVB
(c) a grey siltstone (cf. sites HEB and WK)
(d) 9.5 m of clay and sandstone
(e) 0.4 m of pink siltstone containing the pits
(f) 0.2 m of sandy clay
(g) 0.5 m of ferruginous sand which interfingers with (h), below
(h) 0.7 m of coarse grey sand, the lowest part of the channel
(i) approximately 1 m of clay, locally cut into by the channel
(j) Tuff IID, at the top of Bed II.

Observations by Dr R. L. Hay at JK itself and at TK, a large gully less than 1 km west of JK, indicate that the former river channel exposed at JK continued through TK and further west to FLK (Fig. 2.1). The channel is characterised by coarse sediments at the base, grading into finer-grained sands and silts in the higher levels, as the water became quieter. Thus, the artefact-bearing sediments at JK West and East are at a low level in the channel, while the siltstone with the pits represents a later phase with quieter water action.

All geomagnetic samples from the channel deposits, including those from the base to the clay overlying the pink siltstone, have reversed polarity. On the available evidence, and taking into account the figure of 700,000 years for the close of the Matuyama Reversed Epoch, the age of Bed III has been estimated to be from 1.1 to 0.8 m.y.; the age of the JK site would, therefore, lie somewhere within this period.

The trenches

All excavations carried out by the writer were at JK West. The first trenches to be dug were at the west side of the exposure in the gully and consisted of the following: a Trial Trench 6 m long and 2 m wide, oriented 115° east; Trench I in the central part of the exposure 4 m long and 1 m wide, oriented 30° west, in the area excavated by Dr Kleindienst and cut into the deepest part of the channel; Trench II consisted of a 4 m extension to the north of the Trial Trench; Trench III was situated on the eastern side of the gully and was an extension to the north of a trench excavated by Dr Kleindienst – it measured 4 m in length, aligned 18° west and was approximately 1 to 3 m wide, following the line of the erosion slope. Trench IV was on the west side of the gully and measured 5 m by 1.5 m, aligned 60° east. Trench V, in which the pits were first noted, measured 8 m by 3 m. It was parallel to a trench excavated by Dr Kleindienst in which the remains of O.H.34 were found and was oriented approximately southwest–northeast. Trench VI followed the alignment of a trench dug by Dr Kleindienst which had not reached the base of the channel. Trench VII linked the Trial Trench, Trench II and Trench V.

Stratigraphy of the artefact-bearing level

The exposures in the JK West gully, which drains to the south, have cut into the old river channel roughly at right angles to it. The highest archaeological level encountered during excavation was at the surface of the pink siltstone containing the pits. Some 130 flakes and chips and 150 fragments of bone were recovered from the clay filling one of the larger pits, while a few scattered bone chips and debitage fragments were found in other pits and on the surface of the siltstone. The siltstone itself contained few remains. It was underlain by bedded ferruginous sands with clay lenses. Artefacts and faunal remains occurred throughout these sands which varied in depth in different parts of the channel. The basal channel filling of coarse sand interfingered in places with the overlying fine-grained ferruginous sand.

All trenches were excavated in 10 cm spits but in view of the bedding planes these are probably of little significance. In describing the archaeological and faunal material the spits have been disregarded since it was impossible to correlate accurately the spits in different trenches. The specimens have therefore been divided into the following stratigraphic levels:

(a) coarse sand, the basal filling of the channel, generally grey in colour, the lowest artefact-bearing level
(b) fine-grained sand with clay lenses, generally red in colour, sometimes interfingers with grey sand
(c) the pink siltstone with pits
(d) clay above the pink siltstone, the highest artefact-bearing horizon.

Owing to interfingering, the subdivision between the coarse grey sand and the overlying fine-grained red sand is not precise and a few specimens have probably been placed in the wrong context. However, since parts of what appears to be a single elephant have been recovered from both levels and 20 m apart it is probably of little value to attempt precise subdivisions.

(a) The lower coarse grey sand

The greater part of the artefacts and faunal material from this level were obtained from Trench III on the east side of the JK West gully.

The trenches dug on the west side and in the central area yielded relatively little material. There is no record available as to the distribution of the remains found by Dr Kleindienst.

The industry (3,503 specimens)
These comprise eighty-one tools, eighty-one utilised pieces and 3,341 debitage.

Tools

Bifaces (12 specimens)   There are one cleaver, four handaxes, three unfinished examples, two tip ends and two median fragments.

The cleaver is made on an end-struck lava flake measuring 186 × 98 × 58 mm. The cleaver edge is straight and 62 mm wide. It is formed by the intersection of the primary flake scar with an area of weathered surface which also occurs towards the butt. One rather crude handaxe is made of trachyandesite. It is plano-convex in cross section. The convex face has been trimmed all round the circumference but the reverse, flat surface has only been trimmed at the butt and along one lateral edge (192 × 102 × 52 mm). Another handaxe is made on a basalt end flake. Only three small trimming scars are present on the primary surfaces; the dorsal face consists mostly of cortex (173 × 103 × 47 mm).

Only one of the finished handaxes is of quartzite. It is on a tabular fragment trimmed round the entire circumference. The extreme tip is missing. Present length 166 mm, width 90 mm, thickness 42 mm. There is also a small, broad ovate in unusually sharp condition. Nearly all the trimming scars are stepped and the general appearance is crude (97 × 74 × 34 mm). Two of the unfinished specimens are of quartzite. One consists of a fragment of tabular material with the widest part at one end. It may represent a rough-out for a cleaver (179 × 104 × 67 mm). The second example is an untrimmed end flake measuring 187 × 116 × 49 mm. The third unfinished specimen consists of an unusually large, massive flake of Kelogi gneiss. The tip is broken and three flakes have been detached from one edge, on the primary surface, but there is no further trimming. Present length 205 mm, width 160 mm, thickness 75 mm.

Choppers (7 specimens)   There are six side choppers and one unifacial end chopper. One of the side choppers is made on a basalt cobble and the balance on blocks of quartzite. The basalt specimen exhibits a point in the centre of the working edge. This has been formed by trimming, which is from opposite directions on either side of the point (70 × 110 × 70 mm). Four of the quartzite choppers are made on blocks with flat or weathered surfaces forming the butts. The fifth and smallest example is on a quartzite pebble with a cortex butt (31 × 43 × 31 mm). In all five specimens the working edges are bifacially flaked. Measurements of the remaining four side choppers are 71 × 81 × 53 mm, 55 × 77 × 43 mm, 46 × 57 × 38 mm and 39 × 54 × 40 mm.

The single end chopper is unifacially flaked and is made on a flat cobble of basalt or trachyandesite. A slight pecked hollow or pit is present on one face, measuring 14 × 12 × 0.5 mm. The specimen measures 102 × 76 × 34 mm.

Polyhedrons (3 specimens)   These are angular specimens with three or more bifacially flaked sharp edges. They appear to be deliberately flaked tools and have therefore been classed as polyhedrons although it is not impossible that they are utilised cores. One specimen is of trachyte and two of quartzite. Measurements 78 × 71 × 60 mm, 66 × 60 × 54 mm and 53 × 45 × 46 mm.

Discoids (3 specimens)   All three examples are of quartzite. One is fine-grained and another is made on a cobble with cortical surface retained on one face, it is crudely flaked and angular although discoidal in general form (70 × 66 × 48 mm). The smallest example appears to be broken along one edge, it is also crudely flaked and measures 45 × 44 × 26 mm. The third specimen is slightly oblong but has a cutting edge round the whole circumference; it measures 64 × 51 × 39 mm.

Spheroids/subspheroids (20 specimens) The whole series is of quartzite and consists entirely of angular, subspherical specimens. A few examples retain fairly sharp ridges, but these have usually been reduced by battering. Measurements range from 97 × 93 × 73 mm to 33 × 28 × 23 mm, with an average of 55 × 48 × 41 mm.

Scrapers, heavy-duty (2 specimens) A broken trachyandesite cobble has been steeply trimmed on part of the circumference from a flat, cortical surface. There is a slight projection or 'nose' in the central part of the working edge (58 × 43 × 41 mm). The second example is of quartzite. It is roughly discoidal and part of the circumference has been steeply but crudely trimmed from a flat undersurface to form an irregular working edge (68 × 62 × 42 mm).

Scrapers, light-duty (24 specimens) These can be subdivided into four categories, namely: side, end, discoidal and hollow scrapers. Side scrapers are the most numerous with ten examples. They are all made on broken flakes or other fragments of quartzite. The trimming of the working edges is fairly regular although sometimes there are small indentations, perhaps intentional or due to use. Measurements range from 44 × 72 mm to 15 × 23 mm with an average of 23 × 33 mm. There are four quartzite end scrapers with straight, transverse working edges. In three examples the working edge is the widest part of the tool. Measurements 34 × 32 mm, 22 × 22 mm, 23 × 18 mm and 24 × 23 mm. Seven specimens are roughly discoidal in form although not necessarily trimmed all round the circumference. One is of phonolite and the balance of quartzite. The trimming varies from steep to shallow and is generally even. Measurements range from 49 × 47 mm to 19 × 19 mm, with an average of 29 × 28 mm.

Two quartzite flakes and one of phonolite show well-defined, trimmed notches. One example has two notches, one on either lateral edge of a flake. In another specimen the notch is at the end of a flake so that there is a projecting 'horn' on either side. Overall measurements: 43 × 31 mm, 37 × 26 mm and 33 × 20 mm. Width and depth of the notches: 27 × 5 mm, 17 × 2.5 mm, 12 × 1.5 mm and 16 × 2.5 mm (the last two on the same specimen).

Burin (1 specimen) This is a double-ended angle burin on part of a quartzite flake which has also been trimmed along the lateral edge opposite the burin spalls. A single spall has been removed from each end, both of which are trimmed transversely (42 × 20 mm).

Laterally trimmed flakes (2 specimens) Minimal flaking is present on the lateral edges of two phonolite flakes. One specimen is pointed although trimmed only on one side of the pointed extremity (47 × 28 mm). The second specimen is blunt ended. It has one deeply indented and two shallower flake scars on the primary flake surface. The opposite dorsal face is not retouched (55 × 38 mm).

Outils écaillés (7 specimens) Five examples are double ended and two single ended. With one exception the working edges are broad, unlike the punches found commonly at later sites. Length/breadth measurements range from 34 × 20 mm to 24 × 13 mm, with an average of 28 × 16 mm.

Pitted anvils and hammerstones (2 specimens) Two small cobbles of vesicular basalt have a single pit on one face. In one example the pit measures 16 × 13 × 2.5 mm. The cobble shows slight traces of battering at one end (79 × 56 × 36 mm). The second specimen has been extensively battered at both ends. It measures 69 × 48 × 34 mm and the pit 17 × 15 × 1.5 mm.

Utilised material

Cobbles (10 specimens) These consist of whole or broken cobbles with rough flaking or breakage apparently due to heavy use. They range in size from 100 × 63 × 61 mm to 63 × 52 × 41 mm, with an average of 89 × 63 × 61 mm.

Light-duty flakes, etc. (69 specimens) All these specimens show chipping or blunting of the edges. The majority consist of flakes or broken flakes but there are also some thin fragments of tabular

ANALYSIS OF THE INDUSTRIES FROM JK LOWER
COARSE GREY SAND

	Numbers	Percentages
Tools	81	2.3
Utilised material	81	2.3
Debitage	3341	95.3
	3503	99.9
Tools		
Bifaces	12	14.8
Choppers	7	8.6
Polyhedrons	3	3.7
Discoids	3	3.7
Spheroids/subspheroids	20	24.7
Scrapers, heavy-duty	2	2.5
Scrapers, light-duty	24	29.6
Burin	1	1.2
Laterally trimmed flakes	2	2.5
Outils écaillés	7	8.6
	81	99.9
Utilised material		
Pitted anvils/hammerstones	2	2.6
Cobbles	10	12.3
Light-duty flakes	69	85.1
	81	100.0
Debitage		
Whole flakes	253	7.6
Broken flakes	1568	46.9
Core fragments	1519	45.5
Core	1	0.0
	3341	100.0

quartzite. Only eight examples are of lava. The utilised edges can be roughly subdivided into convex, notched and straight. Convex-edged specimens amount to twenty. They are all on broken flakes or core fragments. The chipping of the edges is usually from one direction only. Length/breadth measurements range from 43 × 31 mm to 16 × 15 mm with an average of 23 × 19 mm. Twenty-three examples show chipping on edges which are approximately straight. The chipping is both coarse and fine and is sometimes present on both faces. Measurements range from 74 × 54 mm to 15 × 13 mm, with an average of 30 × 23 mm.

Specimens with notched or hollowed chipped edges amount to twenty-one. The notches vary from wide and shallow to narrow, v-shaped indentations. They range in width/depth from 27 × 2.5 to 10 × 0.5 mm with an average of 13 × 1.9 mm. Overall measurements of the specimens range from 45 × 43 mm to 19 × 19 mm with an average of 36 × 24 mm.

Debitage (3,341 specimens)

These comprise 253 whole flakes, 1,568 broken flakes, 1,519 core fragments and one core.

Whole flakes End flakes are the most numerous with 145 specimens, against 108 side flakes. Quartzite predominates in both categories, lava flakes amounting to only thirty-nine specimens. Fifty-nine quartzite side flakes are divergent and eight convergent. In the end flakes there are 104 divergent, thirty-seven convergent and four roughly parallel sided. Twenty-five lava flakes are end struck and fourteen side struck. They are mostly divergent and include one phonolite re-sharpening flake apparently struck from the edge of a handaxe.

The quartzite end flakes range in length/breadth from 70 × 48 mm to 14 × 13 mm, with an average of 29 × 22 mm. The lava series ranges from 68 × 65 mm to 19 × 14 mm, with an average of 32 × 25 mm. In the quartzite side flakes the range is from 51 × 60 mm to 14 × 16 mm, with an average of 22 × 27 mm. The lava side flakes range from 38 × 56 mm to 13 × 15 mm, with an average of 22 × 29 mm.

Broken flakes and core fragments Broken flakes and core fragments occur in almost equal proportions and in both quartzite is by far the most common material with 94 per cent in broken flakes and 93 per cent in core fragments. In the whole flakes it is somewhat lower with 83 per cent. In a number of Bed IV industries lava predominates in the whole and broken flakes although core fragments are predominantly of quartzite.

Core This is part of a quartzite cobble which still retains an area of cortex. It appears to be a small core from which flakes have been detached in two directions from a single striking platform (47 × 42 × 41 mm).

In addition to the above material there are seven blocks of quartzite and one of phonolite which are virtually unmodified and may represent unused raw material (they are not included in the debitage).

(b) The fine-grained ferruginous sand

This level was excavated in the Trial Trench and in Trenches II to VII; the greatest depth was found in the Trial Trench and in Trench V where it reached a thickness of approximately 1 m. In some areas the red sands were cross bedded and a number of small channels also occurred. Many of the artefacts, bones and teeth from this level are heavily abraded while others are in fresh condition.

The industry (2,190 specimens)
This consists of sixty-one tools, fifty-four utilised pieces and 2,075 debitage.

Tools

Bifaces (13 specimens) There are six complete specimens, three without tips and four butt ends. The complete specimens are unusually variable, they include a large, well-made elongate ovate of nephelinite, a massive quartzite pick, a small quartzite ovate and three small, broad specimens with unusually thick, untrimmed butts, resembling proto-handaxes. The nephelinite ovate measures $221 \times 99 \times 55$ mm and is weathered; the small quartzite ovate is in fresh condition and measures $79 \times 56 \times 26$ mm. The quartzite pick is made on a triangular fragment of tabular quartzite and measures $181 \times 130 \times 73$ mm. One lateral edge has been regularly trimmed on both faces whilst the opposite edge shows minimal trimming, except at the tip, which has been retouched to a sharp point from both sides. The butt is thick and consists of a flat surface lying at right angles to the upper and lower faces of the tool. The three small specimens with thick butts are in sharp condition, including one made on a basalt cobble. They measure $64 \times 70 \times 49$ mm, $56 \times 47 \times 38$ mm and $35 \times 35 \times 23$ mm. Among the three specimens lacking tips there are two small quartzite examples in which the original lengths cannot have been more than 58 and 85 mm. They are in sharp condition, in contrast to the basalt specimen which is heavily abraded.

Choppers (9 specimens) There are seven side choppers including one miniature and one double-edged specimen, two end choppers which are virtually rectangular, but slightly longer than wide and therefore rank as end choppers. Three lava side choppers are on cobbles and one on an elongate block of phonolite in which the working edge is unusually obtuse, with an angle of approximately 90°. Measurements for these four specimens range from $105 \times 117 \times 71$ mm to $54 \times 64 \times 40$ mm, with an average of $71 \times 87 \times 55$ mm. They are all bifacially flaked and one bears a blunt point approximately in the centre of the working edge. The double-edged specimen is of quartzite with both edges battered and blunted. The entire surface is also abraded ($66 \times 95 \times 46$ mm). The miniature example is also of quartzite; it is bifacially flaked with a point in the centre of the working edge and is in fresh condition ($36 \times 43 \times 22$ mm). Both end choppers are made on cobbles, one of lava and one of quartzite. They are bifacially flaked. Measurements $84 \times 70 \times 63$ mm and $81 \times 79 \times 53$ mm.

Discoids (6 specimens) Three of these tools are diminutive and two of the larger specimens exhibit a short projection or small, blunt point on the circumference, which appears to have been retained deliberately. In all other respects these tools are typical discoids. Five examples are of quartzite and one of basalt, the latter being more abraded than the quartzite specimens. Sizes range from $72 \times 64 \times 45$ mm to $31 \times 29 \times 20$ mm, with an average of $50 \times 47 \times 31$ mm.

Spheroids/subspheroids (14 specimens) There is one almost exactly spherical specimen. It is pecked over the entire surface and all projections and traces of trimming scars have been removed. It is probably made of nephelinite ($74 \times 72 \times 66$ mm). The remaining thirteen specimens are subspherical. They are of quartzite and are angular, although a few have had the major

M. D. LEAKEY
ANALYSIS OF THE INDUSTRIES FROM JK LOWER COARSE GREY SAND

	Quartzite	Fine-grained quartzite	Gneiss, feldspar, etc.	Phonolite	Nephelinite	Trachyte	Basalt/trachyandesite	Lava, indeterminate	Totals
Heavy-duty tools									
Bifaces, choppers, polyhedrons, discoids, spheroids/subspheroids, scrapers	34	1	1	2	—	1	7	1	47
Light-duty tools									
Scrapers, burins, laterally trimmed flakes, awls, *outils écaillés*, punches	28	2	—	4	—	—	—	—	34
Utilised material, heavy-duty									
Pitted anvils and hammerstones, cobbles and blocks	—	—	1	—	1	—	9	1	12
Utilised material, light-duty									
Flakes, etc.	59	2	—	7	—	—	1	—	69
Debitage									
Whole flakes	212	2	—	14	2	1	22	—	253
Broken flakes	1486	7	1	23	—	2	49	—	1568
Core fragments	1491	7	—	4	5	—	12	—	1519
Cores	1	—	—	—	—	—	—	—	1
	3311	21	3	54	8	4	100	2	3503

projections largely removed. Sizes range from 73 × 67 × 68 mm to 27 × 25 × 23 mm, with an average of 45 × 42 × 38 mm.

Scrapers, heavy-duty (2 specimens) One example is made on a basalt cobble steeply trimmed along one edge. Some additional flaking is also present on another edge. The butt and both upper and lower faces consist of cortex (70 × 83 × 52 mm). The second example consists of a double-edged side scraper made on part of an unusually thick slab of green-tinted quartzite. The working edges have been trimmed from opposite faces and both extremities are vertical, consisting of natural cleavage planes. Width from edge to edge 90 mm, length 98 mm.

Scrapers, light-duty (11 specimens) All these tools are irregular in shape, but six specimens can be classed as side scrapers, three as end scrapers and two as hollow scrapers. Five of the eleven specimens are of phonolite, an unusually high proportion for tools of this material at JK. In the side scrapers the working edges are generally irregular but the majority show signs of use. With the exception of a single core fragment they are made on flakes. Length/breadth measurements range from 27 × 50 mm to 21 × 25 mm, with an average of 27 × 41 mm. Two of the three end scrapers, made on fragments of tabular quartzite, exhibit distinct 'noses' at the working ends (53 × 34 mm and 49 × 28 mm). The third example is almost discoidal but has only been trimmed on part of the circumference (21 × 20 mm). The two hollow scrapers are on flakes measuring 45 × 34 mm and 50 × 37 mm respectively, with notches 17 × 1.5 mm and 16 × 2 mm.

Awls (2 specimens) Both examples are of quartzite. They exhibit exceptionally sharp points with fine retouch on either side. Measurements 50 × 43 mm and 24 × 30 mm.

Laterally trimmed flake (1 specimen) A broken flake of fine-grained quartzite has been retouched on both lateral edges. The tip is rounded (35 × 31 mm).

Outil écaillé (1 specimen) There is a single small example of fine-grained quartzite. Only one end shows scaled utilisation (17 × 14 mm).

Punches (2 specimens) Both examples are of quartzite. They are single-ended and show a minimum of crushing at one end (36 × 19 mm and 26 × 17 mm).

Utilised material

Pitted anvils and hammerstones (11 specimens) There are five examples with single pits, two with a pit on either face and four with multiple pits. There are no twin pits in this series and, with the exception of one specimen with multiple pits, the pits are not so deep nor so well defined as in the specimens from Bed IV.

(a) *Single pits.* All five examples are on basalt or trachyandesite cobbles. They range from 127 × 96 × 68 mm to 74 × 63 × 44 mm with an average of 87 × 70 × 50 mm. The pits range from 23 × 18 × 4 mm to 7 × 6 × 0.5 mm with an average length/breadth/depth of 16 × 12 × 1.2 mm. It is noticeable that only one specimen shows the battered utilisation commonly found on the pitted hammerstones from Bed IV.

(b) *Pits on opposite faces.* Both these specimens consist of basalt or trachyandesite cobbles. They measure 104 × 80 × 75 mm and 85 × 70 × 40 mm respectively. The pits in the larger specimen measure 25 × 15 × 5 mm and 18 × 11 × 3 mm but the pitted surfaces in the smaller specimen consist only of a series of shallow indentations, too indefinite to be measured.

(c) *Multiple pits.* These consist of three basalt or trachyandesite cobbles and one block of quartzite. The largest lava cobble exhibits four pits and both the others three each. The quartzite example has four, or possibly five pits. The largest specimen is a flattened rectangular cobble with a pit on each of its four faces. Both extremities are heavily abraded. It measures 138 × 99 × 58 mm and the pits measure 52 × 38 × 9 mm, 30 × 26 × 55 mm, 21 × 21 × 4 mm and 24 × 18 × 5 mm. The remaining two lava examples measure 76 × 75 × 56 mm and 63 × 43 × 34 mm. The pits on both are similar in size and average 17 × 10 × 2.3 mm. The quartzite specimen (84 × 72 × 69 mm) shows battered utilisation along one edge, similar to that seen on anvils from Bed II. It has four well-defined and one shallow pit. Average measurements of the five pits are 21 × 16 × 2.7 mm.

The average weights of the specimens in the three categories are as follows:

(a) single pits 665 g
(b) pits on opposite faces 525 g
(c) multiple pits 575 g.

Hammerstones (3 specimens) These consist of basalt or trachyandesite cobbles with battering and bruising at the extremities. They measure 79 × 64 × 48 mm, 72 × 60 × 40 mm and 63 × 55 × 37 mm.

Cobbles and blocks (20 specimens) There are twelve basalt or trachyandesite cobbles, two of trachyte and one of an unidentified lava, which have been crudely flaked and battered but show little of the bruised utilisation seen on hammerstones. The measurements range from 186 × 115 × 59 mm to 55 × 34 × 40 mm, with an average of 96 × 73 × 55 mm. The four blocks are all of quartzite and exhibit battering and blunting on the edges. One example also has hammerstone type of utilisation. Measurements range from 66 × 59 × 52 mm to 55 × 47 × 37 mm, with an average of 62 × 56 × 40 mm.

Light-duty utilised flakes, etc. (20 specimens) These consist mostly of quartzite flakes and other small fragments with blunted and chipped edges. They follow no particular pattern and the chipping is generally irregular. Two flakes, probably of trachyandesite, are exceptionally abraded and are probably derived. Omitting these two specimens, which are larger than any others in the series, the length/breadth measurements range from 51 × 43 to 30 × 16 mm, with an average of 39 × 29 mm.

Debitage (2,075 specimens)

Whole flakes These comprise 127 specimens, of which 101 are quartzite and twenty-six lava. End flakes are the most numerous, with seventy-four examples. One hundred and three are divergent, sixteen convergent and eight parallel-sided. The quartzite end flakes range in length/breadth from 72 × 44 mm to 16 × 11 mm, with an average of 31 × 23 mm. Lava end flakes range from 54 × 38 mm to 26 × 21 mm with an average of 42 × 30 mm. Quartzite side flakes range from 57 × 69 mm to 15 × 21 mm with an average of 25 × 32 mm, while the lava side flakes range from 63 × 88 mm to 10 × 13 mm with an average of 25 × 35 mm. The average figures for the lava flakes are lower than usual owing to the presence of a number of small, thin phonolite flakes, probably derived from the refined trimming often seen on handaxes made from this material.

Broken flakes These total 1,117 specimens in which quartzite is the most common material with 1,010 pieces.

Core fragments These comprise 821 specimens; all are of quartzite except for nine pieces of lava.

Cores There are ten specimens consisting of six blocks of quartzite, a piece of basalt, two lava cobbles and part of a small boulder. One lava cobble is unusual in this context since it is a prepared core resembling those of the Middle Stone Age. The cobble has been broken transversely retaining the cortex surface on one face, whilst the opposite face has been radially trimmed, with the exception of the fracture. A relatively large flake has then been struck, using the broken surface as a platform (69 × 65 × 40 mm). The small lava boulder is flat in cross section and has had flakes removed from both faces along one edge, thus resembling a gigantic side chopper (189 × 125 × 79 mm). The largest scar is of a side flake measuring 64 × 116 mm. The remainder of the cores are formless. They range in size from 105 × 72 × 43 mm to 44 × 39 × 28 mm with an average of 69 × 56 × 48 mm. In addition to the above artefacts there are thirteen modified blocks of tabular quartzite which appear to have been brought in as raw material but not used. The largest piece measures 102 mm in maximum diameter.

ANALYSIS OF THE INDUSTRIES FROM JK FINE-GRAINED FERRUGINOUS SAND

	Numbers	Percentages
Tools	61	2.8
Utilised material	54	2.5
Debitage	2075	94.7
	2190	100.0
Tools		
Bifaces	13	21.3
Choppers	9	14.8
Polyhedrons	—	—
Discoids	6	9.8
Spheroids/subspheroids	14	23.0
Scrapers, heavy-duty	2	3.3
Scrapers, light-duty	11	18.0
Awls	2	3.3
Laterally trimmed flakes	1	1.6
Outils écaillé	1	1.6
Punches	2	3.3
	61	100.0
Utilised material		
Pitted anvils/hammerstones	11	20.4
Hammerstones	3	5.6
Cobbles and blocks	20	37.0
Light-duty flakes	20	37.0
	54	100.0
Debitage		
Whole flakes	127	6.1
Broken flakes	1117	53.8
Core fragments	821	39.6
Cores	10	0.5
	2075	100.0

(c) The pink siltstone

A total of fifteen artefacts was recovered from the pink siltstone in Trench V. The artefacts consist of the following: a crude lava side scraper made on a basalt flake (54 × 79 mm), a small side scraper made on the lateral edge of a quartzite

BED III. SITE JK

ANALYSIS OF THE INDUSTRIES FROM JK FINE-GRAINED FERRUGINOUS SAND

	Quartzite	Fine-grained quartzite	Gneiss, feldspar, etc.	Phonolite	Nephelinite	Trachyte	Basalt/trachyandesite	Lava, indeterminate	Totals
Heavy-duty tools									
Bifaces, choppers, discoids, spheroids and subspheroids, scrapers	32	1	—	1	2	1	7	—	44
Light-duty tools									
Scrapers, awl, laterally trimmed flake, *outil écaillé*, punches	7	5	—	5	—	—	—	—	17
Utilised material, heavy-duty									
Pitted anvils and hammerstones, cobbles and blocks	5	—	—	—	—	2	26	1	34
Utilised material, light-duty									
Flakes, etc.	14	1	—	3	—	—	2	—	20
Debitage									
Whole flakes	96	5	—	15	—	1	10	—	127
Broken flakes	1000	10	1	80	—	3	23	—	1117
Core fragments	811	1	—	5	1	1	2	—	821
Cores	6	—	—	1	—	—	3	—	10
	1971	23	1	110	3	8	73	1	2190

flake (18 × 27 mm), a discoidal lava cobble measuring 78 × 78 × 78 mm with a slight pit on one face – the pit is ill defined but measures approximately 20 × 11 × 0.5 mm; a utilised block of quartzite and a lava cobble from which some flakes have been detached; one whole and six broken quartzite flakes and two core fragments.

(d) Clay above the pink siltstone

Leaving aside the few bone fragments and artefacts that were found on the surface of the pink siltstone in the area of the pits and in the fillings of the pits, which will be described in the relevant section, the only artefacts recovered from this level were from the Trial Trench. In this trench the siltstone had been partially cut through by a small channel, in which the artefacts occurred. It was not possible to determine the horizon from which they were derived.

The artefacts consist of 176 specimens comprising one chopper, one crude discoid, two subspheroids, two utilised flakes, one hammerstone, ten complete flakes, 119 broken flakes and forty core fragments. Quartzite is the most common material but there are also a few pieces of phonolite among the debitage. The single chopper is a side chopper with a sharp, bifacially flaked working edge. The butt has been considerably battered (62 × 78 × 53 mm). The discoid is made of trachyte. It is flat on one face and steeply trimmed on the circumference of the convex face. It is more heavily abraded than any of the quartzite tools (64 × 54 × 35 mm). The two subspheroids are angular but projecting ridges have been battered to some extent. Measurements are 95 × 90 × 77 mm and 60 × 56 × 51 mm. Both the utilised flakes are under 35 mm in length. One has been chipped along a convex, scraper-like edge while the second has a thick, rather pointed

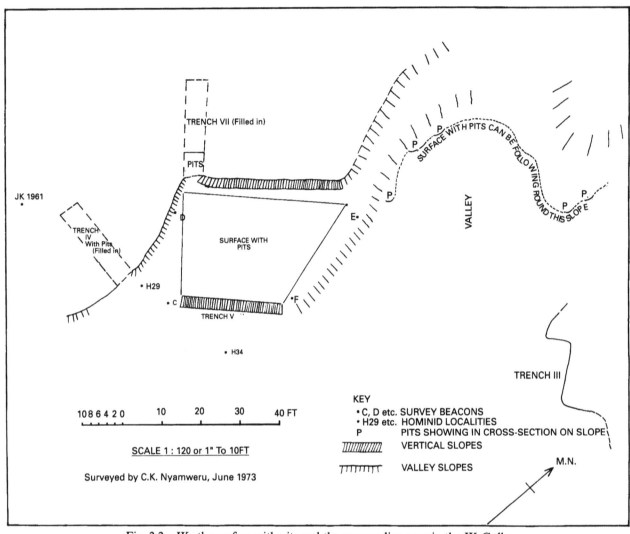

Fig. 2.2 JK: the surface with pits and the surrounding area in the JK Gully

extremity that shows crushed utilisation similar to that seen on punches.

The pits and furrows

Four pits belonging to this complex were first noted during 1971 in the course of excavating a trench adjacent to one dug by Dr M. R. Kleindienst in 1961, where she recovered the hominid femur and tibia of O.H.34. The 1971 trench (Trench V) was parallel to and to the northwest of the previous excavation and was dug in the hope that further remains of O.H.34 might come to light. None was found and the trench yielded little faunal or lithic material of significance except in the lower part of the coarse grey sand at the bottom of the channel, where there were numerous quartzite flakes and chips.

Excavation of the upper part of the pink siltstone was proceeding when a number of hollows filled with the overlying grey clay were observed (see Fig. 2.2; Pl. 2). Most noticeable were four almost exactly circular depressions together with a number of smaller holes. When the small holes were cleared of infilling clay they appeared to be artiodactyl footprints and included several examples approximately 25 cm long and 40 cm wide perhaps made by hippopotamus. After the filling of the large circular depressions had been removed and the surface cleared it was noted that some of the tracks crossed the interiors of the depressions and had therefore been made subsequently.

The four large depressions varied in diameter from 85 × 72 cm to 45 × 46 cm and were between 11 and 20 cm deep but must originally have been deeper since the rims and upper parts of the walls had been dug away before the pits were recognised. A striking feature of the pits was the intense reddening of the interiors; this gradually faded on exposure but for a time led to the belief that they might have held fire.

When the pits had been removed, after being strengthened with preservative, cast and wrapped in plaster bandages for transportation to the National Museum in Dar es Salaam, Trench V was dug down to the base of the channel. Excavation then proceeded to the northwest in 2 m squares and an area of approximately 560 sq m was eventually uncovered. Narrow, exploratory trenches were also excavated to the west and northwest to determine the extent of the area containing pits. It was found that former erosion, preceding the deposition of the Ndutu Beds, had truncated the siltstone in both these directions. More pits, however, could be seen in section exposed in the side of the gully to the north but none was found further up the gully where the siltstone became much coarser in texture and graded into a sandstone.

When the surface of the siltstone was being cleaned a number of furrows or large grooves came to light, apparently associated with the pits. They crossed the area in various directions often terminating in a pit. Owing to faulting and earth movements that have taken place since Bed III times it is no longer possible to determine the direction of drainage, if they were intended to channel water.

Excavation of the pits and furrows was extremely slow and carried out exclusively with dental instruments, paint brushes and spoons, sometimes even teaspoons, since a proportion of the furrows were too narrow for a wider instrument to be used for removing the excavated soil. The fact that the surface of the siltstone was considerably harder than the overlying clay and that there was a distinct contrast in the colour of the two deposits made it possible to follow the contours of the pits and furrows with a degree of confidence that would not otherwise have been possible. Admixture of the clay and siltstone rarely occurred and the surface of the siltstone between the pits was generally flat; it was apparent that the deposit removed when the pits were dug was no longer present in any quantity. The edges of the furrows, on the other hand, sometimes had small ridges where the deposit had been pushed upwards. Only the larger and more distinct pits and furrows will be described here. As will be seen from the contoured plan (Fig. 2.3, see page viii) and photographic mosaic (Pl. 2) there were also many small pits and holes probably made by large ungulates moving across the siltstone when the surface was damp and soft. Excluding the pits found initially in Trench V, in which the depth is uncertain since the surface of the siltstone was damaged during excavation, there are twenty-seven clearly recognisable and well-preserved pits that are described here. Some are single but others are in groups within large, irregular hollows that contain several individual pits. The pits within the hollows seem unlikely to be contemporaneous since they cut into one another, but it was impossible to determine exactly the order of priority beyond the fact that the most complete examples are the most recent.

In the best-preserved pits, groups of short, parallel and approximately horizontal grooves were noted on the walls, generally in sets of four and each set separated from the next by a small vertical ridge, suggestive of finger marks made by scraping the sides of the pits by hand when they were damp. In other pits there were pecked marks generally 1 to 2 cm in diameter, some of which had a short groove on one side such as would occur if a pointed stick were used for digging and had sometimes slipped sideways.

An attempt to make a contoured map of the site was first undertaken by Miss C. Cannon and Mr R. Jurmain but they unfortunately omitted to establish a datum or fixed level so that the relative elevations of different parts of the area and depth of the pits was not recorded. Mr R. I. M. Campbell then kindly made the photographic mosaic shown in Pl. 2 and Dr Celia Nyamweru, Professor of Geography at Kenyatta University College, Nairobi, surveyed the area for the contoured map (Fig. 2.3). The methods she evolved to meet the unusual requirements of this survey were published in the *Journal of Field Archaeology* and extracts are reprinted here in Appendix B. Her field plan was later reduced and redrafted for publication by courtesy of the Cartographic Department of the National Geographic Society, Washington DC. Dr Nyamweru found it most practical to carry out the survey in feet and tenths of feet since her assistants were unfamiliar with metric units. In order to conform with her plan the measurements of the pits and furrows given below are in the same units although elsewhere in the volume the measurements are metric. In measuring the pits the maximum diameter has been recorded with the transverse diameter as far as possible at right angles to it.

Pit 1

An incomplete pit truncated to the south by the edge of Trench V. The northeast–southwest diameter is 2.4 ft and depth 0.3 ft. The interior is rough with a number of small pecked indentations. On the northern side of the pit there is an oblong depression that for a time was considered possibly to be a small human footprint. It measures 135 mm in length and 62 mm in maximum width, across what would have been the ball of the foot. The 'heel' is towards the rim of the pit with the 'toes' pointing inwards to the centre; the impression is very shallow here, but becomes deep at the 'heel'. A slight longitudinal ridge is present in the centre of the depression with a broad shallow groove on either side, both of equal depth, features not compatible with the majority of human footprints in which the lateral edge receives more weight and is more deeply depressed than the medial. No further impressions were discovered that could be considered to resemble human footprints and the single example described above appears to be of doubtful validity; superficially it resembles a child's footprint but it lacks the essential criterion of normal weight distribution. Moreover, the depth to which the 'heel' has been impressed suggests a weight much greater than is likely for a child with a foot of comparable size.

Pit 2

This pit has also been cut into by the edge of Trench V but is of particular interest in view of certain features being unusually well preserved. There are clearly defined horizontal grooves round the sides in sets of four, a relatively deep furrow that leads directly into the pit from the west and a utilised triangular fragment of quartzite lying on the bottom. The maximum width now preserved is 1.8 ft and the depth to the floor of the pit is 0.7 ft. There is a small hole a few centimetres in diameter at the bottom of the pit (Pl. 3). Similar holes occur in several other pits but no explanation can be offered as to their significance. The grooves occur round the sides of the pit and not on the bottom. They are in parallel sets of four, each set measuring 62 to 65 mm across with small ridges between the grooves. The sets are not continuous and nine groups can be seen. Each groove terminates in a small raised lump and these together form a series of knobbly ridges down the sides of the pit (Pl. 3). Experiments have demonstrated that a similar pattern of grooves and ridges can be duplicated when the muddy walls of a hollow are systematically scraped off with the fingers. Comparable grooves are made and the knobbly vertical ridges are formed when the fingers are lifted off after each handful of mud has been removed. Random scraping would not, of course, result in such a consistent pattern.

The furrow that leads into this pit from the west is particularly well preserved. It is approximately 5 ft long in its deepest part, where it is 0.4 ft deep, but continues west for another 2 ft, where it becomes shallower. The sides are nearly vertical and the base is flat; a cross section exposed where it is cut through by Trench V shows that there is no trace of a fissure in the bottom. A branch furrow that leads from it to the south was probably connected with one of the pits in Trench V.

Pits 3, 4, 5, 6, 7

These five pits are enclosed within one large rather irregular hollow that in overall shape is roughly quadrangular with a medial constriction of the walls on opposite sides. It measures 7.4 ft in length southeast–northwest by 4.7 ft at the widest part and narrows to 2.5 ft where the sides are incurved. Pits 3 and 4 are still partly obscured by a baulk of clay filling that was left in place. The southeast–northwest diameters therefore cannot be measured. Pit 3 measures 2.5 ft northeast-southwest and is 0.9 ft deep. The southwest wall is almost vertical and undercut at the southern end for a distance of 0.4 ft. The opposite wall slopes gradually. Pit 4, to the southwest of Pit 3 measures 2.2 ft northeast–southwest and is 0.6 ft deep; it abuts on to Pit 5 (0.9 ft deep), which is central to the group, but the contact of the two pits is obscured by the clay baulk. Pits 5 and 6 are at the northwest end of the large hollow and are separated from the other pits by the medial constriction of the sides. These two pits form a pair of equal depth and almost equal size although Pit 7 has steeper walls, undercut in two areas on the west side. The ridges separating these two pits from one another and from Pit 5 are sharp and narrow and it evident that the most complete of the existing pits, such as No. 6, have been dug into and partially destroyed, but it is difficult to determine the sequence with any certainty. Yet another pit seems to have been present within the hollow to the northwest, but is now almost entirely destroyed by Pits 6 and 7. Pit 6 measures 3 ft (southeast–northwest) by 2.4 ft and is 1.2 ft deep; Pit 7 is 2 ft in diameter (southeast-northwest), and is 1.9 ft wide, and 1.1 ft deep. The interiors of all seven pits are rough with small cavities; groups of horizontal grooves are also present on the sides of Pits 5 and 7. A steep-sided furrow leads from Pit 6, to the east. It is 0.4 ft at the deepest part and is now rather irregular since the edges appear to have crumbled off in places.

Pit 8

This is a small pit in which the walls are either undercut or almost vertical. It measures 1.9 ft (east–west) by 1.6 ft and is 0.8 ft deep. The interior is rough and a furrow leads off it to the west; it is quite deep where it cuts into the edge of the pit but is blocked a little further on and could not be followed for any distance.

Pits 9 and 10

These consist of twin pits within a single oblong hollow measuring 4.3 ft north–south by 3 ft east–west. The two pits are separated by a sharp ridge projecting from the walls on either side and extending across the bottom of the hollow. Both pits are 0.8 ft deep with 'pecked' sides and bottoms. The northern and western walls of Pit 10 are almost vertical and undercut in two areas. A rather broad furrow 0.2 ft deep connects these two pits with Pit 11 which lies 2.5 ft to the west. Two small symmetrical holes are present to the north outside the rim of Pit 10. They are 0.5 ft in diameter and 0.6 ft apart. A ring of limestone also occurred to the south of Pit 9 (cross hatched on Fig. 2.3). This was probably formed below ground level within a pit and is not contemporaneous.

Pit 11

A rather irregular-shaped small pit 2.2 ft north–south by 1.9 ft east–west and 0.6 ft deep. It is connected by furrows to Pit 9 and 10 to the east and to Pit 12 to the west. Two further furrows lead off it to the south and northwest, both of which cut across the two convergent furrows forming a long V orientated to the southeast (Pl. 4).

Pit 12

A small, almost exactly circular pit 1.7 ft in diameter and 0.6 ft in depth linked to Pit 11 by one well-preserved straight furrow 0.2 ft deep and approximately 0.3 ft wide. A second furrow not now connected with any pit, that is also well preserved, measures 0.2 ft deep and is 0.3 to 0.4 ft wide. This is also cut across the long convergent grooves and was made subsequent to them.

Pits 13 and 14

Two small pits within an oblong hollow with a median constriction on either side. The hollow measures 2.2 ft in length (east–west) and 1.3 ft across Pit 13, which is the widest part, narrowing to 0.8 ft where the sides curve inwards between the pits. Pit 13 is 0.4 ft deep, steep sided to the north and with a gradually sloping wall to the south. Pit 14 is the same depth, but steep sided and largely undercut except where it abuts on Pit 13. A broad furrow leads off Pit 13 to the east and probably once connected it to Pit 7 although there is now an obstruction, possibly of material removed from another pit.

Pit 15

At irregular-shaped pit with almost vertical or undercut walls and with shallower extensions to the east and west. The pit measures 2.5 ft northeast–southwest and the rims of the extensions are 3.9 ft from edge to edge. The transverse diameter of the pit is 2.2 ft and the depth 1.2 ft. The interior is rough and 'pecked'.

Pits 16 and 17

Double pits within the same hollow, separated by a low ridge on the bottom and on the walls on either side. The hollow measures 3.3 ft east–west and 2.4 ft in width across Pit 16, the larger of the two; it is 1 ft deep with steep sides and undercut in one area. Pit 17 is undercut on the greater part of its outer circumference and is 0.4 ft deep. Measurements are 1.4 ft (northeast–southwest) by 1.3 ft from the northwest to the southeast edge to where it abuts on to Pit 16. A furrow 0.2 ft deep leads from Pit 16 to a pit now filled with a circular deposit of limestone 1.8 ft in diameter. A second furrow, also 0.2 ft deep, connects Pit 17 to a small pit on the periphery of the large hollow which is to the northwest and contains multiple pits as well as the main pit, 21.

Pit 18

Part of the western side of this pit lies outside the excavated area so that the northeast–southwest diameter cannot be measured, but it appears to have been circular with a southeast–northwest diameter of 2 ft and a depth of 0.5 ft.

Pit 19

This is the largest single pit uncovered. Other hollows of comparable size contain double or multiple pits. It measures 4.4 ft east–west, 3.5 ft in transverse diameter and is 1.6 ft deep. The northwest wall is steep and undercut in two places and the whole interior is rough.

Pit 20

This is an oblong hollow that extends to the north beyond the excavated area. The maximum length is now approximately 4 ft north–south but was originally greater. The pit is 2 ft wide and 0.5 ft deep with a small cavity at the bottom similar to that in Pit 2.

Pit 21

A large irregular-shaped hollow, now containing only one complete pit but appearing formerly to have contained at least five, cut into one another. This complex of pits is now difficult to interpret since the hollow was filled with water and damaged during a freak rainstorm when the guttering on the protective building collapsed. The northern side of the hollow extends beyond the edge of the excavated area but now measures 6 ft southeast–northwest, 5 ft across and 1 ft in depth. The west side of the hollow is steep and partly undercut. A nearly complete small pit measuring 1.4 ft across extends to the south-east of the main hollow and a furrow 0.2 to 0.3 ft deep connects it to Pits 16 and 17. (Only part of the hollow containing the complex of pits that includes Pit 21 is shown in the mosaic photograph.)

Pit 22

Another large irregular hollow containing a complex of four or five incomplete pits. It measures 6.6 ft (northeast–southwest) and 4 ft in maximum width. The sides are steep and undercut in several places and the undercutting of the northeast wall is in two levels forming a step. Series of horizontal grooves are well preserved on several of the pit walls particularly in one pit that appears to have been the most recently excavated. This is 0.8 ft deep but the deepest pit, in the southern part of the hollow, is 1.7 ft deep.

Pit 23

The central pit is approximately circular but there is a shallower extension to the south so that the hollow as a whole is oblong. It is 7.3 ft long (north–south) by 3.6 ft wide (east–west) and 1.7 ft deep. The edges of the main pit are steep and undercut in three areas. Peck marks and grooves are present on the walls and there is a small cavity in the bottom similar to that in Pit 2. The filling of this pit yielded a substantial number of bone fragments and artefacts, all of which are quite small. The total recovered consists of 148 faunal remains and forty-four artefacts, as follows: indeterminate bone fragments 132, catfish bones 4, reptilian bones 7, bovid molar 1, indeterminate tooth fragments 3, suid tooth fragment 1. The artefacts consist of forty quartzite chips and flakes and three of phonolite. There was also one lava cobble.

Pit 24

This is a small circular pit 1 ft long, 0.9 ft wide and 0.3 ft deep with a well-marked furrow leading off it to the south. The walls are steep but not undercut.

Pit 25

A small steep-sided pit in which the walls are undercut for more than half the circumference. It measures 2.2 ft east–west by 1.8 ft north–south and is 0.6 ft deep.

Pits 26 and 27

Dual pits within one oblong hollow which measures 3.2 ft east–west by 1.7 ft north–south. Pit 26 is 0.4 ft deep, steep sided and undercut on the northern side. Pit 27 is to the west and is 0.3 ft deep with more gently sloping sides except to the north, where the wall is steeper and partly undercut. One other pit can be seen at the extreme northern corner of the photographic mosaic but was not surveyed by Dr Nyamweru. It is small and quite deep, with a rough interior and has three well-preserved furrows radiating from it.

The twenty-seven pits uncovered within the area excavated consist of twelve single pits, four double pits within single hollows and three groups of multiple pits also within single large hollows. There are, in addition, numbers of small, irregular holes that seem likely to have been made by animals walking over the surface when it was damp. In multiple groups of pits many have cut into previous pits and partially destroyed them so that only the more complete, and presumably most recent, can be measured. These vary in depth from 1.7 ft to 0.3 ft with an average for the 27 examples of 0.7 ft.

Many of the furrows have undergone considerable damage before being buried beneath the clay. Some have been filled with reworked siltstone that appears to have been derived from digging further pits, and in others, when the sides are nearly vertical, the edges have crumbled off and fallen inwards. A few, however, are still in a good state of preservation and remain relatively undamaged. Among these are three furrows at the southern edge of the excavated area. One connects with Pit 2, and the other two join together to form a single furrow, not truncated by the edge of Trench V where an excellent cross section is exposed. The sides are almost vertical, with sharp edges at the top, while the bottom is flat or slightly hollowed out. There is no trace of a fissure running downwards.

The second group of particularly well-preserved furrows is at the southeast corner of the excavation. It exhibits clear evidence of two periods of activity since two short furrows, orientated east–west, cut across a pair 10 ft long that are 3 ft apart at the northern end and converge to the southeast where they probably joined or led into a pit but are now cut off by the present erosion slope. Both the pair of convergent furrows and the two that cut across them have raised lips along the edges in certain areas, where the siltstone has been pushed upwards. Parts of many shorter furrows are also clearcut and where preserved are generally connected to a pit. Two instances were noted of furrows continuing below the surface of the siltstone as small tunnels. This may have been due to deliberate tunnelling, such as was noted in the salt workings at the Magado Crater or to siltstone being piled above the furrows when they had become partially silted up.

There are indications, seen more clearly on the ground than in the contoured map, that a ditch or furrow, wider and deeper than any other, crossed the area from northeast to southwest. It is now interrupted by Pits 6 and 7 which appear to have been dug subsequently and destroyed part of it. A reconstruction of the site at any one point of time is not possible since it is evident that the pits and furrows now revealed are almost certainly not all of the same period. There are numerous instances of partial pits that have been dug into and damaged by others dug subsequently. This also applies in the case of furrows that cut across others (Pl. 4). It is evident, too, that the volume of siltstone once contained in the pits is no longer present. The surface between the pits was virtually flat with only minor differences in elevation. No mounds of siltstone were found, which would have been quite considerable in view of the size of some pits, had the excavated siltstone remained on the sites, even allowing for flattening by subsequent earth pressure. In fact earth pressure does not appear to have been responsible for deformation of the surface, in view of the well-preserved raised ridges of siltstone along the edges of some furrows.

Many tentative explanations have been suggested for this site since its discovery and I have to thank colleagues and friends for searching the literature and enquiring among living primitive peoples for possible analogies. Natural phenomena such as root casts of bulbous plants, for example *Adenium* sp., have been postulated to account for the pits and it has been suggested that the horizontal, shallow roots of *Acacia* such as *A. tortilis* might have caused the furrows. Examination of *Adenium* and other plants and *A. tortilis* roots systems on the Serengeti indicates that they leave traces unlike the pits and furrows at JK. It has also been postulated that the pits were made by animals in search of salt and it has been pointed out that antelopes scraping off salty earth with their lower incisors leave grooves, but these do not compare at all closely with the grooves on the walls of the JK pits; nor do animals undercut the sides of salt licks at ground level, although elephants are capable of excavating even caves with their tusks. It has also been suggested that the furrows are natural fissures in the siltstone but this explanation is clearly invalid since the grooves are not V-shaped but flat bottomed and never exhibit cracks running downwards. In fact the only analogy so far known that appears to correspond to most of the features to be seen in the JK pits and furrows complex are present day salt workings such as those in the central Sahara and at Magado Crater in Kenya where salt is collected by evaporation from alkaline water.

In order to test whether the pits dug into the siltstone were capable of retaining water, several kilos of siltstone were collected, ground to a powder, moistened to the consistency of damp clay and shaped into thick bowls. These were sun dried and afterwards filled with water. A certain amount of water was absorbed by the siltstone but once it became saturated the bowls held water quite satisfactorily provided they were thick enough.

Nine samples of the sediments from JK were submitted for analysis to Professor P. A. Robbins, head of the Chemistry Department, University of Nairobi. Although the present chemical composition of the JK deposits does not necessarily correspond with that of Bed III times, the results were suggestive. The nine samples were taken from different levels and different localities in the area of the pits as listed in Table 2.1. The results show that all the samples are strongly alkaline and therefore contain a high content of soluble salts, demonstrated by the qualitative analyses. However, the sample from a pit filling is

BED III. SITE JK

Table 2.1 *Soil samples from Olduvai: the JK Pits and vicinity*

Sample No. (ML)	Sample description (ML)	Sample Description (PAR)	Aq. extract colour	Qualitative ion conc.			Chloride ion		Sodium ion	
				ce	PO_4'''	SO_4''	m.moles/kg	ppm	m.moles/kg	ppm
1	Base of small pit in Trench V	Grey grit with low content of very colloidal red clay	?	?	?	?	68.5	2430	52.2	1200
2	Base of pink sandstone away from pits	Pinkish-grey grit with low clay content	Yellow	+	+	+	31	1100	111	2550
3	Pink sandstone from N gully, no pits	Pinkish-red very coarse grit, low clay content	Colourless	+	−	+	37	1310	109	2500
4	Grey clay above pits	Pale grey, very high colloidal clay content	Colourless	+ +	+ +	+ +	72.6	2550	187	4300
5	Grey bed below pink siltstone of pits	Grey grit with medium clay content	Colourless	+ +	+	+	119	4220	222	5100
6	Sand 2 ft below pits siltstone	Pinkish grit, low clay content	Colourless	+	−	+	49	1740	178	4100
7	Base of pit in gully	Very coarse red brown grit, low clay content	Pale yellow	+ +	−	+ +	60.3	2140	163	3750
8	Filling of pit from which sample 7 was taken	Grey grit with low clay content	Yellow	+ + +	−	+ + +	266	9420	844	19400
9	Pits siltstone, from S end of bed, away from pits	Grey grit with medium content of pink clay	Pale yellow	+	−	+ +	36.4	1290	193	4450

exceptionally high (no. 8). This may be significant but, as has been pointed out by Professor Robbins, it is probably that there may have been a redistribution of soluble material like sodium chloride by ground movement since Pleistocene times. Today, the sedimentary deposits at Olduvai are strongly alkaline and in the northern part of the main gorge trona is obtained by the Masai from seepage areas at the bottom of the Gorge. They depend entirely on evaporation of the alkaline water from the surface of the ground and do not improve the yield by digging pits or by any other means.

In all cases where the walls of the JK pits are well preserved and not weathered to a crumbly surface they exhibit either sets of small shallow grooves or else indentations resembling pick marks. The sets of grooves are found on the walls of the pits but not on the bottom, but in three pits there are small circular holes resembling solution cavities, such as that in Pit 2 shown in Pl. 3. The sets of grooves are also most distinct in this pit, where they can be seen to occur in sets of four with the grooves within each set parallel to one another and of equal length, ending in small vertical ridges. The grooves are smoothly concave and the intervening ridges also rounded, never sharp. Similar grooves can be seen on the walls of many other pits but are less clearly defined. As noted, they suggest finger grooves made by scraping the walls of the pits by hand when the siltstone was damp.

The indentations resembling pick marks tend to be less constant in size than the grooves but no more than resulted from experimental digging with a pointed stick approximately 1 inch in diameter, the size and depth of the hole depending on the amount of force exerted and the consistency of the soil being dug.

Characteristic features of the furrows are the nearly vertical sides and flat or rounded bases; none is fissured at the bottom. In some cases the rims have raised lips suggesting that the deposit was pushed upwards when the furrows were formed. The furrows are multidirectional and the majority connect with pits; one small pit in the northern end of the site, shown in Pl. 2 but not included on the contoured plan, has seven furrows radiating from its circumference. It is evident that not all the furrows are of the same period; there are a number of instances where one or more transect others (Pl. 4) and some are blocked by reworked siltstone.

Although natural causes are surely responsible

for some of the features at JK, such as the small holes probably made by animals that would have frequented the site if it were rich in salt, others seem to indicate artificial agency. For example, the fact that the interiors of the pits are not smooth but exhibit either grooves or indentations resembling pick marks demonstrates that they are not pot-holes. But perhaps the most significant factor is the removal of the siltstone extracted from the pits, which must have amounted to a considerable volume of material. The possibility that it was swept away by flood water is unlikely in view of the well-preserved sharp rims of the pits that would inevitably have become abraded and damaged had this occurred. Moreover, the pits would have been filled with redeposited siltstone instead of the grey clay they contained, which also overlay the entire surface of the siltstone.

Any interpretation for the JK pits and furrows complex must be speculative and depends to a great extent upon whether the features displayed at this site are regarded as natural or artificial, but it is the writer's opinion that the combination of features, none of which occurs elsewhere at Olduvai, indicates that they are most likely to be artificial. If this is accepted, then by present-day analogy, the most plausible interpretation would seem to be that the pits were used for collecting salt by evaporation of alkaline water channelled into them by means of the furrows. The grooves and pick-like marks within the pits would thus have been caused by removal of the residual salt, as at the two sites now in active use, described below.

Following the discovery of the JK pits and furrows, quite extensive inquiries were made to obtain information that might contribute to a tentative interpretation. In order to cover as wide a field as possible among the more primitive peoples of Kenya who might be able to provide present-day analogies, a questionnaire was circulated to a large number of bush schools through the good offices of Mrs Jean Sassoon. The children were asked whether any pits or furrows were dug by their people and, if so, for what purpose. The replies were disappointing and did not contribute any relevant information although small irrigation channels were described from several districts. The only present-day activities I have been able to discover that relate to the JK site are the salt workings in northern Kenya and the central Sahara.

Magado Crater salt workings

The present-day salt workings in Kenya are in a small crater known as Magado where trona is being worked commercially by the Meru people (Pls 5–7). The crater lies 0° 37′ N by 30° 2′ E; it is rather more than 1.5 km in diameter and 150 to 200 m deep, with a small soda lake in the bottom that supplies strongly alkaline water for the trona workings. The floor of the crater consists of clay, and fresh water seeps out from a slightly higher level at certain areas round the sides. Cattle are watered regularly from these springs and it is recorded that one of the early travellers, W. H. Brown, obtained water from them in 1904.

My attention was first called to the trona workings at Magado by my son Philip who flew over the area in a light aircraft and was impressed by the apparent similarity to the pits and runnels at JK. I was able to visit the crater in 1985 by courtesy of Mr Robert Lowis and his safari clients, Mr and Mrs J. Hume. We encountered some antagonism from the trona workers during our visit which made it impossible to do more than make general observations.

The present trona workings occupy about half the crater floor and are adjacent to the soda lake from which alkaline water is channelled into the pits by ditches and smaller runnels. Although the general appearance seems to be confused, a definite system of irrigation exists, as can be seen from the aerial photo (Pl. 5). Ditches 0.5 to 1 m deep and up to 0.3 m wide at the bottom, dug into the clay floor, lead water from the lake into the working area (Pls. 6, 7). Smaller ditches and furrows connect with the main drainage system and carry water into pits 7 or 8 ft in diameter that serve as reservoirs. They are both larger and deeper than the pits from which trona is collected. The latter are circular, generally 0.6 to 1 m in diameter and are connected either to runnels or to the reservoir pits by small inlets that are blocked with clay when sufficient water has trickled in, usually to a depth of 15 to 25 cm. The small pits are generally in groups of a dozen or so, but one group of forty-six was seen. Several instances were noted of overlapping pits, where a later one was dug into an earlier, abandoned pit, possibly because this no longer held water. Each group of collecting pits appeared to belong to a separate owner or family unit although the main irrigation systems were communal.

We were told that two or three months may elapse before the water in the collecting pits has evaporated leaving a crust on the bottom of yellowish-white relatively pure trona. This is removed first and regarded as top-quality material. The sides and bottoms of the pits are then scraped by means of a broken coconut shell, a piece of bent tin or a panga, the all-purpose machete-like tool of East Africa. Marks left by the tools used for scraping can be seen quite clearly and remain visible until such time as the pits are refilled with water. The trona-impregnated clay scraped from the pits is piled up into conical heaps on level ground nearby and allowed to drain in the sun. It is then wrapped in banana leaves and tied into oblong packages weighing 5 kg or more and carried by women to the crater rim whence it is transported by donkeys for sale in other parts of the district and used for cooking as well as for mixing with tobacco to make snuff.

Narrow footpaths give access to the different parts of the workings, generally along the top of one of the banks enclosing an irrigation channel. Occasionally small tunnels are dug through the banks on which there are footpaths in order to lead off water and avoid a ditch across the pathway that would cause an obstruction.

L. A. J. Williams states that the late Mrs Joy Adamson sent a sample of the prepared trona to the Mines and Geological Department, Nairobi in 1955, where a qualitative chemical analysis was carried out. It was noted that: 'The sample contains water-soluble alkali salts including sodium and potassium carbonate, bicarbonates, chlorides and a small amount of phosphates. The alkali metal is mainly sodium.' A further sample was collected by L. A. J. Williams from one of the pits in the crater floor, near the lake. An analysis showed the following:

	%
Na_2CO_3	8.06
Na_2HCO_3	15.46
Na_2SO_4	0.68
$NaCl$	5.08

Balance insoluble in water.

Anal. W. P. Horne

The above figures are stated to be approximate only and all are likely to be low owing to the persistence of suspension and great difficulty in filtering during the analysis. The insoluble portion of the analysis is evidently a fine clay.

A further example of extracting salt by evaporation of alkaline water has been recorded at Teguida-n-Tisent in the central Sahara. A published photograph shows that the pits are shallow, set close together and approximately 0.6 m in diameter. They appear very similar to those observed in the Magado Crater. It seems likely that in addition to these two recorded instances salt has been extracted elsewhere by this method in areas where conditions are suitable and where commercial salt is either too costly or too difficult to obtain.

3

THE BASE OF BED IV.
WK HIPPO CLIFF, PDK TRENCH IV,
WK LOWER CHANNEL

M. D. LEAKEY

The three sites, WK Hippo Cliff, PDK IV and WK Lower Channel, are grouped together since they are approximately at the same stratigraphic level. WK Hippo Cliff and WK Lower Channel are within a conglomerate at the Beds III–IV interface and at PDK IV the artefacts lay on the surface of the conglomerate and extended into it for a depth of 10 cm. All three sites can be assumed to antedate Tuff IVA although there is no direct association. The sites in Lower Bed IV (Part 4) postdate Tuff IVA and are stratigraphically between Tuffs IVA and IVB.

WK Hippo Cliff (Wayland's Korongo)

In the course of exploring Beds III and IV during 1970 Philip Leakey noted a number of hippopotamus bones eroding from the III–IV interface at a locality on the south side of the gorge between WK and WK East. This site became known as WK Hippo Cliff. It was excavated between May and August 1970 when a trench was dug measuring 5.2 m long (east–west) and 3 m wide. The greater part of a hippo skeleton was recovered although most of the bones of the forelegs were missing and had probably eroded out many years ago. A number of artefacts and a few bones and teeth of other mammals, as well as crocodile and fish remains, were found with the skeleton.

The deposit in which the bones occurred consisted of a conglomerate filling a former river or stream channel. It contained several hundreds of small pebbles as well as about fifty cobbles, all of which appeared to be of natural origin. Some disturbance of the hippo skeleton had taken place, as may be seen from Fig. 3.1, but the extent to which the bones had been displaced was similar to that seen at the present time on the Serengeti plains when a large animal has been eaten by predators and scavengers. The cranium lay 1 m distant from the mandible and approximately 25 cm higher in the deposit. The top had been damaged but otherwise it is reasonably well preserved. It is of exceptionally large size and exhibits the raised orbits and other features characteristic of *H. gorgops*. The mandible was also well preserved, although somewhat crushed, and lay in an inverted position with the tooth row downwards.

A number of other skeletons of large mammals associated with tools have been found at Olduvai, but these have always been in clays, under conditions indicating that the animals had died as a result of becoming engulfed in mud, either accidentally or as a result of being driven into the swamps and butchered by humans. This is not the case for the WK hippopotamus, which lay in a pebbly conglomerate in a former water course and must, therefore, have died by other means. Moreover, owing to the nature of the deposit, it was impossible to determine whether this was a butchery site where the artefacts were contemporary and genuinely associated with the skeleton or whether they were washed into the same river bed by floods, in common with the fragmentary remains of other animals, all of which could have been derived from a nearby site. The presence of teeth representing a juvenile and a second adult hippopotamus may perhaps mean that this river was a hippo habitat and that the association of artefacts was accidental.

BASE OF BED IV

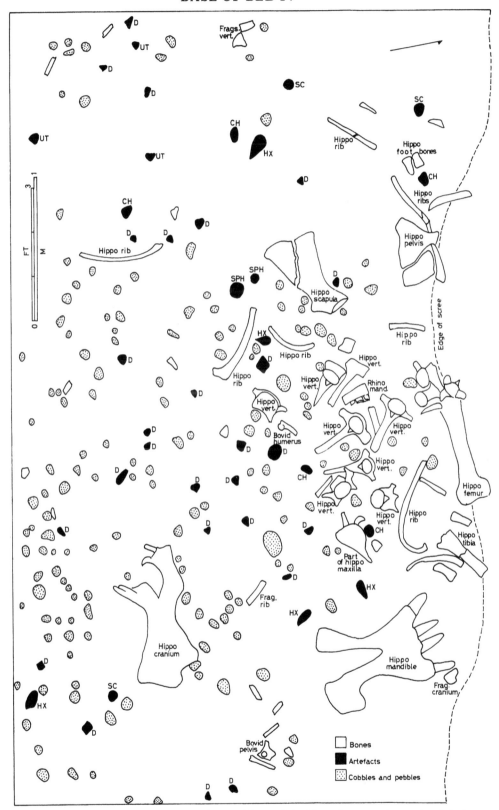

Fig. 3.1 WK Hippo Cliff: plan showing positions of the hippo bones and associated finds

ANALYSIS OF THE INDUSTRIES FROM WK HIPPO CLIFF

	Numbers	Percentages
Tools	14	27.5
Utilised material	5	9.8
Debitage	32	62.7
	51	100.00
Tools		
Bifaces	5	35.7
Choppers	5	35.7
Spheroids/subspheroids	2	14.3
Scraper, heavy-duty	1	7.1
Scraper, light-duty	1	7.1
	14	99.9
Utilised material		
Light-duty flakes, etc.	5	100.0
Debitage		
Whole flakes	9	28.1
Broken flakes	12	37.5
Core fragments	11	34.4
	32	100.0

The industry

The stone industry found with the hippo skeleton consists of fifty-one specimens, of which fourteen are tools, five utilised material and thirty-two debitage. Quartzite from Naibor Soit is the predominant material, but there is also a small number of lava specimens.

Tools

Bifaces (5 specimens, comprising 3 whole and 2 broken handaxes) The handaxes differ greatly in size and form. Two quartzite and one lava specimen have unusually sharp, slender points and are also in mint condition. Two other lava handaxes are weathered and abraded. The five specimens are as follows:

(a) ?Basalt. A small, slender, elongated ovate, plano-convex in cross section, indicating that it was probably made on a flake, although no part of the primary surface has remained (89 × 43 × 31 mm).

(b) Quartzite, probably tabular. An unusual specimen in that the tip is sharp and finely retouched, whereas the remainder of the implement is crudely flaked although it does not appear to be unfinished. The lateral edges near the tip are trimmed and sharp but near the butt and at the butt itself they consist of natural

ANALYSIS OF THE INDUSTRIES FROM WK HIPPO CLIFF

	Quartzite	Fine-grained quartzite	Gneiss, Feldspar, etc.	Phonolite	Nephelinite	Trachyte	Basalt/trachyandesite	Lava, indeterminate	Totals
Heavy-duty tools									
Bifaces, choppers, spheroids/subspheroids, scraper	9	1	—	—	—	—	3	—	13
Light-duty tools									
Scraper	1	—	—	—	—	—	—	—	1
Utilised material, light-duty									
Flakes, etc.	2	1	—	1	—	—	1	—	5
Debitage									
Whole flakes, broken flakes and core fragments	27	1	—	—	1	—	3	—	32
	39	3	—	1	1	—	7	—	51

cleavage planes at right angles to the upper and lower faces of the tool (98 × 56 × 33 mm).

(c) Made on a piece of tabular quartzite 26 mm thick. The upper and lower faces consist of the natural cleavage planes, but the circumference has been finely trimmed all round and has an exceptionally sharp edge. The tip is slenderly pointed and quite undamaged (183 × 90 × 36 mm).

(d) ?Basalt. The tip is missing and has broken off diagonally. One face shows cortical surface and the tool appears to have been made on a flake struck from a boulder. It is in fresh condition. Present measurements 126 × 93 × 47 mm.

(e) ?Trachyandesite. A small, flat, weathered specimen with the tip broken obliquely. Present measurements 93 × 69 × 29 mm.

Choppers (5 specimens)   All are made of quartzite and consist of side choppers with relatively wide, curved working edges; they closely resemble the quartzite side choppers that characterise the industry from site BK in Upper Bed II. The working edges have been bifacially flaked and have been chipped and blunted by use. This wear is particularly pronounced in one example where the entire working edge has been battered to a rounded surface. The measurements are as follows: 51 × 78 × 49 mm, 60 × 80 × 48 mm, 56 × 81 × 43 mm, 64 × 100 × 46 mm, 62 × 91 × 42 mm.

Spheroids/subspheroids (2 specimens)   The largest example is very weathered but appears to be made from fine-grained quartzite. It has been shaped over the whole surface and is symmetrical although the surface is uneven and not smooth. Measurements 116 × 104 × 43 mm. The second specimen is subspherical, measuring 58 × 52 × 43 mm. Projecting parts have been reduced by battering, but the tool has not been shaped in the manner of the preceding specimen. (It may be noted that these two spheroids were found lying together and that no others occurred at the site.)

Scraper, heavy-duty (1 specimen)   A roughly discoidal fragment of tabular quartzite flaked on opposite sides of the circumference from reverse directions. The edges are uneven and blunted (62 × 61 × 30 mm).

Scraper, light-duty (1 specimen)   Made on an end-struck flake measuring 37 × 30 mm. Part of the tip and one lateral edge have been unevenly trimmed.

Utilised material

Light-duty flakes and other fragments (5 specimens)   These five specimens exhibit chipping and blunting on the edges which appears to be due to use. Length/breadth measurements range from 68 × 57 mm to 29 × 23 mm.

Debitage

Whole flakes (9 specimens)   These consist of six end and three side flakes. They are irregular but most have relatively large, thick striking platforms. The end-struck flakes range in size from 83 × 54 mm to 33 × 26 mm and the side-struck flakes from 57 × 61 mm to 37 × 50 mm.

Broken flakes (12 specimens)   These include both butt and tip ends.

Core fragments (11 specimens)   There is one fragment of tabular quartzite measuring 70 × 61 × 46 mm, a heavily abraded piece of nephelinite and sundry other fragments. It is noticeable that the small angular fragments of quartzite that are so abundant in the WK main channel and at the nearby sites of WK East are not present.

PDK Trench IV (Peter Davies Korongo)

The PDK Gully lies on the south side of the gorge immediately northeast of the WK East sites and separated from them only by a narrow stretch of grass and scrub, sloping down into the gorge. Excavations were first undertaken in this gully during 1960 when Dr A. J. Sutcliffe of the British

Museum of Natural History excavated a partial hippopotamus skeleton from Bed I deposits at the mouth of the gully. The site PDK IV lies at the head of the gully and was first recorded in 1970 when Philip Leakey noted bifacial tools eroding from the base of Bed IV. The site of PDK I–III is also at the head of the gully, but to the southwest of PDK IV, on the eastern side of the slope. PDK, in fact, is one of a number of localities in the WK, WK East and PDK area of the gorge where artefacts can be seen eroding from basal Bed IV. In other localities the overburden was too great to permit excavation since Bed IV and overlying deposits had formed an almost vertical cliff approximately 7.5 m high. Natural erosion had partially reduced the overlying beds at PDK IV, including Tuff IVB, but an appreciable amount of overburden still had to be removed before the artefact-bearing level was reached. In the southeast part of the trench Tuff IVB had been partially removed by erosion but was still present in the southwest portion and in the adjacent cliff face, at a height of 2.5 m above the level of artefacts.

There is no direct relationship between this site and Tuff IVA, but it appears to antedate the deposition of the tuff since this occurs as a residual block in the nearby gully of WK East, although it is not present in the PDK gully. The block of Tuff IVA lies at a height of 1.5 m above the Beds III–IV contact and therefore at a higher level than the artefacts in PDK, Trench IV. Thus it would seem that the artefacts appear to be the earliest excavated in Bed IV and antedate the lowest levels at HEB and HEB East, all of which are subsequent to Tuff IVA.

The majority of the artefacts lay on the surface of a hard conglomerate 30 cm thick at the Beds III–IV interface, underlain by the eroded surface of Bed III. A few small flakes also occurred in the upper 10 cm of the conglomerate. This deposit extended beyond the limits of the trench and appeared to be more widespread than the conglomerate often found within former stream channels. The artefacts were not orientated in any particular direction and, with the exception of three choppers and a spheroid, were in exceptionally fresh condition. Two cleavers and several flakes were found lying edge-on, but their positions seem more likely to be due to the uneven underlying surface than to human agency. No evidence of battering or other utilisation on the upper edges could be discerned, such as was present on bone fragments embedded vertically at site MNK, in Bed II.

The area excavated at PDK Trench IV amounted approximately to 51 sq m. This yielded a total of 297 artefacts and some sparse faunal remains. The density of artefacts and bones is low compared with other sites in Bed IV and the area excavated may represent the peripheral part of a site or one used only for a short while.

The industry

Subdivision of artefacts on the basis of physical condition is not usually desirable, but in the material under review there is such a marked contrast between the mint-fresh series and the four rolled and weathered specimens that it is probably justifiable to treat them separately and consider the latter to be derived. The series of artefacts in fresh condition consists of 297 specimens, comprising thirty-eight tools, seventeen utilised pieces and 242 debitage.

Tools

Bifaces (15 handaxes, 6 cleavers) The fifteen complete handaxes are made from a variety of materials, quartzite, nephelinite, basalt and trachyandesite. Perhaps for this reason they are not standardised and vary in both shape and degree of retouch. Although a few are flaked entirely over both faces most of the lava specimens retain part of a primary flake scar. Even when this has been largely trimmed off, areas of cortical surface often remain on the dorsal aspect of the tools and it is likely that they were made from large flakes struck from boulders. Five of the six cleavers are made from flakes of nephelinite struck from boulders. They are particularly elegant and show that the makers had full control of this intractable material. Two examples were found lying on one lateral edge in nearly vertical positions. The upper and lower edges are chipped,

BASE OF BED IV

ANALYSIS OF THE INDUSTRIES FROM PDK TRENCH IV

	Numbers	Percentages
Tools	38	12.8
Utilised material	17	5.7
Debitage	242	81.5
	297	100.0
Tools		
Bifaces	21	55.3
Polyhedrons	4	10.5
Spheroids/subspheroids	3	7.9
Scrapers, light-duty	9	23.7
Punch	1	2.6
	38	100.0
Utilised material		
Pitted anvils/hammerstones	2	11.8
Cobbles	6	35.3
Light-duty flakes, etc.	9	52.9
	17	100.0
Debitage		
Whole flakes	29	12.0
Broken flakes	64	26.4
Core fragments	144	59.5
Cores (?)	5	2.1
	242	100.0

but no more so than in other specimens that were lying horizontally. (Measurements of these bifaces will be found in chapters 8 and 9.)

Polyhedrons (4 specimens) One is of quartzite and three of trachyandesite. They are very similar in size and particularly angular. With the exception of one lava specimen the edges are sharp, although there is noticeable chipping and wear. Length/breadth/thickness measurements range from $67 \times 44 \times 45$ mm to $47 \times 47 \times 35$ mm, with an average of $57 \times 46 \times 42$ mm.

Spheroids/subspheroids (3 specimens) These three small quartzite spheroids are angular and of approximately the same size, with an average of $47 \times 43 \times 40$ mm.

Scrapers, light-duty (9 specimens) All these tools are made from quartzite. Four are made on tabular fragments and the balance on irregular-shaped pieces or broken flakes. Seven are side scrapers, one an end scraper and one a hollow scraper. They conform to no particular pattern and it is evident that the pieces of quartzite were not expressly shaped prior to the chipping of the scraping edges. The side scrapers range in length and breadth from 38×55 mm to 20×24 mm, with an average of 32×39 mm.

Punch (1 specimen) There is one quartzite punch 43 mm long. One end is crushed and blunted and a longitudinal flake scar resembling a burin spall scar is present on one lateral edge. The opposite end is shattered.

Utilised material

Pitted anvils/hammerstones (2 specimens) A lava cobble and a heavily rolled end chopper measuring $100 \times 93 \times 42$ mm have pitted hollows. The chopper has pitting on both the upper and lower faces. On one side the pitting is shallow and dispersed over an area of 37×35 mm whilst on the opposite side it consists of a single pecked hollow 30×28 mm in diameter and 3 mm deep. The second specimen is an irregular-shaped cobble, measuring $101 \times 80 \times 77$ mm, with three hollows, one of which is possibly natural. The largest hollow is elongate and measures 47×25 mm in diameter and 5 mm in depth. The remaining two hollows are smaller and very irregular.

Cobbles (6 specimens) These lava cobbles show rough flaking and battering. One is of trachyte and five of basalt or trachyandesite. They range in size from $94 \times 68 \times 35$ mm to $58 \times 48 \times 46$ mm, with an average of $77 \times 55 \times 36$ mm.

Light-duty flakes, etc. (9 specimens) Nine flakes and other fragments of quartzite show chipping and blunting along the edges. The chipping is irregular. These specimens range in size from a broken flake 67 mm long and 51 mm wide to a fragment measuring 24×20 mm.

Debitage (29 whole flakes, 64 broken flakes and 149 sundry fragments)

In the complete flakes twenty-one are end struck and eight side struck. Four of the end flakes and three of the side flakes are typical large 'handaxe trimming' flakes. One of the large side flakes was

ANALYSIS OF THE INDUSTRIES FROM PDK TRENCH IV

	Quartzite	Fine-grained quartzite	Gneiss, feldspar, etc.	Phonolite	Nephelinite	Trachyte	Basalt/trachyandesite	Lava, indeterminate	Totals
Heavy-duty tools									
Bifaces, polyhedrons, spheroids/subspheroids	7	—	—	—	6	1	14	—	28
Light-duty tools									
Scrapers, punch	9	1	—	—	—	—	—	—	10
Utilised materials									
Heavy-duty, pitted anvils, cobbles	—	—	—	—	—	1	1	6	8
Light-duty, flakes, etc.	8	1	—	—	—	—	—	—	9
Debitage									
Whole flakes	13	—	—	1	2	2	5	6	29
Broken flakes	47	1	—	2	1	—	4	9	64
Core fragments	138	1	—	1	1	—	—	3	144
Cores (?)	5	—	—	—	—	—	—	—	5
	227	4	—	4	10	4	24	24	297

found edge-on, but as in the case of the two cleavers found in a similar position, there is no evidence of more extensive utilisation on the edge that was uppermost. Divergent flakes are the most common, with only four convergent and no parallel-sided specimens. Quartzite is the most common material among both whole and broken flakes. It is noticeable that there are only one whole and two broken flakes of nephelinite, whereas the manufacture of the one handaxe and six cleavers made from this material would require a minimum of forty-eight trimming flakes, not counting minor chips along the edges. Among the sundry fragments are five relatively large pieces of tabular quartzite that may be cores. The largest measures 116 × 80 × 43 mm.

The rolled artefacts

These consist of a damaged chopper that may have been pointed originally, two end choppers and a discoid. One of the end choppers is pitted, and the second has also possibly been used as an anvil.

WK Lower Channel

A description of site WK will be found on pp. 75–6 in the section describing the Upper Channel, the most important archaeological level.

The industry

The industry from the Lower Channel consists of 535 specimens, comprising ten tools, fifteen utilised pieces and 510 debitage. A number of pebbles and cobbles found at the same level are not included since they appeared to be natural components of the conglomerate.

Tools

Bifaces (5 specimens)   These tools are not described by Dr Roe on account of the small size of the sample. They consist of one finished and one unfinished handaxe of Engelosen phonolite, one unfinished basalt tip which appears to have broken during manufacture, an asymmetrical but probably finished quartzite handaxe and a mass-

BASE OF BED IV

ANALYSIS OF THE INDUSTRIES FROM WK LOWER CHANNEL

	Numbers	Percentages
Tools	10	1.9
Utilised material	15	2.8
Debitage	510	95.3
	535	100.0
Tools		
Bifaces	5	50.0
Choppers	3	30.0
Light-duty scrapers	2	20.0
	10	100.0
Utilised material		
Hammerstones	3	20.0
Cobbles and blocks	6	40.0
Light-duty flakes	6	40.0
	15	100.0
Debitage		
Whole flakes	32	6.3
Broken flakes	281	55.1
Core fragments	197	38.6
	510	100.0

ive tabular quartzite 'rough out' measuring 172 × 116 × 86 mm. The finished phonolite specimen lacks the extreme tip but is otherwise intact, although weathered. Part of one face consists of a natural cortical surface. Present measurements are 128 × 82 × 40 mm. The quartzite specimen is made on an end-struck flake and does not appear to be of the tabular material, but of a light-brown quartzite that occurs on an inselberg adjacent to the source of the tabular material. The tip is skewed relative to the overall length and the trimming is irregular (126 × 98 × 46 mm).

Choppers (3 specimens) These consist of: a small, bifacially flaked side chopper, made on a pebble, in which the butt is formed by cortical surface (39 × 44 × 24 mm); a basalt side chopper, also bifacially flaked, with a wide working edge extending round both ends (48 × 69 × 42 mm); and a crudely flaked trachyte side chopper in which the butt consists of either weathered surface or cortex (59 × 99 × 44 mm).

Scrapers, light-duty (2 specimens) A flake of fine-grained quartzite, measuring 30 × 30 mm, has been trimmed to a rounded end scraper. The second specimen consists of a thin fragment of tabular quartzite, sub-triangular in shape, in which one edge has been trimmed (39 × 33 mm).

Utilised material

Hammerstones (3 specimens) One cobble of indeterminate lava measuring 59 × 58 × 51 mm and one roughly cuboid fragment of quartzite measuring 77 × 65 × 51 mm have been battered and bruised. A third specimen consisting of an oblong cobble of trachyandesite has been heavily battered at one end and along one edge, causing some flakes to be removed. Three small depressions on one face may be either artificial pitting or natural (89 × 71 × 51 mm).

Cobbles and blocks (6 specimens) One fragment of quartzite, one of trachyte, three cobbles of trachyandesite or basalt and one possibly of welded tuff show evidence of battering and heavy utilisation. Sizes range from 80 × 69 × 53 mm to 78 × 65 × 55 mm, with an average of 77 × 66 × 55 mm.

Light-duty flakes, etc. (6 specimens) These consist of a broken phonolite flake and five fragments of quartzite, all of which exhibit chipping along the edges. The length/breadth measurements range from 48 × 39 mm to 28 × 16 mm, with an average of 40 × 29 mm.

Debitage (510 specimens)

These comprise thirty-two whole flakes, 281 broken flakes and 197 sundry angular fragments (these proportions are unusual, since in most Bed IV industries angular fragments greatly outnumber the broken flakes). Twenty-three of the complete flakes are end struck and nine side struck. They are mostly thick in cross section with pronounced bulbs and only a few specimens appear to be handaxe trimming flakes. Seven are convergent, twelve divergent and four approximately parallel-sided. The end flakes range in length/breadth from 89 × 46 mm to 20 × 16 mm, with an average of 47 × 31 mm and the side flakes from 39 × 60 mm to 16 × 25 mm, with an average of 26 × 39 mm.

ANALYSIS OF THE INDUSTRIES FROM WK LOWER CHANNEL

	Quartzite	Fine-grained quartzite	Gneiss or granite	Phonolite	Nephelinite	Trachyte	Trachyandesite or basalt	Lava, indeterminate	Totals
Heavy-duty tools									
Bifaces, choppers	2	—	1	2	—	1	2	—	8
Light-duty tools									
Scrapers	1	1	—	—	—	—	—	—	2
Utilised material									
Hammerstones	1	—	—	—	—	—	1	1	3
Cobbles and blocks	1	—	—	—	—	1	3	1	6
L/D flakes and other fragments	5	—	—	1	—	—	—	—	6
Debitage									
Whole flakes	23	—	—	3	—	—	6	—	32
Broken flakes	267	—	—	8	—	—	6	—	281
Core fragments	195	—	1	—	—	—	1	—	197
	495	1	2	14	—	2	19	2	535

4

LOWER BED IV.
HEB EAST, HEB AND HEB WEST, WK INTERMEDIATE CHANNEL

M. D. LEAKEY

The sites in Lower Bed IV are situated stratigraphically above Tuff IVA and below IVB. They consist of HEB East, HEB, HEB West Levels 4 to 2b and the Intermediate Channel at WK.

HEB East, HEB and HEB West (Heberer's Gullies)

The sites designated HEB lie on the south side of the gorge, near the museum. HEB East is reached by a footpath from the museum down over the edge of the gorge, and HEB is approximately 80 m to the west, along the edge of the gorge; HEB West is adjacent to it but further west. All except one of the levels exposed at these sites are within Lower Bed IV, below the siltstone referred to earlier. By inference they are also later than Tuff IVA since clasts of this tuff occur within the deposit beneath the lowest archaeological level at HEB West (Level 4). Some artefacts were also recovered from a limited area above the grey siltstone and were therefore in Upper Bed IV; they are described on pp. 113–15.

At HEB East only one archaeological level was present. Artefacts and faunal material occurred in an unusually dense concentration in a small channel 1.5 m above the Beds III–IV contact. Level 4 at HEB and HEB West is at a height of 1.2 m and appears to be either a continuation of the same channel or, more likely, a related channel of the same braided stream. Cobbles and pebbles in the channels in both areas were of similar rocks and mostly derived from Lemagrut and Sadiman.

The exceedingly abraded condition of some artefacts and fresh appearance of others in the HEB East conglomerate and in Level 4 at HEB and HEB West suggest that these may be mixed assemblages. The extent of abrasion, however, does appear to depend to some extent on the hardness of the rocks: phonolite and trachyandesite specimens are generally more heavily abraded than those of quartzite or trachyte.

HEB East

This site was first noted by Mr Kamoya Kimeu in 1969. Excavations were undertaken during June and July of that year and consisted of a single trench 5.5 m long (east–west) by 4.5 m wide that was cut into the deposits at the level from which the artefacts were eroding (Fig. 4.1). Digging was carried out in 10 cm spits and the remains in each spit were plotted. The implementiferous horizon proved to be 7.5 m below the base of the Masek Beds and was overlain by sterile deposits and an accumulation of recent rubble.

The artefacts were found in a single level 40 cm thick in the eastern end of the trench, where the lower 10 to 15 cm consisted of a conglomerate within a narrow channel (Figs 4.2–4.6). In the higher part of the deposit, 15 to 20 cm above the channel, the artefacts and faunal remains were sparsely scattered over the entire area of the trench. The channel was orientated approximately north–south and had a maximum width of 2 m. It occupied an area of approximately 16 sq m, narrowing to the south and truncated to the north by the present erosion slope. Specimens were densely concentrated throughout its length and those in the southern part of the channel were at a higher level than those at the northern end

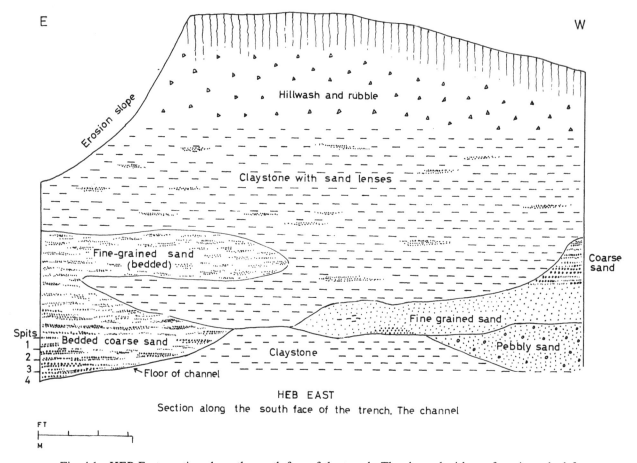

Fig. 4.1 HEB East: section along the south face of the trench. The channel with artefacts is to the left

although separated by only a few metres, suggesting that the gradient may have been quite steep. The deposit below the channel generally consisted of clay, although in certain areas it cut through sands. The filling was coarse to fine-grained sand, usually horizontally bedded, except in the deepest part where the sediments dipped and followed the slope of the channel bottom. The deposits overlying the artefact level were generally clayey with local lenses of sand, often current-bedded.

Cobbles associated with the artefacts in the channel are so common that it must be assumed that they were components of the conglomerate, although a few may have been manuports. Precambrian material of natural origin was not present, with the exception of two small pebbles out of a total of 355 unmodified cobbles and pebbles. Faunal remains were mostly very fragmentary and also scarce in relation to the number of artefacts. They included ten large freshwater bivalves, some of which were embedded vertically, partly open, with the sharp edges of the shells uppermost. Bovidae are the most numerous mammalian group and a well-preserved frontlet with horn cores of *Antidorcas recki* was found in the lower part of the channel. A large bovid and one of medium size also appear to be represented, but apart from the frontlet the remains are too fragmentary for identification.

The industry

This comprises 971 specimens, consisting of 194 tools, 98 utilised pieces and 679 debitage. The raw materials include Naibor Soit quartzite, phonolite, trachyte, basalt and trachyandesite. Surface specimens have been included in the following analysis since the deposits overlying the imple-

LOWER BED IV

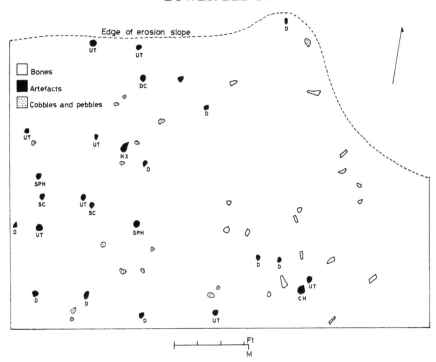

Fig. 4.2 HEB East: plan of finds above the channel (Spit 1)

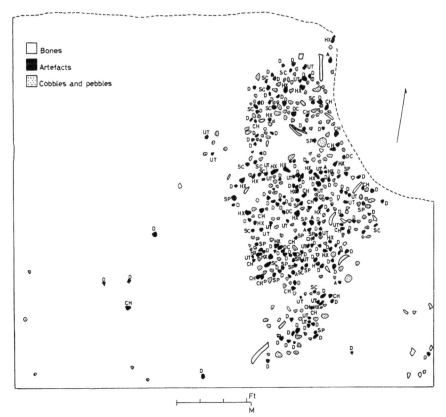

Fig. 4.3 HEB East: plan of finds in the upper part of the channel (Spit 2)

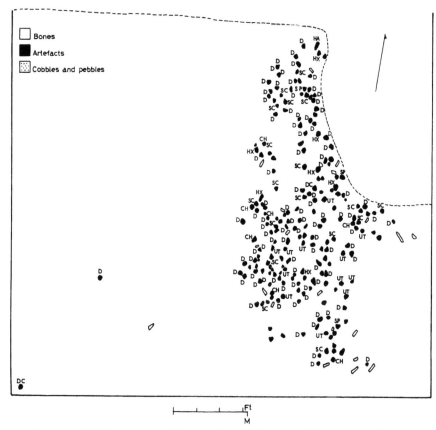

Fig. 4.4 HEB East: plan of finds in the middle part of the channel (Spit 3)

mentiferous level are barren, so that it is unlikely that extraneous material would be incorporated.

Tools

Bifaces (35 complete handaxes, 22 broken or unfinished specimens and 2 cleavers) Quartzite is the most common material, but a substantial number of specimens are also made from Engelosen phonolite and trachyandesite. The comparable use of all three materials for making handaxes is unusual since at other sites there is generally a definite preference for one particular rock. The complete handaxes show a considerable variation in size and technology, resembling in this respect the bifaces from the Developed Oldowan levels in Upper Bed II at SHK, BK and TK. (The series is analysed by Drs Roe and Callow in chapters 8 and 9.)

Choppers (20 specimens) These consist of nineteen side and one end chopper. In common with the bifaces, the whole series is variable in size and technology and there is also a marked difference in the degree of abrasion. Six of the side choppers are made on lava cobbles. The working edges are bifacially flaked and jagged, while the butts consist of rounded cortical surface. The quartzite specimens are made on blocks, flaked over most of the surface, and the butt ends generally show some degree of battering. The length/breadth/thickness measurements range from $110 \times 116 \times 60$ mm to $40 \times 45 \times 54$ mm. The single end chopper is made on a trachyandesite cobble. It is unifacial, flaked only from the cortex surface ($88 \times 75 \times 67$ mm).

Polyhedrons (18 specimens) The most common material is Naibor Soit quartzite, but there are also three examples of phonolite, one of nephelinite and four of trachyandesite or basalt. The utilisation of the cutting edges has resulted in

LOWER BED IV

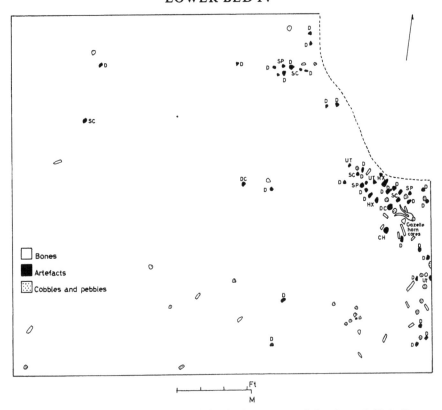

Fig. 4.5 HEB East: plan of finds in the lower part of the channel (Spit 4)

numbers of small step flakes being removed, besides battering and crushing that has completely blunted the edges so that some specimens resemble subspheroids. The range in size is from 103 × 88 × 62 mm to 44 × 33 × 30 mm, with an average of 74 × 65 × 57 mm.

Discoids (11 specimens) Although this is only a small series, seven different raw materials are represented, namely Naibor Soit quartzite, fine-grained quartzite, phonolite, nephelinite, trachyte, basalt and trachyandesite. There is considerable variation in size, from one large trachyte specimen weighing 1.4 kg to a quartzite specimen weighing only 50 g. With the exception of one example made on a cobble, in which a small area of cortical surface has remained, these tools have been radially trimmed on the whole circumference. In the majority, the trimming is bifacial, resulting in a biconvex cross section, but two examples are trimmed round the edge on one face only, from a flat under-surface, so that the cross section is plano-convex. The range in size is from 125 × 125 × 83 mm to 38 × 36 × 23 mm, with an average of 76 × 73 × 48 mm.

Spheroids/subspheroids (23 specimens) All are of quartzite except for one faceted trachyandesite example. Only four specimens are symmetrically rounded and smooth over the entire surface although in the majority the projecting ridges have been partially reduced. The series ranges in size from 87 × 80 × 73 mm to 39 × 35 × 35 mm, with an average of 60 × 53 × 45 mm.

Scrapers, heavy-duty (15 specimens) With the exception of two trachyandesite examples these tools are made from quartzite, five on tabular fragments, one on a flake and the balance on blocks. There are five side scrapers, seven in which the scraping edge is at one end of an oblong block, one hollow and two discoidal specimens. In the side scrapers the edges are both convex and nearly straight, while the trimming is usually

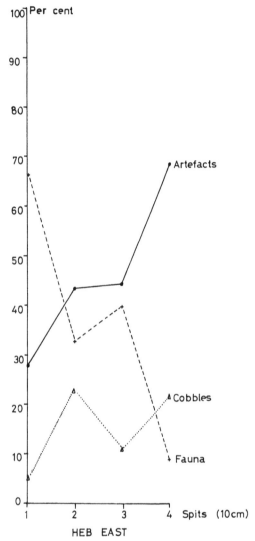

Fig. 4.6 HEB East: graph to show proportionate occurrences of artefacts, faunal remains and cobbles at different levels in the channel

steep and rather irregular. In two examples the edge has been entirely blunted. The side scrapers range in size from 60 × 114 × 27 mm to 64 × 75 × 31 mm. In the seven examples made on oblong blocks the scraping edges are convex and steeply trimmed but exhibit the same degree of irregularity. The range in size is from 81 × 61 × 53 mm to 54 × 45 × 39 mm. The single hollow scraper is made on a slab of quartzite measuring 101 × 44 × 30 mm. The notch is situated on one lateral edge and is 33 mm long and 6 mm deep. One of the two discoidal specimens is the largest in the series. It measures 109 × 107 × 54 mm. It is plano-convex in cross section and appears to be made on part of a large flake. The greater part of the circumference has been steeply trimmed from the flat under-surface and is also blunted by use. The second specimen is also on a piece of tabular quartzite that measures 72 × 58 × 34 mm and has likewise been steeply trimmed on the greater part of the circumference.

Scrapers, light-duty (34 specimens) These comprise side, end and hollow scrapers. With the exception of one hollow scraper of trachyandesite the series is of quartzite, including four made from the fine-grained variety.

Side. These are the most common type, with twenty-three examples. Two are made on flakes, two on pieces of tabular quartzite and the remainder on irregular fragments or broken flakes. The edges are generally curved and the trimming is rather irregular. The series includes four diminutive, double-edged examples, trimmed on opposite edges, the average length/breadth for the four being 17 × 27 mm. In two of the larger specimens, made on flakes, the working edges have been trimmed from the dorsal faces. The range in length/breadth for the whole series is from 46 × 61 mm to 18 × 23 mm with an average of 31 × 38 mm.

End. There are nine specimens with working edges at the tips of end flakes or other oblong fragments. The edges are generally convex and the trimming sometimes extends for a short distance along either lateral edge, forming a nosed scraper. The range in size is from 49 × 41 mm to 25 × 24 mm, with an average of 36 × 32 mm.

Hollow. There are only two examples and both are on flakes. The hollows are trimmed from the primary surface and measure 24 × 6 mm and 21 × 5 mm respectively.

Awls (3 specimens) These three specimens are all of quartzite. The points are formed by a trimmed notch on either side. Measurements are 40 × 30 mm, 27 × 32 mm and 28 × 22 mm.

Laterally trimmed flakes (5 specimens) Five quartzite flakes show some degree of trimming along the lateral edges but are quite variable in form. Length/breadth measurements range from 38 × 28 mm to 27 × 17 mm, with an average of 30 × 20 mm.

Outils écaillés (4 specimens)   Three examples are single ended and one is double ended. They exhibit the usual shallow, scaled flaking along the working edges. The range in length/breadth is from 38 × 34 mm to 23 × 21 mm.

Sundry tools (2 specimens)   Two small bifacially flaked tools do not conform to any of the above categories. They are triangular, with pointed tips and wide butt ends. It is possible that they represent atypical miniature bifaces but in method of flaking and general form they are unlike the small bifaces that occur in the Developed Oldowan at other sites. Measurements 36 × 38 × 20 mm and 36 × 35 × 16 mm.

Utilised material

Pitted anvils/hammerstones (3 specimens)   These consist of three lava cobbles pitted on opposite faces. The largest measures 123 × 83 × 57 mm. It has twin hollows on one face and triple on the opposite side. The hollows are adjacent and the two sets occupy areas of 36 × 25 mm and 35 × 22 mm. The central parts of the hollows are 6 mm deep. The weathering of the pitted areas differs noticeably from that on the surface of the cobble, being dark grey, while the surface is light grey and slightly reddened. In the second example (90 × 60 × 54 mm) there is a single hollow on one face and twin hollows on the opposite face. The single hollow measures 23 × 19 mm and is 6 mm deep, while the two adjoining hollows together measure 37 × 20 mm and have a maximum depth of 5 mm. The interiors are pitted and rough, but have weathered to the same colour as the surface of the stone. The third example consists of a circular cobble with flat upper and lower faces, each of which has an irregular hollow formed by a number of small, overlapping pits. These two areas measure approximately 35 × 25 mm and 26 × 17 mm.

Anvils (7 specimens)   One high-backed lava specimen is utilised on the circumference of a flat base. There are also three irregular blocks and three pieces of tabular quartzite, all of which show utilisation along edges that are approximately 90°. The range in size is from 81 × 89 × 49 mm to 61 × 56 × 50 mm.

Hammerstone (1 specimen)   This is an oval cobble of trachyandesite that shows bruising and battering at both ends. It measures 80 × 63 × 46 mm.

Cobbles (18 specimens)   Most of these specimens are of trachyandesite. There is no evidence of shaping and the flakes that have been removed appear to have resulted only from use. The cobbles range in size from 154 × 115 × 47 mm to 41 × 36 × 20 mm, with an average of 70 × 58 × 35 mm.

Light-duty flakes, etc. (69 specimens)   These consist of whole and broken flakes and other fragments with utilisation of the edges. Some of the larger core fragments also show battering of parts of the surface, such as is seen in hammerstones. The flakes can be subdivided into those with approximately straight, convex or notched edges. The series with straight edges amounts to twenty-nine specimens; except for six examples, all are broken. The type of utilisation varies greatly, from fine, even wear to irregular coarse chipping that has considerably damaged the edges. In the series with convex edges (twelve specimens) the type of chipping is also variable. It is confined to one face and, in the case of flakes, is present on the dorsal surface only. The notches in the third group (fifteen specimens) are smaller and less well defined than in the hollow scrapers and they do not appear to have been hollowed out intentionally. They vary in width from 15 to 8 mm and all are shallow. The whole series of utilised flakes ranges in size from 86 × 70 mm to 22 × 20 mm, with an average of 36 × 26 mm. The remaining thirteen specimens are irregularly shaped core fragments.

ANALYSIS OF THE INDUSTRIES FROM HEB EAST

	Numbers	Percentages
Tools	194	20.0
Utilised material	98	10.1
Debitage	679	69.9
	971	100.0
Tools		
Bifaces	59	30.4
Choppers	20	10.3
Polyhedrons	18	9.3
Discoids	11	5.7
Spheroids/subspheroids	23	11.9
Scrapers, heavy-duty	15	7.7
Scrapers, light-duty	34	17.5
Awls	3	1.6
Laterally trimmed flakes	5	2.6
Outils écaillés	4	2.0
Sundry tools	2	1.0
	194	100.0
Utilised material		
Pitted anvils/hammerstones	3	3.1
Anvils	7	7.1
Hammerstone	1	1.0
Cobbles	18	18.4
Light-duty flakes, etc.	69	70.4
	98	100.0
Debitage		
Whole flakes	116	17.1
Broken flakes	421	62.0
Core fragments	134	19.7
Cores	8	1.2
	679	100.0

Debitage (679 specimens)
The debitage comprises 116 whole flakes, 421 broken flakes, 134 sundry fragments and eight cores.

Whole flakes   Naibor Soit quartzite is the most common material followed by trachyandesite. There are also a few flakes of Engelosen phonolite, trachyte and basalt. Sixty-two per cent of the flakes are end struck. Both convergent and divergent flakes are present and the whole series tends to be irregular. The end flakes vary in length/breadth from 83 × 77 mm to 15 × 12 mm, with an average of 42 × 32 mm, and the side flakes from 84 × 85 mm to 20 × 21 mm, with an average of 41 × 51 mm.

Broken flakes   The proportions of the various raw materials are similar to those of the whole flakes. Bulbar ends are not so common as tip ends and medial sections.

Sundry fragments   These consist of blocks and irregular fragments, some of which exhibit a few random negative flake scars.

Cores   These comprise a cuboid block of quartzo-feldspathic gneiss, a fragment of trachyte and six blocks of quartzite. In most of the specimens flakes have been detached from a number of different directions. The cores range in size from 106 × 95 × 81 mm to 67 × 60 × 51 mm, with an average of 81 × 72 × 55 mm.

HEB and HEB West

The first trenches were dug at HEB during 1962 by the late Dr John Waechter of the London Institute of Archaeology. More extensive excavations were carried out by the writer in 1969, 1970 and 1971 when investigation of Beds III and IV was in progress. The existence of several Acheulean industries at different levels was revealed but their relationship to other Acheulean sites in Bed IV was not established until 1972 when Dr F. H. Brown of the University of Utah was visiting Olduvai to collect palaeomagnetic samples. He recognised the importance of a grey siltstone that occurs at these sites and elsewhere in Bed IV for correlating archaeological levels at various localities. More detailed work by Dr Hay revealed that in certain areas, for example at JK, there are two siltstones. But it has been established that one or both always occur in Lower Bed IV and are not present in Upper IV, above Tuff IVB.

The excavations at HEB by Dr Waechter consisted of two trenches 5 × 2 and 6 × 4 m, at right angles to one another (Figs 4.7 and 4.8, see pg. viii). Two implementiferous levels were found, the lowest being the stream channel already referred to (Level 4), which continues as far west as HEB West, with a higher level (Level 3) present only at HEB. When work on the site was resumed by the writer in 1969 a small trench was excavated

LOWER BED IV

ANALYSIS OF THE INDUSTRIES FROM HEB EAST

	Quartzite	Fine-grained quartzite	Gneiss, feldspar, etc.	Phonolite	Nephelinite	Trachyte	Basalt/trachyandesite	Lava, indeterminate	Totals
Heavy-duty tools Bifaces, choppers, polyhedrons, discoids, spheroids/subspheroids, scrapers	80	2	1	16	2	5	40	—	146
Light-duty tools Scrapers, awls, laterally trimmed flakes, *outils écaillés*, sundry tools	42	5	—	—	—	—	1	—	48
Utilised material, heavy-duty Pitted anvils, anvils, hammerstones, cobbles	6	—	—	1	—	—	22	—	29
Utilised material, light-duty Flakes, etc.	49	6	1	8	—	2	3	—	69
Debitage Whole flakes	64	—	—	11	—	2	38	1	116
Broken flakes	377	3	—	12	—	2	27	—	421
Core fragments	127	—	2	1	1	—	3	—	134
Cores	6	—	1	—	—	1	—	—	8
	751	16	5	49	3	12	134	1	971

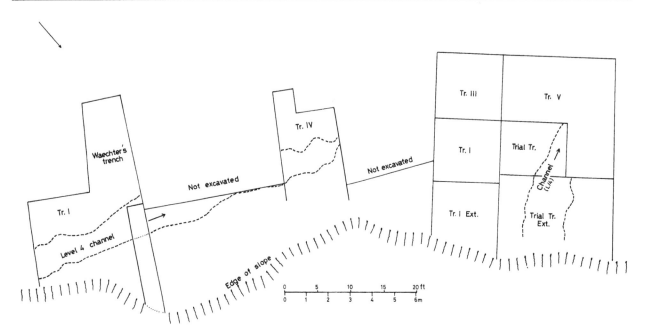

Fig. 4.7 HEB and HEB West: plan of excavations and channels in Level 4

at HEB to confirm the positions of the two artefact-bearing levels and to expose fresh sections through the deposits; the western part of the site was excavated more extensively. A trial trench dug by Dr Waechter in this area had failed to locate the source of a number of rather crude bifaces that had eroded on to the slope. The 1969 trenches revealed that they were derived from one or more of three artefact-bearing levels stratigraphically higher than Levels 3 and 4.

The sequence of deposits at HEB and HEB West, in descending order, was as follows.

HEB		*HEB West*	
(a)	1.8 m of Masek Beds	(a)	40 cm of sand
		(b)	approx. 1 m of claystone, merging into 75 cm of sand at base
		(c)	30–50 cm of dark grey claystone
		(d)	approx. 50 cm of sand
		(e)	10–20 cm of claystone, *Archaeological Level 1*
(b)	5 m of red-brown clay with lime-filled cracks and root casts, particularly in the upper part	(f)	30–60 cm of grey siltstone
(c)	1 m of reddish-black sand	(g)	10–30 cm of coarse quartz sand
(d)	20–50 cm of grey siltstone	(h)	10–60 cm of claystone
(e)	approx. 1.5 m of brown clay with sandy lenses containing *Archaeological Level 3*, 45 cm from the base	(i)	50 cm of sand with *Archaeological Level 2a* at the base
(f)	10–20 cm of conglomerate in small channel, *Archaeological Level 4*	(j)	20–60 cm of fine-grained quartz sand, *Archaeological Level 2b*
(g)	channel 1 m deep cut into (h); filling of cross-bedded red sand, coarser at base, with clasts of Tuff IVA	(k)	10–20 cm conglomerate in small channel, *Archaeological Level 4*
(h)	1 m of dark brown clay with sand lenses, cut into by channel (g)	(l)	cross-bedded red sand with clay lenses, maximum thickness 2 m, cut into by channel (k)
(i)	1 m red-brown claystone, top of Bed III	(m)	red-brown claystone, top of Bed III

The archaeological levels at the three localities, HEB East, HEB and HEB West can be correlated as follows:

HEB East	*HEB*	*HEB West*
Missing	*Missing*	*Level* 1
Missing	*Missing*	*Level* 2a
Missing	*Missing*	*Level* 2b
Missing	*Level* 3	*Missing*
?*Level* 4	*Level* 4	*Level* 4

The grey siltstone and the conglomerate of Level 4 can be identified at both the sites but the uppermost part of Bed IV, overlying all the archaeological levels, as well as the Masek Beds, both of which occur at HEB, have been removed by erosion to the west. The stratigraphy below the siltstone is also more complicated at HEB West, and the three artefact-bearing horizons (Levels 1, 2a and 2b) that occur at this site are not present at HEB. Conversely, Level 3 of HEB is missing at HEB West. Sizeable fragments of Tuff IVA are to be seen at both sites in the deposits underlying Level 4, indicating that all the archaeological levels postdate the deposition of this tuff.

In analysing the cultural and faunal material

from these sites, each of the artefact-bearing levels has been treated separately but in the faunal analysis all the material has been pooled and attributed to lower Bed IV.

HEB Level 4. The channel

This level consists of a small, shallow stream channel marked by a spread of pebble conglomerate approximately 3 m wide extending from HEB to HEB West. At HEB it is at a depth of 7 m below the base of the Masek Beds. In general the channel is aligned approximately southeast–northwest and appears to have drained to the northwest, but slight bends occur, such as that uncovered at HEB West (Fig. 4.7). Some artefacts and fossils were found in the conglomerate; they are in various stages of abrasion, the heavy-duty classes of tools being the most affected. In view of the differences in physical condition, this assemblage may not be homogeneous. The greatest number of artefacts and fossils were recovered from an area at HEB West where the stream bed curved to the southwest and it seems likely that they were caught up at the bend. In this area, too, the base of the channel, beneath the conglomerate, was a brilliant orange-red when first uncovered but the colour quickly faded after exposure.

The components of the conglomerate were largely pebbles and cobbles of basalt and trachyandesite from Lemagrut with a smaller proportion of nephelinite from Sadiman. There was a suggestion that some oblong artefacts and a few long-bone fragments were orientated either parallel to the course of the channel (120° west) or at right angles to it, but for the most part the alignment appeared to be haphazard.

The industry
This consists of 1110 specimens comprising 122 tools, 75 utilised pieces and 913 debitage.

Tools

Bifaces (21 specimens) These tools have been analysed by Drs Roe and Callow, in chapters 8 and 9. There are fourteen complete, three broken and one probably unfinished handaxe as well as three cleavers. Three specimens are of quartzite, one of nephelinite, six of trachyte, three of phonolite and eight of trachyandesite or basalt. Except for the single unfinished handaxe all the bifaces appear to be finished tools. In some of the handaxes the tips have been trimmed to sharp points and the butts have either been flaked all round or else retain only a small area of cortex. Both upper and lower surfaces have usually been extensively flaked and the scar count is relatively high. In other specimens the flaking is less intensive and the scars tend to be deeper. The variation in size, form and technique of manufacture confirms the belief suggested by the difference in physical condition of the assemblage as a whole that it is probably mixed. The three cleavers vary considerably in size but are similar in form, with straight transverse edges, rather narrow in relation to the width of the specimens.

Choppers (11 specimens) All are made of quartzite except for one heavily rolled trachyandesite specimen that may be derived. It is a unifacial end chopper from which only a minimum of flakes has been detached. One of the quartzite choppers is double edged and the balance are side choppers, including an unusually large specimen ($98 \times 124 \times 74$ mm) with a wide, curved working edge extending round both ends as well as along the principal edge. The remaining specimens are considerably smaller, varying between $53 \times 68 \times 37$ mm and $34 \times 48 \times 27$ mm, with an average of $45 \times 53 \times 34$ mm.

Polyhedrons (4 specimens) Two examples are of quartzite and two of trachyandesite or basalt. A number of ridges are present on all four specimens formed by multi-directional flaking; they are usually chipped and blunted. One of lava and one of quartzite are heavily abraded while the other two are in fresh condition. Measurements range from $82 \times 68 \times 48$ mm to $41 \times 37 \times 33$ mm, with an average of $65 \times 48 \times 46$ mm.

Discoids (8 specimens) Five are of quartzite,

one of nephelinite and two of trachyandesite or basalt. Two of the lava specimens are exceedingly rolled; they are both larger and flatter than the rest of the series and measure 81 × 73 × 29 mm and 59 × 57 × 21 mm respectively. The five quartzite specimens are unusually small, one being only 32 mm in maximum diameter. Cross sections may be either biconvex or plano-convex. The remaining discoids are either in fresh condition or only moderately abraded, they range in size from 56 × 50 × 30 mm to 32 × 31 × 19 mm, with an average of 39 × 36 × 24 mm.

Spheroids/subspheroids (11 specimens) All these tools are made of quartzite. They are angular and small, the largest specimen having a mean diameter of only 43 mm. Although they are crudely flaked and generally asymmetrical, most of the projecting parts have been battered and reduced to some extent. Sizes range from 51 × 45 × 35 mm to 30 × 30 × 24 mm, with an average of 40 × 36 × 30 mm.

Scrapers, heavy-duty (5 specimens) Four are of quartzite and one of basalt or trachyandesite. Two of the quartzite specimens are massive and steeply trimmed from flat under-surfaces. One is roughly circular although only one edge is trimmed, and measures 108 × 105 × 49 mm. The second consists of a piece of 'sugary' quartzite evenly trimmed to a convex working edge along one side (64 × 138 × 35 mm). The third, made on a rather thick fragment of tabular quartzite, also has an evenly trimmed, but shorter and convex working edge (59 × 54 × 35 mm). The remaining two are hollow scrapers. One is a triangular fragment of tabular quartzite on which there are two notches, one 40 mm wide and 7 mm deep and the second 20 mm wide and 3 mm deep (97 × 76 × 22 mm). The second hollow scraper is on a roughly discoidal fragment of basalt or trachyandesite with a notch 22 mm wide and 5 mm deep (67 × 64 × 26 mm).

Scrapers, light-duty (44 specimens) Four are of fine-grained quartzite, one of trachyte, one of phonolite, one of trachyandesite or basalt and the remainder of Naibor Soit quartzite. Side scrapers are the most common form, amounting to twenty-five specimens. There are also seven end, three nosed, three hollow and six roughly discoidal examples. Only six are made on flakes, the majority being on sundry fragments, including pieces of tabular quartzite. In the side scrapers the working edges vary from almost straight to convex. Most are rather uneven with small projections and indentations. Blunting and chipping is usually present. Length/breadth measurements range from 52 × 62 mm to 16 × 21 mm, with an average of 25 × 36 mm. Similarly, the working edges of the end scrapers vary from straight to convex and are also uneven. One example is made on a flake of fine-grained pink quartzite, of which the source is unknown. The length/breadth measurements range from 55 × 43 mm to 32 × 19 mm, with an average of 34 × 28 mm. The discoidal scrapers are to some extent angular and not symmetrical, but are trimmed on the circumference and are more nearly discoidal than the perimetal scrapers found in the Oldowan and Developed Oldowan of Bed II, some of which are triangular but trimmed round the entire circumference. Three examples are made on tabular pieces of quartzite and one on a split quartzite pebble. In one small specimen the working edges on opposite sides have been trimmed from reverse directions. The length/breadth measurements range from 52 × 50 mm to 19 × 19 mm, with an average of 32 × 30 mm. Three scrapers, that are in fact side scrapers, have pronounced 'noses' in the central part of the working edges. They are evenly trimmed, with a slight notch on either side. Measurements are 27 × 35 mm, 31 × 39 mm and 20 × 24 mm. The largest of the three hollow scrapers, which is made on a lava flake measuring 54 × 57 mm, exhibits a notch 37 mm wide and 6.5 mm deep. This has been trimmed entirely from the primary flake surface and is situated on one lateral edge. The remaining two specimens consist of angular fragments of quartzite measuring 26 × 32 mm and 26 × 34 mm, in which there are notches 10 mm wide and 2 mm deep and 16 mm wide and 2.5 mm

LOWER BED IV

ANALYSIS OF THE INDUSTRIES FROM HEB LEVEL 4

	Numbers	Percentages
Tools	122	11.0
Utilised material	75	6.7
Debitage	913	82.2
	1110	99.9
Tools		
Bifaces	21	17.2
Choppers	11	9.0
Polyhedrons	4	3.3
Discoids	8	6.6
Spheroids/subspheroids	11	9.0
Scrapers, heavy-duty	5	4.1
Scrapers, light-duty	44	36.1
Burin	1	0.8
Awls	8	6.5
Laterally trimmed flakes	9	7.4
	122	100.0
Utilised material		
Hammerstone	1	1.3
Cobbles and blocks	10	13.3
Light-duty flakes, etc.	64	85.3
	75	99.9
Debitage		
Whole flakes	94	10.3
Broken flakes	465	50.9
Core fragments	354	38.8
	913	100.0

deep respectively. These have also been trimmed from one direction only.

Burin (1 specimen) There is one well-made quartzite burin. It measures 50 × 38 mm and the working edge is 16 mm wide with multiple spall scars on either side and also traces of utilisation.

Awls (8 specimens) These tools are of quartzite and the points are sharp but thick-set, generally formed by a trimmed notch on either side. In two specimens they are skewed sideways, to the right, but in the balance they are straight and central to the tool. The length/breadth measurements range from 40 × 28 mm to 33 × 23 mm, with an average of 35 × 25 mm.

Laterally trimmed flakes (9 specimens) Unlike the scrapers and awls, which are made on various fragments, these tools are all on whole or broken flakes; one is of phonolite, two of trachyandesite or basalt and six of quartzite. They vary in form and in the extent of trimming but the feature common to all is the presence of retouch on one or both lateral edges. Most are bluntly pointed. Length/breadth measurements range from 76 × 40 mm to 33 × 21 mm, with an average of 45 × 31 mm.

Utilised material

Hammerstone (1 specimen) A quartzite cobble measuring 81 × 75 × 61 mm shows battering and bruising on various parts of the surface.

Cobbles and blocks (10 specimens) Two quartzite cobbles, three of trachyandesite or basalt, one of trachyte and one of nephelinite show battering and flaking at the ends which appears to be the result of heavy use. The specimens are very similar in size and have average measurements of 86 × 53 × 40 mm. There are also three blocks of quartzite in which a number of thick-set edges are battered. Measurements are 98 × 98 × 67 mm, 91 × 89 × 49 mm and 60 × 47 × 45 mm.

Light-duty flakes, etc. (64 specimens) Two nephelinite, five phonolite, one trachyte, nine basalt or trachyandesite and forty-seven quartzite whole or broken flakes show irregular chipping on the edges. Many are considerably abraded, particularly the lava specimens. The chipping sometimes forms notches and appears to be the result of utilisation except in the case of very thin flakes in which it may be due partly to stream action in the conglomerate. Length/breadth measurements range from 65 × 53 mm to 30 × 16 mm, with an average of 39 × 30 mm.

Debitage (913 specimens)

The debitage consists of ninety-four complete flakes, 465 broken flakes and 354 core fragments. There are equal numbers of complete quartzite and lava flakes, although in the broken flakes and core fragments quartzite is far more common than lava. Fifty-five of the complete flakes are end struck and thirty-nine side struck. Only two small flakes are convergent, all others being divergent.

ANALYSIS OF THE INDUSTRIES FROM HEB LEVEL 4

	Quartzite	Fine-grained quartzite	Gneiss, feldspar, etc.	Phonolite	Nephelinite	Trachyte	Basalt/trachyandesite	Lava, indeterminate	Totals
Heavy-duty tools									
Bifaces, choppers, polyhedrons, discoids, spheroids/subspheroids, scrapers	35	—	—	3	2	6	14	—	60
Light-duty tools									
Scrapers, awls, burins, laterally trimmed flakes,	52	4	—	2	—	1	3	—	62
Utilised material, heavy-duty									
Hammerstones, Cobbles and blocks	6	—	—	—	1	1	3	—	11
Utilised material, light-duty									
Flakes, etc.	47	—	—	5	2	1	9	—	64
Debitage									
Whole flakes	47	—	—	1	1	—	45	—	94
Broken flakes	370	—	—	11	2	2	80	—	465
Core fragments	346	—	—	—	—	4	4	—	354
	903	4	—	22	8	15	158	—	1110

End flakes range in size from 85 × 46 mm to 19 × 15 mm, with an average of 41 × 32 mm, and side flakes from 60 × 94 mm to 13 × 15 mm, with an average of 31 × 40 mm. Large flakes such as are present in Levels 2a, 2b and 3 do not occur.

HEB Level 3

As noted earlier this horizon of artefacts and associated faunal remains (Level 3) occurred only at HEB and did not extend westwards to HEB West. It consisted of 20 to 30 cm of greyish-red sand, sloping gently down to the east. Where excavated, it extended for a distance of 7.5 m east–west. The north–south dimension can only be estimated since the present erosion slope cuts into the deposit, and it was there that the level of tools was exposed in the first place. Some flakes and bone fragments can still be seen projecting from the face of the north–south trench excavated by Dr Waechter, indicating that the artefact level probably extends further to the south.

One notable feature of the artefacts from this level is the abundance of green phonolite from Engelosen volcano, 9 km to the north. Not only are most of the bifaces made from this material but there are also plentiful trimming flakes, large unretouched flakes and substantial blocks that seem to represent raw material brought to the site but never used. No similar abundance of the phonolite is known elsewhere, even at sites closer to the source. The refinement and high quality of workmanship in the handaxes and cleavers is also most striking. They are not equalled from any other site in Bed IV, the only comparable standard of excellence being in the quartzite bifaces from the Masek Beds at FLK and in those from HK and TK Fish Gully, both of which are in disturbed deposits of more recent date.

Flaked and utilised bones were recovered from other levels excavated at HEB and HEB West, but two of the most remarkable were recovered from Level 3. These are a handaxe made from elephant bone and half an elephant pelvis in which the acetabulum has been extensively battered and damaged, probably as a result of use as a mortar. Descriptions of these tools will be found in Appendix A. Handaxes and some cleavers form an unusually high percentage among the tools, and light-duty tools are scarce, with only eight out of the total of 108.

The industry

This consists of 905 specimens, comprising 108 tools, forty-one utilised pieces and 756 debitage.

Tools

Bifaces (81 specimens; Pl. 8) These consist of thirty-six complete handaxes and five cleavers, twenty-five unfinished and fifteen broken specimens. They are described by Drs Roe and Callow in chapters 8 and 9. Forty-eight examples are made of Engelosen phonolite, thirty of quartzite and only three from trachyandesite or basalt, in contrast to the series from Levels 2a and 2b at HEB West, in which these last two are the most common materials. The trimming of the bifaces, particularly of the handaxes, is elaborate with unusually flat, shallow scars; the tips are slender and the cutting edges virtually straight. The handaxes vary considerably in size but the phonolite series shows a consistent preference for elongated oval or lanceolate forms. In the quartzite series the tips are also generally finely trimmed but in some examples there is an untrimmed area at the butt ends, consisting of a natural cleavage plane in the tabular quartzite. Owing to the intensive flaking of both faces it is difficult to determine whether the phonolite tools were made on flakes, but the fact that many tend to be plano-convex in cross section suggests that they are on flakes rather than 'cores'. The elaborate trimming of these handaxes and cleavers gives an unusually high average scar count. Among the broken handaxes are two split longitudinally, but transverse fractures are more common. There are six tip ends, four butt ends, two with only the tips missing and one fragment which has broken off from a lateral edge. Some of the twenty-five examples classed as unfinished are partly trimmed and appear to have been discarded owing to faulty material.

Choppers (2 specimens) Both examples are quartzite side choppers with sharp and very jagged working edges, bifacially trimmed. In one specimen the working edge extends for about two-thirds of the circumference and the butt end has also been battered ($72 \times 78 \times 61$ mm). The second chopper is very similar but has a relatively shorter working edge ($72 \times 81 \times 62$ mm).

Discoids (4 specimens) These are all of quartzite and include a thick, massive specimen that might be classed as an angular spheroid were it not for the presence of a radially trimmed cutting edge on the circumference. There is also one rather oblong example among the smaller specimens. Length/breadth/thickness measurements range from $93 \times 90 \times 68$ mm to $51 \times 50 \times 31$ mm, with an average of $62 \times 57 \times 40$ mm.

Spheroids/subspheroids (10 specimens) This series is very variable in size. It includes one large, symmetrical specimen in which the entire surface has been worked over and smoothly rounded. In two large quartzite examples the projecting parts have also been battered and rounded off, but a phonolite example is particularly angular. Two of the smaller quartzite spheroids are broken. The range in size is from $99 \times 96 \times 82$ mm to $31 \times 29 \times 21$ mm, with an average of $71 \times 66 \times 53$ mm.

Scrapers, heavy-duty (3 specimens) The largest example is made on a slab of quartzite measuring $145 \times 87 \times 55$ mm. It is an end scraper, evenly trimmed to a steep, curved working edge, an unusual form among heavy-duty scrapers from Bed IV. There is also a smaller end scraper (63×56 mm) similarly made on a tabular piece of quartzite. The third example consists of a phonolite side scraper made on a flake, with a long, carefully trimmed convex working edge (71×93 mm).

ANALYSIS OF THE INDUSTRIES FROM HEB LEVEL 3

	Numbers	Percentages
Tools	108	11.9
Utilised material	41	4.5
Debitage	756	83.5
	905	99.9
Tools		
Bifaces	81	75.0
Choppers	2	1.9
Discoids	4	3.7
Spheroids/subspheroids	10	9.3
Scrapers, heavy-duty	3	2.7
Scrapers, light-duty	7	6.5
Outil écaillé	1	0.9
	108	100.0
Utilised material		
Pitted anvils/hammerstones	3	7.3
Hammerstone	1	2.4
Cobbles and blocks	17	41.5
Light-duty flakes	20	48.7
	41	99.9
Debitage		
Whole flakes	181	23.9
Broken flakes	471	62.3
Core fragments	102	13.5
Cores	2	0.3
	756	100.0

Scrapers, light-duty (7 specimens) There are six side scrapers with nearly straight or slightly convex working edges. One is made on a phonolite flake and the balance on fragments of quartzite. The range in size is from 40 × 58 mm to 23 × 31 mm, with an average of 29 × 43 mm. The remaining specimen is an end scraper made on a broken quartzite flake. The working edge is steeply trimmed and forms the broadest part of the tool, which tapers to a narrow butt (64 × 43 mm).

Outil écaillé (1 specimen) This consists of part of a fine-grained quartzite pebble with cortex on one face. It is narrow relative to its length (34 × 16 mm) and both ends show scaled utilisation.

Utilised material

Pitted anvils/hammerstones (3 specimens) Three lava cobbles show small pecked hollows. In two examples there are single hollows, but the third has two pits, one on either face. The measurements of the cobbles and of their respective pits are as follows: (a) 83 × 73 × 38 mm with a pit 23 × 10 mm in diameter and 2 mm deep; (b) 82 × 70 × 53 mm with a pit 17 × 12 mm in diameter and 2.5 mm deep; (c) 81 × 61 × 40 mm with two pits 14 × 11 mm in diameter and 2 mm deep and 16 × 11 mm in diameter and 1.5 mm deep.

Hammerstone (1 specimen) A roughly discoidal cobble of trachyandesite has been battered on the greater part of the circumference (75 × 70 × 51 mm).

Cobbles and blocks (17 specimens) Six broken lava cobbles and eleven blocks of quartzite have been roughly flaked and battered, apparently by use. They range in size from 107 × 81 × 59 mm to 51 × 50 × 40 mm.

Light-duty flakes, etc. (20 specimens) These consist of irregular fragments of quartzite in which one or more edges show chipping and blunting, sometimes forming slight notches. They range in length/breadth from 67 × 43 mm to 28 × 23 mm, with an average of 45 × 32 mm.

Debitage

This consists of 181 complete flakes, 471 broken flakes, 102 sundry fragments and two cores.

Complete flakes These are almost entirely of Engelosen phonolite and quartzite, with only thirteen of other materials. In the quartzite series, end flakes outnumber side flakes, with 53 as against 37 but in the phonolite flakes the proportions are reversed, with 52 side flakes and 35 end flakes. Many phonolite flakes are exceedingly thin and have wide, flat bulbs of percussion, typical of cylinder hammer technique. Some flakes are of considerable size, clearly too large for normal handaxe trimming flakes, and are probably unused material brought from the quarry site. With the exception of three side and five end flakes all the phonolite specimens are divergent. In the quartzite series fourteen are convergent and the balance divergent. The

LOWER BED IV

ANALYSIS OF THE INDUSTRIES FROM HEB LEVEL 3

	Quartzite	Fine-grained quartzite	Gneiss, feldspar, etc.	Phonolite	Nephelinite	Trachyte	Basalt/trachyandesite	Lava, indeterminate	Totals
Heavy-duty tools									
Bifaces, choppers, discoids, spheroids/subspheroids, scrapers	45	—	?1	51	—	—	3	—	100
Light-duty tools									
Scrapers, *outil écaillé*	6	1	—	1	—	—	—	—	8
Utilised material, heavy-duty									
Pitted anvils/hammerstones, hammerstones, cobbles and blocks	11	—	—	—	—	—	10	—	21
Utilised material, light-duty									
Flakes, etc.	17	3	—	—	—	—	—	—	20
Debitage									
Whole flakes	90	—	—	86	—	—	5	—	181
Broken flakes	323	10	—	134	—	—	4	—	471
Core fragments	81	—	1	16	—	—	3	1	102
Core	1	—	—	1	—	—	—	—	2
	574	14	2	289	—	—	25	1	905

phonolite end flakes range in length/breadth from 189 × 67 mm to 19 × 14 mm, with an average of 68 × 43 mm. The range in size for the quartzite end flakes is from 97 × 61 mm to 26 × 24 mm, with an average of 48 × 37 mm. In the side flakes the phonolite specimens are also larger than those of quartzite. They range in length/breadth from 102 × 134 mm to 26 × 40 mm, with an average of 44 × 59 mm, whereas the quartzite flakes range from 98 × 116 mm to 25 × 26 mm, with an average of 39 × 49 mm.

Handaxe re-sharpening flakes (5 specimens) These phonolite flakes clearly have been detached from the edges of handaxes. The striking platforms consist of the bifacially trimmed and utilised original edges. Four are side flakes and one an end flake which has been detached longitudinally, probably from the butt end of a handaxe; this measures 90 × 26 mm while the side flakes range from 56 × 69 mm to 27 × 39 mm.

Cores There are two specimens that appear to be simple cores. One is irregular, of Engelosen phonolite, and has had flakes removed from three platforms (72 × 60 × 39 mm). The second example is of quartzite and has been flaked radially round most of the circumference (58 × 47 × 35 mm).

The 1969 HEB excavations

The trial trench dug by the writer at this site in 1969 measured 3.5 × 3.5 m and adjoined the eastern end of the east–west trench dug by Dr Waechter in 1962. Artefacts and fossil bones were found in reddish-grey sand throughout a depth of 1.7 m. The highest artefact-bearing level lay 3 m below the base of the Masek Beds, and the horizon that yielded the large number of bifaces in 1962 (Level 3) 1.4 m lower. Some artefacts and bones were also found in the 30 cm of sandy clay between this level and the channel conglomerate

(Level 4). The concentration of artefacts in Level 3 extended only in the northwest corner of the 1969 excavation. Elsewhere at that level there was no marked accumulation but some artefacts were found in the deposits above and below Level 3. The trench was dug in 10 cm spits but the number of artefacts from any particular level was too small to be significant and they have therefore been pooled for analysis. The few tools found in Level 3 can be placed in the sequence, but the position of those from above it in relation to Levels 2a and 2b at HEB West is unknown.

The industry from above Level 3

This consists of 1143 specimens, comprising twenty-two tools, thirty-seven utilised pieces and 1084 debitage.

Tools

Bifaces (9 specimens) These consist of four complete quartzite handaxes, one of trachyandesite, lacking the tip, one quartzite tip end, one broken phonolite handaxe and one quartzite cleaver. A long, triangular phonolite flake (148 × 93 × 23 mm) has also been classed as a biface although it has only a minimum of trimming at the tip and butt end. This specimen and one of the quartzite handaxes were collected from the erosion slope but are included with the *in situ* material since they can be assumed only to have eroded from this level. The trachyandesite handaxe is the only weathered specimen; all the rest, including the cleaver, are in exceptionally fresh condition with very sharp edges. The smallest of the quartzite handaxes (112 × 71 × 28 mm) was found edge-on, in a vertical position. The upper edge is considerably more chipped and battered than the edge that lay downwards. Three of the complete handaxes made from Naibor Soit quartzite show great refinement of trimming, similar to that of the phonolite series from Level 3. All three are elongate ovates. Another, made from micaceous quartzite, is crude and may be unfinished although perhaps no better results could be achieved from this coarse, granular material. The broken phonolite handaxe has lost its tip and has

ANALYSIS OF THE INDUSTRIES FROM HEB 1969 EXCAVATIONS

	Numbers	Percentages
Tools	22	1.9
Utilised material	37	3.2
Debitage	1084	94.8
	1143	99.9
Tools		
Bifaces	9	40.9
Choppers	2	9.1
Polyhedron	1	4.5
Discoids	4	18.2
Scraper, heavy-duty	1	4.5
Scrapers, light-duty	5	22.7
	22	99.9
Utilised material		
Hammerstone	1	2.7
Cobbles and blocks	11	29.7
Light-duty flakes, etc.	25	67.5
	37	99.9
Debitage		
Whole flakes	95	8.7
Broken flakes	864	79.7
Core fragments	125	11.5
	1084	99.9

also fractured longitudinally down one side so that almost all the edge has broken off. Trimming scars truncated by this fracture indicate that the edge was originally bifacially flaked, like the opposite edge, although the tool now resembles a blunt-backed knife. The cleaver has an oblique transverse edge. It is made on a piece of tabular quartzite. One lateral edge consists of the natural cleavage plane, while the opposite edge is bifacially flaked (116 × 77 × 40 mm). The complete handaxes range in size from 180 × 93 × 40 mm to 112 × 71 × 28 mm.

Choppers (2 specimens) Both these choppers are unifacial side choppers. They are made on cobbles of trachyandesite which have been trimmed along one edge and are heavily battered and blunted. One measures 65 × 58 × 41 mm and the second 92 × 89 × 60 mm.

Polyhedron (1 specimen) This has six short, intersecting edges, most of which have been chipped by use (78 × 71 × 51 mm).

LOWER BED IV

ANALYSIS OF THE INDUSTRIES FROM HEB 1969 EXCAVATIONS

	Quartzite	Fine-grained quartzite	Gneiss, feldspar, etc.	Phonolite	Nephelinite	Trachyte	Basalt/trachyandesite	Lava, indeterminate	Totals
Heavy-duty tools									
Bifaces, choppers, polyhedrons, discoids, scrapers	10	1	—	2	—	—	4	—	17
Light-duty tools									
Scrapers	2	2	—	1	—	—	—	—	5
Utilised material, heavy-duty									
Hammerstone, cobbles and blocks	8	—	—	—	—	—	2	2	12
Utilised material, light-duty									
Flakes, etc.	17	4	—	4	—	—	—	—	25
Debitage									
Whole flakes	45	1	—	44	—	1	4	—	95
Broken flakes	680	3	—	163	—	1	17	—	864
Core fragments	125	—	—	—	—	—	—	—	125
	887	11	—	214	—	2	27	2	1143

Discoids (4 specimens) There is one large, thick specimen flaked entirely over both faces as well as radially round the circumference; it measures 95 × 85 × 71 mm. Two small quartzite examples are slightly oblong. They show considerable wear on the circumference and measure 43 × 42 × 27 mm and 44 × 36 × 22 mm respectively. The fourth example is less well made. Although the working edge extends round the entire circumference, it is uneven and jagged (72 × 74 × 38 mm).

Scrapers, heavy-duty (1 specimen) This is a side scraper made on a piece of tabular quartzite. The butt consists of a vertical cleavage plane and the working edge has been steeply trimmed from another cleavage plane to a symmetrical curve. It measures 47 × 72 mm.

Scrapers, light-duty (5 specimens) These are all side scrapers. They are made on whole or broken flakes in which the working edges are straight or slightly convex. The trimming is mostly shallow and the edges rather uneven except for one phonolite specimen which is regularly trimmed. Length/breadth measurements range from 50 × 64 mm to 16 × 30 mm, with an average of 38 × 49 mm.

Utilised material

Hammerstone (1 specimen) A large quartzite pebble measuring 50 × 50 × 40 mm has been heavily battered and chipped on all projecting parts.

Cobbles and blocks (11 specimens) Seven quartzite blocks and four broken lava cobbles show battering and crude flaking. They range in size from 91 × 90 × 42 mm to 45 × 36 × 27 mm, with an average of 78 × 54 × 41 mm.

Light-duty flakes, etc. (25 specimens) These are whole or broken flakes and other fragments with chipping on one or more edges. This is generally haphazard and in a few examples there are crushed notches. Length/breadth measurements range from 57 × 35 mm to 21 × 21 mm, with an average of 36 × 26 mm.

Debitage

This consists of ninety-five complete flakes, 864 broken flakes and 125 sundry fragments.

Among the complete flakes there are almost equal numbers of quartzite and phonolite (forty-five and forty-four specimens of each) with only six of other materials. Fifty-five are end struck, and the balance are side struck. A number of small and exceedingly thin flakes occur in both the quartzite and phonolite series that appear to have been detached in the course of trimming bifaces. The phonolite end flakes range in length/breadth from 105 × 64 mm to 13 × 12 mm, with an average of 39 × 28 mm, and the quartzite examples from 67 × 33 to 17 × 12 mm with an average of 33 × 25 mm. The average size of the phonolite side flakes is also greater than the quartzite, with 37 × 52 mm in contrast to 28 × 35 mm.

HEB West Level 2b

This level underlies Level 2a and is separated from it by about 40 cm of sand. The sand is relatively poor in remains and therefore has been regarded as dividing Levels 2a and 2b. However, there appears to be no stratigraphic break in the sequence and it is likely that these two levels are not separated by any substantial interval of time. The artefacts and fossil bones in Level 2b were dispersed through about 50 cm of sand and were not concentrated at any particular horizon. An exception was a series of bifaces in mint condition that occurred together at a single level in an irregular strip about 60 cm wide. This crossed the area of the trench diagonally, approximately from north to south. The disposition of these tools indicated water action, although they can only have been transported for a short distance, in view of their fresh condition (Fig. 4.9). As in the overlying Level 2a, trachyandesite and basalt from Lemagrut are the most common materials for bifaces, but in Level 2b there are some of quartzite, unlike the higher level where no quartzite specimens occurred. There is also similarity in technique of manufacture and in the prevalence of apparently unfinished 'rough-outs'.

The industry

This consists of 1363 specimens, comprising 113 tools, forty-two utilised pieces and 1208 debitage.

Tools

Bifaces (57 specimens; Pl. 9) These consist of thirty-one complete handaxes, fifteen cleavers, four unfinished and seven broken specimens.

The handaxes resemble those from Level 2a in that the majority are crude with an unfinished appearance, but there are also a few examples that appear to be finished as well as two unusual trihedral specimens. A number are sharply pointed, sometimes with the point skewed off-centre, others have rather broader points and some are rounded at the tips and very similar to the series from Level 2a. Eight specimens are made on large end flakes, four on side flakes and six on slabs of quartzite, while twelve are flaked entirely over both faces so that it is impossible to determine whether they are made on flakes or cores. Among both handaxes and cleavers, basalt and trachyandesite are the most common materials. In addition there are also two of tabular quartzite, two of trachyte and one of nephelinite. (It should be noted that none of the bifacial tools from this level is made from Engelosen phonolite although a few flakes of this material occur among the debitage.) Seven cleavers are made on side flakes, three on end flakes and five are indeterminate. The cleaver edges are variable, both wide and narrow, straight or oblique and in some specimens jagged and irregular. They are usually very thin and unlikely to have withstood rough wear. (The series is described by Drs Roe and Callow, in Chapters 8 and 9.)

Choppers (8 specimens) There are five side choppers, one end, one pointed and one double-edged. Quartzite is the most common material, but the largest specimen, which is pointed, is made on a cobble of trachyandesite. It is a massive

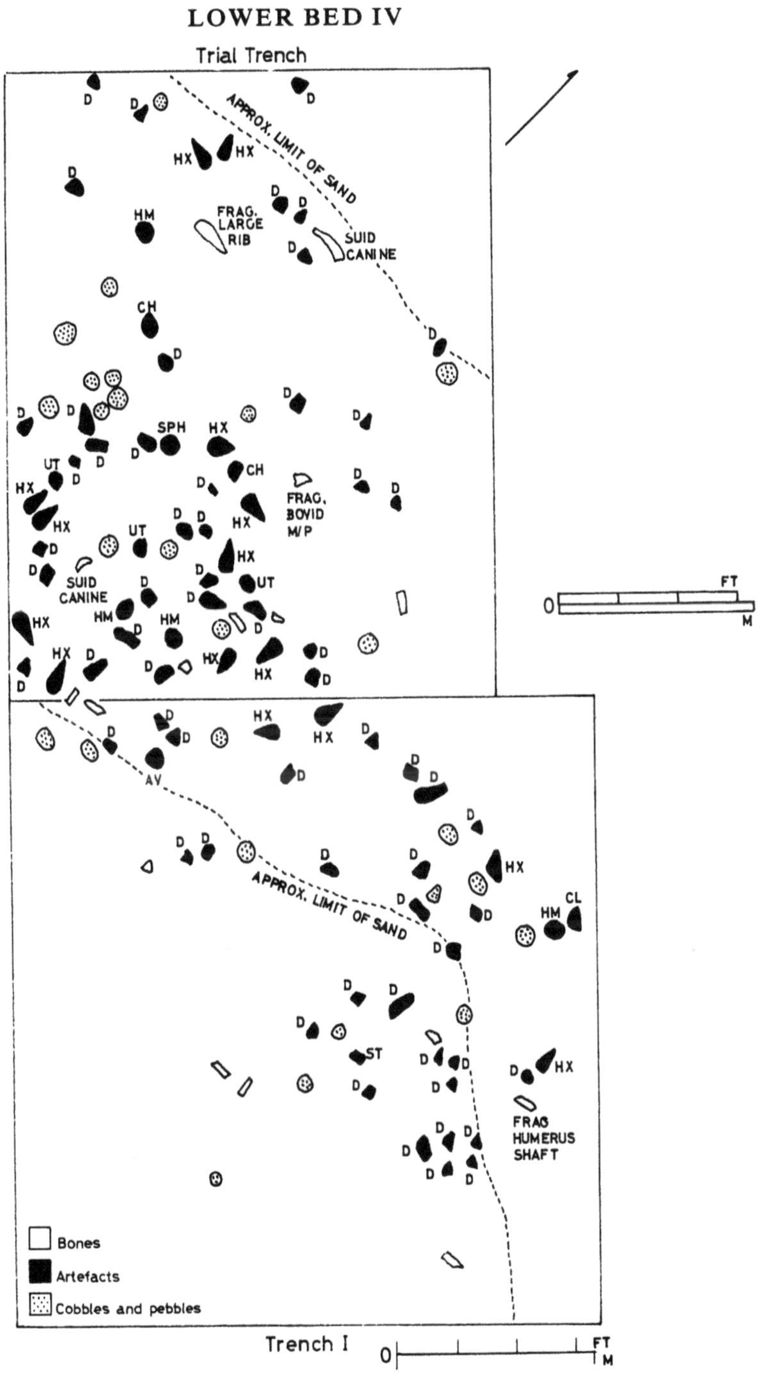

Fig. 4.9 HEB West: plan of finds associated with the sand lens

tool weighing 1.2 kg and has a pronounced projection on the working edge. The butt is cortex and shows three shallow pits which appear to be artificial. Measurements are 96 × 127 × 80 mm. The side choppers are all of quartzite. The butts are flaked or battered and the working edges are bifacially flaked and also show some traces of utilisation. Sizes range from 93 × 110 × 68 mm to 46 × 61 × 40 mm, with an average of 69 × 80 × 48 mm. The end chopper is made on a trachyandesite cobble. It is slightly rolled and some pitting is visible on the cortex surface while the butt shows hammerstone utilisation (92 × 76 × 48 mm). The double-edged chopper is oblong with a working edge at either end; these are bifacially trimmed and show extensive chipping from utilisation (124 × 83 × 61 mm).

Polyhedrons (5 specimens) These tools are classed as polyhedrons by definition since each has three cutting edges but they are not typical specimens. In each case there are two main edges while the third is subsidiary. The largest specimen is of lava and also shows anvil type of utilisation. One example is considerably abraded. The measurements range from 82 × 82 × 61 mm to 47 × 42 × 40 mm, with an average of 66 × 54 × 47 mm.

Discoid (1 specimen) The single discoid is made on part of a lava cobble and is plano-convex in cross section, with cortex remaining on the convex face. It is radially trimmed round the whole circumference (74 × 65 × 45 mm).

Spheroids/subspheroids (9 specimens) All these tools are angular and faceted). The largest and most symmetrical is made from Engelosen phonolite, a material seldom used for spheroids. This specimen measures 87 × 89 × 75 mm and weighs 600 g. The average size for the remainder of the series is 57 × 54 × 46 mm.

Scrapers, heavy-duty (7 specimens) Two examples are on large trachyandesite flakes steeply trimmed along the whole length of the right lateral edge to form side scrapers. These are unusual tools in an industry from Olduvai, in which heavy-duty scrapers are generally made on slabs and blocks of quartzite similar to five in the present series. Both the side scrapers are evenly trimmed to slightly convex working edges. The length/breadth measurements are 123 × 90 mm and 128 × 92 mm. The remaining scrapers consist of a steeply trimmed end scraper of fine grained quartzite (46 × 39 mm), a double-edged side scraper made on a slab of quartzite which has been trimmed on both lateral edges from opposite directions (131 × 88 mm), and a triangular slab of quartzite (125 × 78 × 31 mm) with a hollow scraper on one edge. This notch measures 40 mm in width and is 6 mm deep. There is also a steeply rimmed discoidal scraper with a very uneven working edge (63 × 60 mm) and a rather crude quartzite side scraper (55 × 90 mm).

Scrapers, light-duty (18 specimens) These consist of ten side scrapers, two discoidal, five hollow and one small end scraper. One discoidal specimen is made on a trachyte flake; the balance are on broken flakes and other fragments of quartzite. The side scrapers vary from a steeply trimmed example, made on a slab of quartzite, that measures 45 × 62 mm, to a small, narrow specimen measuring 16 × 33 mm. The trimming on the edges is variable and is unusually even and regular in some examples. The average length/breadth measurements for the side scrapers are 32 × 47 mm. The five hollow scrapers are similarly made on sundry pieces of quartzite varying in size from 68 × 45 mm to 21 × 22 mm with the notches always trimmed from the flat lower surface. They vary in length and depth from 34 × 6 mm to 16 × 2 mm with an average of 21 × 2.8 mm. The three remaining specimens consist of a small end scraper (24 × 14 mm) and two discoidal examples (50 × 61 mm and 53 × 42 mm).

Laterally trimmed flakes (3 specimens) These consist of three irregular, broken flakes which have been trimmed along both lateral edges. In one phonolite specimen there is also a notch on either edge. Length/breadth measurements are 51 × 26 mm, 33 × 33 mm and 27 × 19 mm.

Outils écaillés (5 specimens) These are all of quartzite and four are double ended. They exhibit the scaling and crushing of the edges that is typical of these tools. The length/breadth measurements vary from 52 × 22 mm to 22 × 16 mm, with an average of 29 × 19 mm.

Utilised material

Pitted anvils/hammerstones (4 specimens) Four lava cobbles exhibit one or more pecked hollows and in two examples the ends have also been battered by use. Most of the hollows consist of quite small pits, no more than a few millimetres in diameter, but in the smallest specimen there is a well-defined hollow 18 to 15 mm in diameter and 1.5 mm deep. The cobbles vary in size from 92 × 78 × 46 mm to 72 × 56 × 28 mm, with an average of 91 × 68 × 41 mm.

Anvils (3 specimens) In these three specimens there is one flat surface, roughly circular in shape, that has been battered round the circumference. The sizes of all three are very similar and the average measurements are 79 × 70 × 63 mm.

Hammerstones (6 specimens) These consist of cobbles that show unusually heavy battering, not only at the ends of oblong specimens but also on the greater part of the surface. One example also has two small pits. The two largest specimens weigh 1.6 kg and 1.4 kg and are above the average size for hammerstones. Measurements range from 126 × 111 × 92 mm to 84 × 76 × 63 mm, with an average of 105 × 87 × 72 mm.

Cobbles and blocks (17 specimens) Fourteen cobbles and three blocks of quartzite show some flaking as well as battering. A proportion are also abraded. Sizes range from 105 × 95 × 40 mm to 52 × 38 × 31 mm.

Light-duty flakes, etc. (12 specimens) These consist of broken flakes and sundry fragments with chipping along the edges. This follows no recognisable pattern and appears to have resulted from haphazard utilisation. Sizes ranged from 59 × 52 mm to 23 × 22 mm, with an average of 35 × 26 mm.

Debitage (1,208 specimens)
The debitage comprises ninety-seven complete flakes, 910 broken flakes, 198 sundry angular fragments and three cores.

There are sixty-one end-struck and thirty-six side-struck flakes. Among the end flakes are twelve large specimens, eight of which are over 80 mm long. The average length for the quartzite flakes is slightly greater than for the lava series with 69 mm in contrast to 61 mm. The quartzite flakes are also wider in proportion to length, with an average width of 41 mm, whereas the lava

ANALYSIS OF THE INDUSTRIES FROM HEB WEST LEVEL 2B

	Numbers	Percentages
Tools	113	8.3
Utilised material	42	3.1
Debitage	1208	88.6
	1363	100.0
Tools		
Bifaces	57	50.4
Choppers	8	7.1
Polyhedrons	5	4.4
Discoid	1	0.9
Spheroids/subspheroids	9	7.9
Scrapers, heavy-duty	7	6.2
Scrapers, light-duty	18	15.9
Laterally trimmed flakes	3	2.7
Outils écaillés	5	4.4
	113	99.9
Utilised material		
Pitted anvils/hammerstones	4	9.5
Anvils	3	7.1
Hammerstones	6	14.3
Cobbles and blocks	17	40.5
Light-duty flakes, etc.	12	28.6
	42	100.0
Debitage		
Whole flakes	97	8.0
Broken flakes	910	75.3
Core fragments	198	16.4
Cores	3	0.3
	1208	100.0

flakes have an average of 36 mm. The majority of both end and side flakes are divergent, only five specimens being convergent. There are three blocks of quartzite from which a number of flakes have been detached and which may be considered as cores. These are approximately cuboid in form.

HEB West Level 2a

This horizon was virtually barren of artefacts in the first trenches dug, that is the Trial Trench and Trenches I–III. Subsequently, however, another trench was excavated further south and this revealed a number of bifacial and other tools at a level above the quartz sand that had been termed Level 2, but beneath the horizon known as Level 1. There was thus a level of artefacts to be fitted into the sequence which was stratigraphically between Levels 1 and 2. To meet this situation it

ANALYSIS OF THE INDUSTRIES FROM HEB WEST LEBEL 2B

	Quartzite	Fine-grained quartzite	Gneiss, feldspar, etc.	Phonolite	Nephelinite	Trachyte	Basalt/trachyandesite	Lava, indeterminate	Totals
Heavy-duty tools									
Bifaces, choppers, polyhedrons, discoids, spheroids/subspheroids, scrapers	33	3	—	1	1	9	39	1	87
Light-duty tools									
Scrapers, laterally trimmed flakes, outils écaillés	24	—	—	1	—	1	—	—	26
Utilised material, heavy-duty									
Pitted anvils, hammerstones, anvils, hammerstones, cobbles and blocks	6	—	—	1	—	—	18	5	30
Utilised material, light-duty									
Flakes, etc.	11	—	—	—	—	—	1	—	12
Debitage									
Whole flakes	58	3	—	6	—	—	27	3	97
Broken flakes	800	9	—	34	—	—	67	—	910
Core fragments	192	1	—	4	—	—	1	—	198
Cores	3	—	—	—	—	—	—	—	3
	1127	16	—	47	1	10	153	9	1363

was decided to retain Level 1 for the highest artefact-bearing horizon (above the siltstone) and to term the newly found horizon Level 2a, the original Level 2 being altered to 2b. The two lower implementiferous horizons remained unchanged, that is Level 3 at HEB and Level 4 for the channel conglomerate at both HEB and HEB West. Level 2a consisted of a fine-grained silty sand, approximately 20 to 30 cm thick which was 50 to 60 cm below the grey siltstone. The artefacts were dispersed through the deposit and were not concentrated at any particular level.

The large bifacial tools are in exceptionally sharp condition and evidence of utilisation is uncommon. Trimming is also minimal and most of the handaxes and cleavers have the appearance of factory rough-outs. Although no nearby source of the raw materials is known at present, it seems likely that at the time the rocks may have been conveniently close, perhaps in the form of boulders in rivers draining from Lemagrut, since the majority of tools are made from trachyandesite and basalt originating from that volcano. The fact that the bifaces are generally made on large flakes retaining areas of weathered cortex on the dorsal aspect, suggests that the raw material was probably obtained from boulders.

Whatever the source of the rocks may have been, it is evident that the large bifacial tools were brought to the site ready made since there is only an insignificant number of lava flakes among the debitage but many of quartzite. Furthermore there are two handaxes of a red lava (source unknown) that show a total of thirty-one flake scars but there were no flakes of this material among the debitage. Another feature observed in this group of handaxes and cleavers is the similarity in technique of manufacture, in shape and size and in the order of detaching the trimming flakes which suggests quite strongly that one craftsman or one school of craftsmen was responsible. (A comparable similarity in size and form has also been observed among handaxes from the Masek Beds at FLK, see p. 119.) This series is analysed by Drs Roe and Callow in chapters 8 and 9.)

The industry
This amounts to 726 specimens, consisting of eighty-six tools, forty-five utilised pieces and 595 debitage.

Tools

Bifaces (35 specimens) These comprise seventeen handaxes with sharply pointed tips, seven chisel-ended, three round-ended that are probably very crude rough-outs, five cleavers and three broken tools, and one very heavily rolled specimen which is most certainly derived and is not included in the analysis. There are only a few specimens with refined, flat trimming scars such as occur in Level 3.

Choppers (11 specimens) Five are made on cobbles of basalt or trachyandesite, one on a nephelinite cobble and five on blocks of quartzite. There is one two-edged specimen and one has been so severely damaged by use that the original form can no longer be determined. The remainder are side choppers with cutting edges extending partly round both ends as well as along one lateral edge. The quartzite specimens are in sharp condition but three of the lava choppers are weathered. The side choppers range in length/breadth/thickness from 106 × 109 × 83 mm to 39 × 49 × 23 mm, with an average of 74 × 83 × 52 mm.

Polyhedrons (2 specimens) There are two quartzite polyhedrons. Both are in sharp condition and have four intersecting cutting edges. The measurements are 71 × 60 × 53 mm and 62 × 56 × 52 mm.

Discoids (6 specimens) One is made on part of a trachyandesite cobble and the balance are of quartzite. The series varies considerably in thickness and in form. One specimen is more oblong than discoidal but has been included since there is a cutting edge on the greater part of the circumference. They range in size from 87 × 74 × 50 mm to 49 × 49 × 33 mm, with an average of 65 × 60 × 41 mm.

Spheroids/subspheroids (13 specimens) The whole series is of quartzite. There is one symmetrical specimen measuring 100 × 88 × 78 mm which is considerably bigger than any of the others. Two of the smaller specimens are nearly spherical but the remainder are angular and asymmetrical. Excluding the single large specimen, the series ranges in size from 62 × 61 × 58 mm to 33 × 38 × 33 mm, with an average of 49 × 43 × 39 mm.

Scrapers, heavy-duty (4 specimens) Three are of quartzite and one is made on a trachyte cobble; all are in fresh condition. The trachyte specimen is steeply trimmed from the cortex to a rounded projection or broad 'nose'. Another specimen, made on a piece of tabular quartzite, has also been similarly trimmed from the natural flat cleavage surface to a nosed working edge. A second quartzite specimen is a double-edged side scraper. The fourth specimen has three steeply trimmed edges and on this count might be termed a polyhedron except for the fact that both trimming and wear on each of the edges are from one direction only. The range in length/breadth for these tools is from 94 × 90 mm to 74 × 54 mm.

Scrapers, light-duty (12 specimens) These tools are made on irregular pieces of quartzite and one on a nephelinite flake. The forms of the working edges appear to depend largely on the form of the edge selected for trimming. In two examples there is a suggestion of a projecting nose similar to those in the heavy-duty scrapers. Except for these two specimens and a notched example, the working edges are either straight or only slightly convex. The range in length/breadth is from

ANALYSIS OF THE INDUSTRIES FROM HEB WEST LEVEL 2A

	Numbers	Percentages
Tools	86	11.8
Utilised material	45	6.2
Debitage	595	82.0
	726	100.0
Tools		
Bifaces	35	40.7
Choppers	11	12.8
Polyhedrons	2	2.3
Discoids	6	7.0
Spheroids/subspheroids	13	15.1
Scrapers, heavy-duty	4	4.6
Scrapers, light-duty	12	13.9
Outils écaillés	2	2.3
Punch	1	1.2
	86	99.9
Utilised material		
Pitted anvils/hammerstones	3	6.6
Cobbles	3	6.6
Light-duty flakes, etc.	39	86.7
	45	99.9
Debitage		
Whole flakes	190	31.9
Broken flakes	213	35.8
Core fragments	187	31.4
Cores	5	0.9
	595	100.0

75 × 52 mm to 27 × 31 mm, with an average of 37 × 35 mm.

Outils écaillés (2 specimens) Both examples are of quartzite and both are double ended. They measure 37 × 35 mm and 26 × 25 mm respectively.

Punch (1 specimen) A small, quartzite example with heavy blunting and crushing at both ends (30 × 11 mm).

Utilised material

Pitted anvils/hammerstones (3 specimens) One example consists of an oval cobble, probably of trachyandesite, measuring 123 × 97 × 68 mm. On one face there is a shallow, irregular, pitted hollow 13 mm in diameter. There are also traces of battering on the surface adjacent to the pitted area. The opposite face of the cobble bears one small pit only 4 mm in diameter and another that is oblong, measuring 15 × 5 mm and 1.5 mm in depth. Other slight traces of pitting are also present on this face. The second specimen is oblong and is of vesicular basalt. It measures 79 × 51 × 38 mm and has a hollow 15 mm in diameter and 3.5 mm deep on one face. There is no trace of pitting on the opposite side. The third example consists of a flattened oval cobble of trachyandesite. It measures 124 × 110 × 41 mm and has been flaked on both lateral edges possibly in preparation for making a small biface. One face has a series of small, adjoining pitted hollows, covering an area 20 × 15 mm.

Cobbles (3 specimens) Two whole and one broken cobble of trachyandesite or basalt have been battered and crudely flaked. The complete specimens measure 94 × 70 × 69 mm and 72 × 65 × 51 mm.

Light-duty flakes, etc. (39 specimens) With the exception of one flake and two fragments of trachyandesite, these specimens are of quartzite. They consist of broken flakes, pieces of tabular quartzite and other fragments that have been chipped or blunted along one or more edges. Three examples are slightly notched, but in most cases the utilisation is on slightly convex or straight edges. Length/breadth measurements range from 104 × 45 mm to 20 × 14 mm, with an average of 41 × 30 mm.

Debitage (595 specimens)

This consists of 190 complete flakes, 213 broken flakes, 187 core fragments and five cores.

One hundred and thirty of the complete flakes are end struck and sixty side struck. There are 166 divergent and twenty-four convergent flakes but none which can be classed as parallel sided. Five of the lava side flakes appear to be handaxe trimming flakes. A few specimens are weathered but most are in fresh condition. The bulbs of percussion are variable, but generally rather flat and wide in the handaxe trimming flakes whilst

LOWER BED IV

ANALYSIS OF THE INDUSTRIES FROM HEB WEST LEVEL 2A

	Quartzite	Fine-grained quartzite	Gneiss, feldspar, etc.	Phonolite	Nephelinite	Trachyte	Basalt/trachyandesite	Lava, indeterminate	Totals
Heavy-duty tools Bifaces, choppers, polyhedrons, discoids, spheroids/subspheroids, scrapers	28	—	—	2	1	5	33	2	71
Tools, light-duty Scrapers, *outils écaillés*, punch	13	1	—	—	1	—	—	—	15
Utilised material, heavy-duty Pitted anvils/hammerstones, cobbles	—	—	—	—	—	—	6	—	6
Utilised material, light-duty Flakes, etc.	34	2	—	—	—	—	3	—	39
Debitage									
Whole flakes	155	4	—	2	1	1	27	—	190
Broken flakes	193	—	—	4	—	—	16	—	213
Core fragments	178	1	—	4	—	—	4	—	187
Cores	5	—	—	—	—	—	—	—	5
	606	8	—	12	3	6	89	2	726

those in the quartzite flakes are often shattered. Striking platforms also vary greatly in size, shape and angle. The end-struck flakes vary in length/breadth from 108 × 102 mm to 19 × 15 mm, with an average of 34 × 26 mm. The side-struck flakes range from 73 × 77 mm to 13 × 19 mm, with an average of 31 × 38 mm.

Nearly all the broken flakes are of quartzite, with only nineteen of lava. Among the sundry fragments are four pieces of Engelosen phonolite and four fragments of trachyandesite. The rest of the series consists of angular fragments of quartzite, many from tabular material. There is great variation in size, the largest piece measuring 98 × 70 × 66 mm and the smallest 16 × 12 × 10 mm. There are also five sizeable pieces of quartzite, three of which are tabular, which have had flakes removed and appear to be cores rather than tools. They range in size from 153 × 130 × 86 mm to 74 × 54 × 45 mm, with an average of 109 × 83 × 67 mm.

Artefacts in disturbed context

HEB

Thirty-three artefacts were collected from the surface by the late Dr J. Waechter in 1962, mostly from HEB at the eastern end of the excavated area. However, five bifaces made from lava may have been obtained from HEB West where similar specimens occur *in situ*. A further ten artefacts were collected from the surface at HEB during 1969. The total comprises the following: nineteen bifaces, including six broken; one chopper; four spheroids; two heavy-duty scrapers; seventeen flakes.

HEB West

Five hundred and one artefacts were found in the surface deposit at HEB West and in the disturbed deposit filling the channel in Trench IV. It is likely that the majority are from Levels 2a and 2b but some may also be from Level 1. These tools are not described in detail in view of their uncertain provenance. They consist of the following specimens:

Tools
Bifaces	12, including 3 broken or unfinished
Choppers	6
Polyhedrons	2
Discoids	2
Spheroids/subspheroids	8
Scrapers, heavy-duty	3
Laterally trimmed flake	1

Utilised material
Hammerstones	7
Cobbles and blocks	6
Light-duty flakes, etc.	6

Debitage
Complete flakes	29
Broken flakes	346
Core fragments	73
	501

WK Intermediate Channel

The description of Site WK will be found preceding the analysis of the industry from the Upper Channel on pp. 75–6. The industry from the Intermediate Channel was recovered from a small channel approximately 30 cm deep, lying 2.7 m above the Beds III–IV junction. Only eight bifaces were found and they are not included in the analysis by Drs Roe and Callow.

The industry
The industry consists of 1,651 specimens, comprising forty-six tools, thirty-nine utilised pieces and 1566 debitage.

Tools

Bifaces (8 specimens) The series consists of one complete handaxe, one medial section, two unfinished examples, one butt and three tip ends. Three specimens are of Engelosen phonolite, one of trachyte and four of basalt or trachyandesite. The complete handaxe and two phonolite tips are in fresh condition, but the remainder of the series is rolled. The complete handaxe is made on a flake apparently struck from a trachyandesite boulder, and shows cortex on the butt. It is sharply pointed and the lateral edges are sinuous ($112 \times 71 \times 43$ mm).

Choppers (13 specimens) These consist of eight side and five end choppers. Seven are of trachyandesite or basalt, four of quartzite, one of phonolite and one of nephelinite. Two examples are very crude, with minimal flaking, and have only two or three flakes struck from the working edges. One of these is an end chopper and also has two pitted hollows on the butt. The side choppers include one miniature quartzite example only 32 mm long with a battered, rounded butt. Another quartzite specimen, made from unusually coarse material, has a point in the centre of the working edge, which is wide and curved; it may possibly be the broken tip of a handaxe modified to a chopper. A trachyandesite specimen also has a broad working edge and appears to be made on the butt of a large flake. The side choppers range in size from $71 \times 93 \times 69$ mm to $32 \times 36 \times 28$ mm, with an average of $54 \times 85 \times 38$ mm. The six end choppers are also variable in size, ranging from $121 \times 89 \times 70$ mm to $60 \times 48 \times 50$ mm, with an average of $81 \times 66 \times 51$ mm. All are rolled, particularly one phonolite specimen, and all are made on cobbles or weathered nodules with cortex or weathered surface forming the butts.

Polyhedrons (2 specimens) Both examples are made from trachyandesite or basalt, and both have a number of intersecting edges, most of which appear to have been chipped by use. Measurements are $80 \times 76 \times 70$ mm and $58 \times 56 \times 52$ mm.

Discoids (7 specimens) Four are of quartzite, including one broken specimen, two of trachyandesite or basalt and one of phonolite. It is

LOWER BED IV

ANALYSIS OF THE INDUSTRIES FROM WK INTERMEDIATE CHANNEL

	Numbers	Percentages
Tools	46	2.8
Utilised material	39	2.3
Debitage	1566	94.9
	1651	100.0
Tools		
Bifaces	8	17.4
Choppers	13	28.3
Polyhedrons	2	4.3
Discoids	7	15.2
Spheroids/subspheroids	4	8.7
Heavy-duty scraper	1	2.2
Light-duty scrapers	7	15.2
Outils écaillés	3	6.5
Punch	1	2.2
	46	100.0
Utilised material		
Pitted anvils/hammerstones	3	7.9
Hammerstones	4	10.5
Cobbles and blocks	23	57.9
Light-duty flakes, etc.	9	23.7
	39	100.0
Debitage		
Whole flakes	106	6.8
Broken flakes	645	41.2
Core fragments	815	52.0
	1566	100.0

possible that this specimen was originally a broken section of a handaxe that has been trimmed into a discoid. It is in fairly fresh condition, together with the quartzite example, but both the trachyandesite or basalt discoids are rolled. They are also steeply convex on both faces, but the remainder of the series is flatter in cross section. The complete specimens range in size from $79 \times 71 \times 45$ mm to $48 \times 52 \times 24$ mm, with an average of $59 \times 58 \times 33$ mm.

Spheroids/subspheroids (4 specimens) Three are of quartzite and one of trachyandesite or basalt. They are angular and not exactly spherical, although in each the projecting ridges have been partly reduced. Measurements range from $76 \times 76 \times 60$ mm to $41 \times 41 \times 43$ mm, with an average of $60 \times 54 \times 49$ mm.

Scraper, heavy-duty (1 specimen) One piece of tabular quartzite, sub-rectangular in shape and 21 mm thick, has been trimmed on all four edges, one in an opposite direction to the other three (75×66 mm).

Scrapers, light-duty (7 specimens) These are all made from quartzite. They include five small examples of the usual size and two that are considerably larger, although within the light-duty class since the mean diameters are under 50 mm. They are made on an end flake and on a piece of tabular quartzite and measure $58 \times 44 \times 27$ mm and $50 \times 44 \times 25$ mm respectively. Both have convex, steeply trimmed working edges. Three smaller specimens are miniature side scrapers measuring 20×30 mm, 19×24 mm and 20×29 mm. The two remaining examples are hollow scrapers measuring 35×23 mm and 39×28 mm. The notches measure 24 mm and 15 mm in width and have been trimmed from one direction only.

Outils écaillés (3 specimens) Two examples are double ended and one single ended. One of the double-ended specimens is of phonolite, an unusual material for these tools; the remaining two are of quartzite. Measurements are 32×27 mm, 28×26 mm and 21×22 mm.

Punch (1 specimen) A small quartzite example, measuring 28×13 mm, shows the characteristic utilisation at both ends.

Utilised material

Pitted anvils/hammerstones (3 specimens) All these specimens are of trachyandesite. They show: a single, shallow pitted hollow on one face, measuring 16×9 mm; pitted hollows on opposite faces, measuring approximately 21×18 mm and 21×19 mm; and three ill-defined small pits on one face. Two are oblong cobbles that also have slight bruising at one end. Measurements are $158 \times 151 \times 66$ mm, $131 \times 77 \times 60$ mm and $82 \times 69 \times 51$ mm. As noted above, one end chopper also has pitting on the butt.

Hammerstones (4 specimens) These consist of an oblong cobble of vesicular basalt, battered at both ends (128 × 88 × 76 mm); a discoidal trachyandesite cobble, bruised on the whole circumference (80 × 81 × 40 mm); an irregular cobble of trachyandesite, bruised on two opposite edges (75 × 64 × 52 mm) and an oblong basalt pebble bruised at one end (63 × 48 × 35 mm).

Cobbles and blocks (23 specimens) Sixteen whole and broken cobbles of basalt and trachyandesite, one of trachyte and one of welded tuff exhibit heavy battering and crude flaking which appears to have been incidental to heavy use rather than intentional flaking. Three quartzite blocks and two of feldspar show similar utilisation. Sizes range from 128 × 88 × 62 mm to 60 × 38 × 24 mm, with an average of 80 × 65 × 48 mm.

Light-duty flakes, etc. (9 specimens) These consist of whole and broken flakes and sundry fragments with irregular chipping and blunting along the edges. In two specimens slight notches have been chipped out. They range in length/breadth from 66 × 56 mm to 23 × 20 mm, with an average of 42 × 30 mm.

Debitage

This consists of 1,566 specimens, of which 106 are whole flakes, 645 broken flakes and 815 sundry, angular fragments. Sixty-eight of the complete flakes are end struck and thirty-eight side struck. End flakes range in length/breadth from 87 × 77 mm to 11 × 10 mm, with an average of 43 × 31 mm. Of these, eight are roughly parallel sided, thirteen convergent and forty-seven divergent. The side flakes range in length from 64 × 66 mm to 16 × 18 mm.

ANALYSIS OF THE INDUSTRIES FROM WK INTERMEDIATE CHANNEL

	Quartzite	Fine-grained quartzite	Gneiss or feldspar	Phonolite	Nephelinite	Trachyte	Trachyandesite or basalt	Lava, indeterminate	Totals
Heavy-duty tools									
Bifaces, choppers, polyhedrons, discoids, etc.	12	—	—	5	1	1	16	—	35
Light-duty tools									
Scrapers, *outils écaillés*, punches	9	1	—	1	—	—	—	—	11
Utilised material, heavy-duty									
Pitted anvils/hammerstones, hammerstones, cobbles and blocks	3	—	2F	—	—	1	23	1WT	30
Utilised material, light-duty									
Flakes and other fragments	5	1	—	1	—	—	2	—	9
Debitage									
Whole flakes	63	2	—	10	—	2	29	—	106
Broken flakes	570	8	—	12	4	2	49	—	645
Angular fragments	789	6	1F	5	4	—	10	—	815
Cores	—	—	—	—	—	—	—	—	—
	1451	18	3	34	9	6	129	1	1651

F = Feldspar
WT = Welded Tuff

5

UPPER BED IV.
WK UPPER CHANNEL, WK EAST A AND C, PDK TRENCHES I–III, HEB WEST LEVEL I

M. D. LEAKEY

Site WK

The site WK was first recorded in 1931 when handaxes were noted eroding from Bed IV. It is situated on the south side of the gorge, rather less than 2 km downstream from the confluence of the Main and Side Gorges (Pl. 10 and Fig. 5.1) In 1969 a number of handaxes and cleavers were again observed at this site, on the slope below Bed IV. A trial trench 1 m wide was dug into the lower part of Bed IV, as far down as the Beds III–IV junction. Three artefact-bearing levels were revealed, all of which were in conglomerates or channel fillings (Pl. 11; Figs. 5.2 and 5.3, see page viii).

(a) The lowest level was a pebble conglomerate 50 cm thick at the Beds III–IV junction (see pp. 42–4).
(b) A channel approximately 30 cm deep, 2.7 m above the Beds III–IV junction, will be referred to as the Intermediate Channel (see pp. 72–4).
(c) The Upper Channel was approximately 1.5 m higher. It consisted of an uneven, eroded surface with depressions and a channel 1 m deep in the southwest corner of the excavated area. Artefacts and fossils were concentrated in the depressions and particularly in the deep channel which was densely packed with remains. A few also occurred on the higher parts of the surface, where the hip bone and femur of O.H.28 were also found.

Following the excavation of the trial trench, a series of adjacent trenches was dug which eventually exposed an area of 150.5 sq m of the main channel and occupation surface. Three of the trenches (I, II and V) were dug down to the intermediate level, exposing an area of 30.7 sq m. A single trench (X) was dug at the lowest level and exposed an area of 7.5 sq m.

The sequence of deposits revealed was as follows:

(a) approximately 3.2 m of Ndutu Beds and calcrete
(b) 2 m of aeolian tuff and calcrete of the Masek Beds
(c) 2.5 m of reddish-brown claystone with calcrete veins and nodules
(d) approximately 30 cm of sandy grit
(e) 1 m of sandy claystone
(f) artefact-bearing eroded surface with a channel 1 m deep (the main or Upper Channel)
(g) 1.5 m of siltstone, veined by red claystone and root-marked
(h) intermediate artefact-bearing channel 45 cm deep (see pp. 72–4)
(i) 1.7 m of grey siltstone
(j) 1 m of sandy claystone with coarse lenses
(k) lowest artefact-bearing channel, a pebble conglomerate 30 to 50 cm thick (see pp. 42–4)
(l) Bed III

Tuff IVB, which elsewhere in this area occurs about 3 m above Bed III, was missing at WK and had been cut out by channelling, but blocks of the tuff were contained in the conglomerate at the

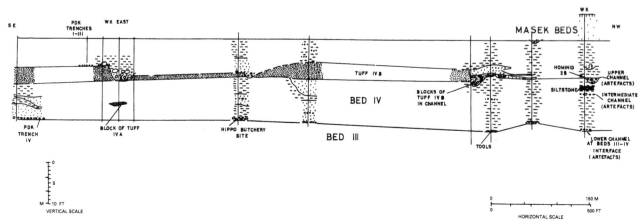

Fig. 5.1 Bed IV: section along the south side of the gorge showing sites PDK Trenches I–III, PDK Trench IV, WK East, the hippo butchery site and WK

base of the main channel (f), so that this archaeological level postdates Tuff IVB.

The intermediate level lies 3 m above the top of Bed III, at the same level as Tuff IVB in places where it has not been removed by channelling. The grey siltstone underlying the intermediate level can be correlated with similar siltstones that occur elsewhere in lower Bed IV and which overlie the principal archaeological levels at HEB and HEB West.

In a preliminary note to *Nature* (Leakey 1971), reporting the discovery of *Homo erectus* remains at WK, R. L. Hay contributed the following note on the geology of Bed IV in the area of WK:

Bed IV ranges from 20–22 feet thick in the vicinity of WK. It sharply overlies Bed III and is overlain by aeolian tuffs of the Masek Beds (formerly known as Bed IVB). Bed IV is 50%–60% claystone and most of the remainder is sandstone and conglomerate, which form lenticular beds. The claystones are usually sandy, root-marked, and locally contain vertical veins of sandy claystone a centimetre thick and a few feet long that represent deep mud cracks filled by sediments during the deposition of Bed IV. Most sandstones contain both volcanic and quartzose basement debris.

A marker tuff (Tuff IVB) extends through most of this area near the middle of Bed IV. It is as much as 7 feet thick, hard and reddish-brown, and is characterised by altered mafic glass and abundant rock fragments. Most of the tuff is thin-bedded or laminated. The tuff is locally eroded away and fragments of it occur in the lowermost sandstones and conglomerates above the tuff. The hominid fossils were found at the base of a sandstone with fragments of tuff, showing that they postdate the tuff, but probably by only a short period of time.

The non-tufaceous deposits of Bed IV are fluvial; sandstones and conglomerates fill the stream channels, and the claystones represent overbank deposits. The WK area lay near the drainage axis of the Bed IV basin which received volcanic detritus from the south and east, and basement detritus from the north and east. Tuff IVB was deposited in a temporary, small, saline playa lake, at a time when the climate was relatively hot and dry.

The industry

The industry from this level consists of 10,904 specimens comprising 429 tools, 428 utilised pieces and 10,047 debitage (see Pls 12–19).

Tools

Bifaces The total of 192 bifacial tools is the largest number excavated during the 1970s and is only exceeded at Olduvai by the assemblage from HK excavated by L. S. B. Leakey during the 1930s. Apart from three specimens made from a lava of which the source is unknown all the bifaces are made from three types of rock: quartzite from Naibor Soit inselberg, phonolite from Engelosen and basalt/trachyandesite from Lemagrut, the last being the most commonly used material. Trachyte is not represented among the main series of handaxes and cleavers, although one of the miniature handaxes described below, that appear to be extraneous, is made from this material. (This series is analysed by Drs Roe and Callow but the following notes may provide some supplementary information.)

Handaxes (78 complete specimens) Five examples are of Engelosen phonolite, twenty-two of quartzite, forty-nine of trachyandesite or basalt

UPPER BED IV

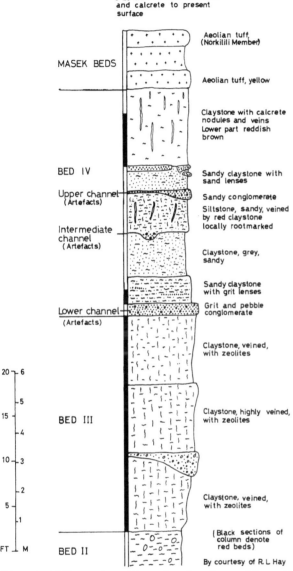

Fig. 5.2 WK: stratigraphic section showing positions of the three channels with artefacts

show very shallow feather flaking on the lateral edges that does not occur on the bifaces made from coarser lavas, although in a few made from quartzite the trimming is almost as refined. Ten of the quartzite specimens exhibit the cleavage planes characteristic of the tabular material from Naibor Soit. In five examples both the upper and lower faces consist of natural cleavage fractures and only the circumference has been flaked. Three other specimens exhibit a cleavage plane on one face only, and in a fourth, made from an unusually thick slab, the butt end is formed by a vertical fracture. All the quartzite handaxes are rather irregular in form with uneven edges and often lack trimming at the butt ends; chipping and blunting from use is often present. Several of the handaxes in the trachyandesite/basalt series show areas of smooth, weathered surface on the dorsal faces or butts indicating that they have been made on flakes struck from boulders. Cortex is present on the butts in six examples and on the dorsal faces in thirteen examples. In three others there is cortex on the butt as well as on one face. The series includes sharply pointed, blunt-ended and chisel-ended examples and others in which the pointed tip is skewed to one side.

In addition to the above specimens there are ten small handaxes, mostly very abraded, that do not appear to be part of the main series but are similar to the crudely made, miniature handaxes found in the Developed Oldowan sites at WK East and PDK. Four of these specimens are of Engelosen phonolite, four of quartzite, one of nephelinite and one of trachyte.

Cleavers (59 specimens) There are fifty-nine complete specimens, of which five are of Engelosen phonolite, twelve of quartzite, forty-one of trachyandesite or basalt and one of a fine-grained lava from an unknown source. The series can be subdivided as follows: two waisted specimens, thirteen oblique-ended, twenty-five square-ended (in nine of which the maximum width is at the cleaver edge) and nineteen semi-rounded or irregular-ended. Of the five specimens made from Engelosen phonolite, three appear to be made on side flakes and two on end flakes. They include

and two of a fine-grained lava of unknown origin from which one cleaver is also made. Except for two of the phonolite specimens that show very refined trimming similar to that on two of the phonolite cleavers, the handaxes have a crude appearance with deeply indented trimming scars. Few specimens show any attempt to remove or reduce the pronounced ridges between the scars. The exceptions are phonolite specimens which

two of the most highly finished cleavers known from Olduvai. One is an oblique specimen with extremely refined, feathered retouch on the dorsal aspect along one lateral edge. This edge is 154 mm long and slightly curved. Eight very flat 'feather' flakes have been removed from it and there is a continuous series of twenty-three small secondary scars. The trimming is so regular that the edge is virtually straight except for one area where it appears to have been damaged by use, although the cleaver edge itself shows only negligible traces of use. On the primary flake surface the bulb and striking platform have been almost entirely trimmed off. The second specimen is waisted and has a splayed cleaver edge which shows a greater degree of wear than the oblique-ended example. Both lateral edges have wide and fairly shallow indentations, situated approximately a quarter of the length from the cleaver edge. One indentation has been bifacially trimmed but that on the opposite side has been trimmed on the dorsal face only. The platform of the side flake on which this tool was made has been entirely trimmed off and none of the original platform remains although the edge on this side is still thick. These two specimens and the handaxe also made from Engelosen phonolite show a much greater refinement of trimming and attention to detail than any bifaces at this site made from other lavas.

The twelve cleavers made from quartzite have probably all been made from tabular material. Seven specimens, in fact, show the typical natural cleavage planes. It seems that when the rock split into fairly thin slabs, not more than 45 mm thick, the cleavers were made by means of trimming the broken edges of the slabs after detaching a diagonal or transverse 'cleaver' flake from one end. Specimens made on thicker slabs of quartzite show either a cleavage plane on one lateral edge, with both upper and lower faces flaked, or on one face only. A further five examples have no trace of cleavage planes; two are made on large side flakes and three on large end flakes or on fragments that might be either natural fractures or parts of broken flakes. The butts of the quartzite cleavers are usually thick and angular with nearly vertical flake scars or broken surfaces. There is some chipping on the cleaver edges, but the lateral edges usually show more extensive chipping and blunting.

The forty-one cleavers of basalt and trachyandesite show a great similarity in manufacture. The whole series appears to have been made on flakes, and the presence of weathered cortex on thirteen specimens indicates that the flakes were probably struck from boulders. In eight examples the cortex has remained in the central part of the dorsal surface while both lateral edges are trimmed. In the remaining five it is present on the butts, which are relatively thick. With one exception these cleavers are made on side flakes. In specimens made either on end or side flakes the striking platforms have been crudely trimmed off by means of a few rather large flakes. Flaking of the primary surface is generally minimal, consisting of a few flakes detached from one or both lateral edges. The chipping and blunting on the edges that appears to be due to wear varies considerably. In some specimens the cleaver edge shows virtually no trace of utilisation; in others there is sporadic light chipping. A few examples show extensive chipping, from one or both directions. More pronounced chipping as well as blunting and small step-flaking are present on the lateral edges of a number of specimens, although others show little evidence of any wear.

Broken and unfinished handaxes and cleavers
The broken bifaces amount to nineteen specimens. They consist of four examples in which the tips are lacking, six tip ends including the square-ended tip of a cleaver, three median sections in which both tips and butts are missing and six butt ends. The raw materials are as follows: Engelosen phonolite three specimens, quartzite nine specimens, trachyandesite or basalt five specimens, feldspathic gneiss one specimen, a broken, but retrimmed butt end and one heavily rolled butt end of green trachyte. The unfinished bifaces comprise thirty-six specimens of which fourteen are of quartzite and twenty-two of trachyandesite or basalt (no other rocks are represented). Seven of the quartzite specimens have the natural cleavage planes of the Naibor Soit material either

on both lateral edges or on one edge or at the butt. The remaining specimens, although made from the same rock, appear to be parts of large flakes or indeterminate fragments.

Picks (2 specimens) Both examples are in fresh condition, sharply pointed, with thick butts; they are very similar in size and form. One is of quartzite and one of trachyandesite. In the former, the butt is formed by a weathered surface and in the latter by cortex. Measurements are 156 × 122 × 83 mm and 128 × 125 × 75 mm.

Choppers (35 specimens) Sixteen are of quartzite from Naibor Soit, two of fine-grained quartzite and seventeen of basalt or trachyandesite. Six of the quartzite specimens are made on pieces of tabular material in which one edge has been bifacially trimmed; one is on a pebble with a cortex butt, and the balance on indeterminate fragments. The quartzite series consists only of bifacially flaked side choppers, some of which have relatively straight working edges, whilst others are markedly convex. One specimen, made from fine-grained quartzite, has a short point in the centre of the working edge, formed partly by a negative flake scar on either side and partly by crushing of the edges. The largest example is extensively battered on the butt and may be a broken hammerstone. The quartzite choppers range in length/breadth/thickness from 76 × 104 × 52 mm to 34 × 37 × 23 mm, with an average of 51 × 64 × 37 mm.

The lava series comprises nine side, seven end and one combined end/side chopper. Three side, six end and the combination chopper are made on cobbles and have cortex on the butts. The remainder are on irregular blocks. The lava series is generally crude and more heavily rolled than the quartzite specimens. All except one of the side choppers are bifacially flaked, although not always along the entire length of working edge. In one example an area of cortex 12 mm wide has remained in the centre. The unifacial chopper has only two flakes removed from one face. The lava side choppers range in length/breadth/thickness from 93 × 122 × 54 mm to 55 × 60 × 32 mm, with an average of 72 × 96 × 50 mm. Two of the seven end choppers are unifacial and five bifacial, although in two specimens only one flake has been removed from either face. Four examples made on cobbles have been battered on the cortex butts and two show small, pitted hollows on one face. Measurements range from 90 × 59 × 54 mm to 60 × 51 × 41 mm. The combination side/end chopper measures 78 × 50 × 42 mm. It is made on an oblong cobble in which one lateral edge has been unifacially flaked from the cortex. One end has also been flaked to a convex working edge from a single negative scar on the opposite face.

The quartzite side choppers are generally smaller than those made from lava, the average measurements for the two groups being 55 × 68 × 40 mm (quartzite) compared to 72 × 90 × 50 mm (lava). Since the end choppers are made only of lava no comparison can be made between the two materials.

Polyhedrons (7 specimens) Two quartzite examples are in fresh condition as well as two made from an unidentified lava; the remaining three, made of basalt of trachyandesite, are considerably abraded. These tools have three or more working edges, generally chipped and blunted besides being abraded. Sizes range from 74 × 69 × 57 mm to 51 × 51 × 37 mm, with an average of 61 × 58 × 44 mm.

Discoids (17 specimens) These tools vary considerably in size and in workmanship. The eleven quartzite examples include five relatively large, rather asymmetrical and crudely flaked specimens as well as three small examples, two of which are made from fine-grained quartzite. These are symmetrical, elliptical in cross section and evenly flaked round the circumference; they resemble small, unstruck tortoise cores. The lava series is more heavily abraded than the quartzite. Part of the circumference of the largest lava discoid has been extensively battered while the central areas of both upper and lower faces appear to have been hollowed by artificial pitting. These two areas measure 35 × 31 mm and 52 × 36 mm respectively. The measurements for the seventeen discoids range from 80 × 80 × 34 mm to 46 × 43 × 25 mm, with an average of 59 × 55 × 32 mm.

Spheroids/subspheroids (25 specimens) These include two broken quartzite examples, battered smooth on the surface, which were probably symmetrical stone balls before they were broken. The remainder of the series are angular and faceted but approximately spherical in general form and with reduction of the projecting parts. Four lava specimens and one of fine-grained purple quartzite are more abraded than the series made from Naibor Soit quartzite. Three of the lava examples are made on cobbles and retain part of the cortex. Sizes range from 77 × 71 × 63 mm to 28 × 25 × 23 mm, with an average of 53 × 50 × 44 mm.

Scrapers, heavy-duty (25 specimens) Eighteen of these tools are made from quartzite, one from Engelosen phonolite and six from basalt or trachyandesite. Seven of the quartzite series are made on pieces of tabular material, either on thin slabs, trimmed along one edge, or on flakes struck from thicker slabs. Three further specimens are end scrapers on large flakes; the remainder are made on indeterminate fragments. Three of the lava examples are on large flakes, two on a broken flake and two on parts of cobbles. The whole series is rather crude, with uneven trimming, but can be subdivided into four categories; end, side, hollow and combined end/side scrapers. There are nine end scrapers, in which the working edges vary from straight to curved. The trimming may be either steep or quite shallow and in some examples the edges have been blunted by use. The end scrapers range in length/breadth from 129 × 101 mm to 71 × 68 mm, with an average of 94 × 78 mm. Eight are side scrapers. Like the end scrapers, they are variable in form, with uneven working edges which may be either straight or curved. In the majority, the edges are both blunted and chipped by use. Length/breadth measurements range from 175 × 126 mm to 62 × 68 mm, with an average of 76 × 90 mm.

In the five combined end/side scrapers the trimming extends along one lateral edge and also across one end. Neither edge is evenly trimmed, except in one example, made on a piece of tabular quartzite 24 mm thick, 119 mm long and 80 mm wide. This has been trimmed across one end and to a concave edge on one side, forming a 'nose' where the two edges meet. Another example is made on a flat cobble 36 mm thick, 82 mm long and 61 mm wide. It has also been trimmed across one end and along one side. The trimming is unusually steep and is nearly vertical. Both upper and lower faces of the cobble show small pitted hollows. The length/breadth measurements for the combined end/side scrapers range from 119 × 83 mm to 67 × 50 mm, with an average of 90 × 60 mm. The single hollow scraper is on a large lava flake measuring 96 × 97 mm. The notch is 34 mm wide and 8 mm deep. It has been chipped and crushed from the primary flake surface only.

Scrapers, light-duty (57 specimens) These tools can be subdivided into five groups, side, end, nosed, hollow and steeply trimmed 'core' scrapers. Most are made of Naibor Soit quartzite, but there are also six of basalt or trachyandesite, two of phonolite, one of trachyte and one of nephelinite. Side scrapers are the most numerous, with twenty-five examples. They vary in degree of trimming and in the nature of the working edges, but most are gently curved. The trimming is generally too irregular to produce an even working edge and in the majority there are small projections that appear to have been created by wear. Length/breadth measurements range from 40 × 93 mm to 16 × 24 mm, with an average of 29 × 41 mm.

End scrapers amount to fourteen specimens. Only three are made on whole flakes, the majority being on broken flakes and two on slabs of quartzite. Only one specimen, made of lava, has an evenly curved working edge. In the remainder the working edges are irregular, as in the side scrapers. Length/breadth measurements range from 57 × 56 mm to 28 × 27 mm, with an average of 37 × 31 mm. Two of the 'nosed' scrapers are of quartzite and one of phonolite. They have well defined, rounded projections on the working edges. These appear to have been trimmed intentionally and are 14 mm, 15 mm and 16 mm wide. Length/breadth measurements are 47 × 39 mm, 41 × 31 mm and 58 × 23 mm. Three of the eight hollow scrapers are made on fragments of Naibor Soit quartzite, two on broken flakes of fine-grained quartzite and three on broken lava flakes. The notches generally show multiple flaking, but in the largest lava specimen the notch is formed by the removal of one large flake, although the edge has subsequently been chipped. In one specimen, a second, convex scraping edge is present opposite the notch. The notches range in width and depth from 36 × 7 mm to 13 × 2.5 mm, with an average of 17 × 3 mm. The overall measurements of the tools range from 78 × 50 mm to 23 × 20 mm, with an average of 37 × 43 mm.

The core scrapers consist of seven specimens. Three are made on pieces of tabular quartzite, two on parts of lava cobbles and two on indeterminate fragments. These tools are thick relative to their overall size and the working edges are steeply trimmed. They vary in form but are generally rounded, with small projections at the intersections of trimming scars. Four are side scrapers and three end scrapers. The length/breadth measurements range from 61 × 45 mm to 44 × 35 mm, with an average of 50 × 44 mm. The average thickness for these seven scrapers is 30 mm in contrast to 14 mm for the side scrapers and 18 mm for the end scrapers described above.

Burins (3 specimens) Two examples are angle burins in which the spalls have been detached from a transverse, trimmed edge. In one specimen, made from fine-grained quartzite, a single spall has been removed from one side. The second angle burin is made from Naibor Soit quartzite and is similar to the first, except that at least two spalls have been removed. The third specimen, also of Naibor Soit quartzite, has a single spall removed on either side of a broken transverse edge. All three burins exhibit wear on the chisel edges. Measurements are 40 × 24 mm, 46 × 21 mm and 41 × 24 mm.

Laterally trimmed flakes (19 specimens) The materials include Naibor Soit quartzite (eight specimens), fine-grained quartzite (seven specimens), phonolite (two specimens), and one each of vein quartz and basalt. In fourteen of these tools the two trimmed edges converge to form small projections which are generally at the tips of flakes, but in three examples they are situated laterally; they are not as robust as the points classified as awls. The retouch is unifacial and only on the dorsal aspects of the flakes, the primary surface being devoid of trimming or any form of chipping. In two cases the extreme tips of the points have been broken off. The remaining five specimens are flakes trimmed along both lateral edges, but lacking points. Measurements range from 66 × 52 mm to 21 × 16 mm, with an average of 38 × 26 mm.

Outils écaillés (33 specimens) There are twenty-two double and eleven single-ended examples. The entire series is made from quartzite and includes an unusually high percentage of fine-grained quartzite (23 per cent). Typical scaled utilisation is present on all these specimens and the edges are generally concavo-convex, but in a few specimens are nearly straight. These appear to be less heavily used and the extent to which the edges are curved may perhaps depend on the degree of wear. Five of the single-ended examples exhibit crushed utilisation on the butt ends, which are quite narrow. This utilisation is similar to that seen on the punches, but to a less marked extent. Both single and double-ended specimens are within the same size range, for which the length/breadth measurements are from 39 × 28 mm to 20 × 16 mm, with an average of 28 × 22 mm.

Punches (14 specimens) All these specimens are of quartzite. They are very similar in appearance, and consist of oblong, thick-set fragments, rectangular or polyhedral in cross section, with crushing at both ends, and generally narrow in relation to their width. Length/breadth measurements range from 52 × 28 mm to 27 × 14 mm, with an average of 33 × 19 mm.

Pitted anvils and hammerstones (107 specimens) The pitted anvils and hammerstones can be subdivided into four categories, based on the number and position of the pits, as follows:

(a) single pits
(b) pits on opposite sides
(c) twin pits
(d) multiple pits.

All except one of the pitted stones can be accommodated within these four categories. The exception is a specimen bearing two sets of twin pits in addition to a number of single pits. This has been included in category (c), in view of the absence of any similar specimens.

The majority of these stones consist of unmodified water-worn cobbles, but there are also a few angular blocks, as well as a number of heavy-duty tools, such as choppers and discoids, which have been re-utilised as anvils or hammerstones and now bear pits, sometimes on the trimming scars. A proportion of the pitted cobbles also exhibit bruising and battering on the extremities, of the type usually seen on hammerstones. Trachyandesite and basalt are the most common rocks and account for ninety-seven specimens, of which forty-two are vesicular. Variation in overall weight and size of the stones is considerable, from 4,400 g (217 × 201 × 121 mm) to 100 g (53 × 44 × 37 mm). Owing to the fact that most of the specimens have not been modified they also vary in form, but the pits are generally situated on relatively wide, flat surfaces, although in the case of multiple pits they are also present on lateral edges. The interiors of the pits are rough and pecked, not worn smooth by abrasive action. A proportion are circular, but a greater number are oblong; others exhibit a broad groove extending from one side of the pit, tapering as it becomes shallower. Besides clearly demarcated and deeply indented pits there are a number of cobbles with small, shallow indentations set close together. The pits range in length/breadth/depth from 80 × 65 × 21 mm to 7 × 6 × 0.5 mm.

There seems little doubt that these specimens are the result of bipolar flaking, particularly of quartzite. It is impossible to determine from the nature of the pits whether they represent anvil or hammerstone utilisation and it can only be assumed that the heavier stones served as anvils and the lighter as hammerstones, to strike the upper end of the 'core' held on the anvil. The ratio between the weight of the stones and size of pits is not consistent, but in some cases rather larger pits are present on the heavier stones. In both single and twin-pitted specimens all the examples are of a size which could be held in the hand comfortably. The average weight in relation to the size of the pits in the four categories (see Fig. 5.4) is as follows:

	weight	mean diameter of pits
(a) single pits	382 g	12.3 mm
(b) pits on opposite sides	500 g	10.0 mm
(c) twin pits	632 g	16.3 mm
(d) multiple pits	798 g	15.9 mm

(a) *Single pits* (28 specimens). These consist of twenty-two unmodified cobbles, five cobbles with battered utilisation at the extremities and one unifacial side chopper. The entire series is of trachyandesite or basalt, twelve being vesicular. The range in weight is from 775 g (116 × 90 × 62 mm) to 110 g (53 × 45 × 36 mm). The pits vary in form, some are shallow and poorly defined and others oblong with a groove. In two oblong grooved examples the pit has a slight raised ridge in the centre, transverse to the length of the groove, but too ill defined for the hollows on either side to be classified as separate pits. The pits in this category range in length/breadth/depth from 48 × 46 × 3 mm to 10 × 8 × 3 mm, with an average of 21 × 15 × 3 mm.

(b) *Pits on opposite sides* (40 specimens). This is the most numerous category and also includes the three largest specimens in the collection, which probably served as anvils. Twenty-three specimens are unmodified cobbles, seven are tools consisting of four choppers, two discoids and one scraper, the remaining ten are cobbles with battered utilisation at the extremities. The rocks comprise trachyandesite and basalt (thirty-seven specimens, of which all are vesicular), green welded tuff (one discoid and one chopper) and quartzite (one discoid). The series ranges in

UPPER BED IV

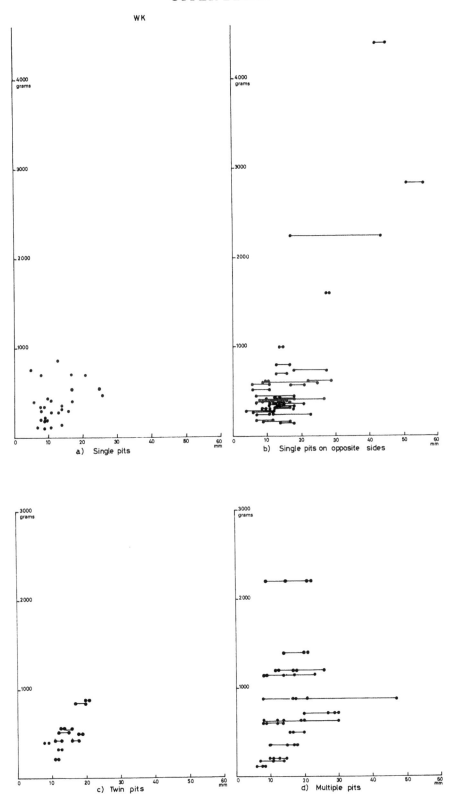

Fig. 5.4 WK: pitted anvils and hammerstones; diagrams to show the size of the pits (mean diameter) in relation to the weight of the stones

weight and size from 4,400 g (217 × 201 × 121 mm) to 175 g (90 × 58 × 25 mm).

The pits on either side of these stones are usually of different sizes. A deep or wide pit may be present on one face, whilst the pit on the opposite face is ill defined and shallow. The three largest specimens are of coarse vesicular lava. In each case the surface bearing the pit has been hollowed out into a shallow depression over the entire area, in addition to the actual pits, which are situated approximately in the centre. The largest specimen, weighing over 4 kg, and two of the smaller ones are so symmetrically circular that they may have been artificially shaped, although there is no visible evidence of such preparation. The seven tools bearing pitted utilisation consist of the following: two crude unifacial end choppers in which pits are present on the upper and lower cortical surfaces; a heavily rolled end chopper with one pit on the cortical surface and a second pit on the opposite face, within one of the trimming scars; another end chopper with well-defined circular pits on both upper and lower faces; two discoids and a scraper made on a discoidal lava cobble also all with centrally placed pits. In all three of these last specimens the pit on one face is relatively deep and well defined, whilst that on the opposite face is shallow and irregular.

(c) *Twin pits* (11 specimens). Eight examples are unmodified cobbles, two are choppers and one is a cobble with battered extremities. Except for two specimens of welded tuff, the series is of trachyandesite or basalt, two being vesicular. The range in weight and size is from 980 g (117 × 86 × 67 mm) to 230 g (81 × 70 × 37 mm).

In nearly all cases the twin pits are oblong and orientated transverse to the long axis of the tool, either at right angles or obliquely. Four examples are pitted on one face only, two have a single pit on the opposite face and one example, made on an oblong cobble, has twin pits on both faces and grooves on both lateral edges. The two pits are always set close together and parallel, separated by an intervening ridge which is generally below the level of the original surface and has been subjected to a certain amount of wear. One of the choppers with twin pits consists of a rolled bifacial side chopper made on an oblong cobble. It bears two pits on the cortex surface, on one face, and a second pair on the opposite face, within the largest of the trimming scars. There are also two additional single pits on the butt. The second pitted chopper is made on a flat cobble in which one lateral edge has been unifacially trimmed. Two adjacent pits are present on one face, one of which is narrow, deep and sharply defined, while the second is shallower and more irregular.

(d) *Multiple pits* (13 specimens). The stones in this series are more irregular in form and include a number of specimens larger than any in categories (a) or (c). In consequence, more surfaces suitable for use were available. The maximum number of pits on a single example is eight, with an average of five for the total, discounting very small indentations. Eleven specimens consist of cobbles or blocks, either unmodified or with pits on broken surfaces that already existed at the time of use. One is a unifacial side chopper which also exhibits heavy battered utilisation at both ends and one a roughly rectangular, steep-sided cobble with pits on the upper and lower surfaces as well as on the circumference. The series ranges in weight from 2,300 g (173 × 115 × 82 mm) to 150 g (107 × 71 × 56 mm). In one specimen, the surface on one side has been entirely hollowed out into an oblong shallow depression measuring 110 × 58 mm, in the centre of which are two slightly deeper indentations. Another specimen, mentioned earlier, bears two sets of twin pits as well as a number of single pits. It has been classified under category (c), but could equally well be included here.

Broken pitted anvils and hammerstones (15 specimens) The broken specimens exhibit no unusual features. So far as it is possible to ascertain, they have been fractured during or after use since in most cases the line of fracture passes through a pit.

In Fig. 5.4 the weight of the hammerstones and anvils has been plotted in relation to the mean diameter of the pits. It will be seen that the single and twinned specimens are fairly constant in weight and in the size of the pits, none exceeding 1,000 g and all the pits lying between 5 and 25 mm in mean diameter. The majority of specimens in categories (b) and (c), that is with pits on opposite sides and multiple pits, also lie within the same range of weight and pit dimensions, but a proportion are considerably heavier and there is also a wider range in size of the pits.

Utilised material

Anvils (7 specimens) In addition to the pitted anvils there are one lava and six quartzite specimens with battered right-angled edges. They are generally cuboid and the battered edges are on the circumference of a flat surface which may be circular or rectangular. Measurements range from $119 \times 84 \times 70$ mm to $67 \times 58 \times 35$ mm, with an average of $87 \times 70 \times 60$ mm.

Hammerstones (9 specimens) These comprise seven basalt or trachyandesite cobbles, one block of quartzite and one of trachyte, all of which have been battered with varying degrees of force. The two largest specimens are oblong and measure 164 mm and 154 mm in length. They have both been used so heavily that a number of large flakes have been knocked from the ends. One of these specimens also appears to have been used in a similar manner to the pitted anvils, since one face is pitted and battered although not deeply enough to form a hollow. Measurements range from $164 \times 100 \times 83$ mm to $67 \times 43 \times 33$ mm, with an average of $99 \times 71 \times 59$ mm.

Cobbles and blocks (58 specimens) Sixteen blocks and one pebble of quartzite, thirty-four cobbles of basalt or trachyandesite, two cobbles of nephelinite, one cobble and three blocks of phonolite and one block of trachyte are broken, battered and flaked by use but do not exhibit the bruising of the surfaces considered characteristic of hammerstones. Measurements range from $139 \times 92 \times 73$ mm to $62 \times 52 \times 27$ mm, with an average of $85 \times 66 \times 45$ mm.

Light-duty flakes, etc. (24 specimens) The majority of these specimens are of Naibor Soit quartzite (199), but there are also thirty-seven of phonolite, eleven of basalt or trachyandesite and one of gneiss. They bear traces of utilisation on the edges, which can be subdivided approximately into five groups: concave (fifty-two specimens), convex (fifty-six specimens), straight or nearly straight (eighty-eight specimens), numerous edges on angular fragments (nineteen specimens), and crushing of the type seen to a more marked extent in *outils écaillés* and punches (thirty-three specimens). The whole series consists of irregular fragments, generally small, and it seems that any convenient piece of stone was used, regardless of shape.

The utilisation on the concave-edged series varies from slight chipping along a widely curved edge to small, rather deep notches that may have been caused by grinding against a hard substance. Most of these specimens consist of broken flakes, but some are angular fragments and seven are flakes in which only the tips are missing. Length/breadth measurements range from 50×45 mm to 16×31 mm, with an average of 28×32 mm. (Length is measured from the utilised edge to the butt and width is the maximum transverse measurement.) The majority of specimens with utilisation on convex edges consist of broken flakes or angular fragments, but eleven complete flakes show this type of wear. It is either shallow chipping on thin edges or rather steep flaking on more thick-set edges. Length/breadth measurements range from 58×83 mm to 17×22 mm, with an average of 25×29 mm. The chipping on approximately straight edges is most often on both sides of relatively thin, sharp edges but there are some examples with thicker edges in which the wear is on one side only. These are sometimes blunted. Broken flakes and angular fragments are far more common than whole flakes, which amount to only four specimens. Length/breadth measurements range from 68×36 mm to 13×18 mm, with an average of 23×31 mm.

The nineteen angular fragments showing utilisation on a number of edges are angular and asymmetrical, but a few are subspherical, although they are too irregular to class as such. The

ANALYSIS OF THE INDUSTRIES FROM WK UPPER CHANNEL

	Numbers	Percentages
Tools	429	3.9
Utilised material	429	3.9
Debitage	10047	92.2
	10905	100.0
Tools		
Bifaces	192	44.7
Choppers	35	8.1
Polyhedrons	7	1.6
Discoids	17	4.0
Spheroids/subspheroids	25	5.8
Scrapers, heavy-duty	25	5.8
Scrapers, light-duty	57	13.3
Burins	3	0.7
Laterally trimmed flakes	19	4.4
Outils écaillés	33	7.7
Punches	14	3.3
Sundry tools (picks)	2	0.5
	429	99.9
Utilised material		
Pitted anvils/hammerstones	107	24.9
Anvils	7	1.6
Hammerstones	9	2.1
Cobbles and blocks	58	13.5
Light-duty flakes, etc.	248	57.9
	429	100.0
Debitage		
Whole flakes	405	4.0
Broken flakes	4357	43.4
Core fragments	5281	52.6
Cores	4	0.0
	10047	100.0

utilisation generally takes the form of crushing and blunting on thick edges. Length/breadth/thickness measurements range from 68 × 51 × 36 mm to 29 × 23 × 17 mm, with an average of 33 × 33 × 24 mm.

The crushed and scaled wear on opposite edges, to be seen on thirty-three specimens, has clearly been caused by the same type of utilisation as in *outils écaillés* and punches but is less pronounced and more irregular. In three cases the pressure exerted at the ends has detached flakes from either side that simulate burin spalls. These specimens range in length/breadth from 43 × 32 mm to 22 × 16 mm, with an average of 28 × 20 mm.

Debitage This comprises 405 complete flakes, 4,357 broken flakes, 5,281 angular fragments and four cores.

It will be seen from the accompanying table that almost equal numbers of complete flakes are made from lava and Naibor Soit quartzite, but quartzite is by far the most common material in the broken flakes and angular fragments. Among the complete flakes, 266 are end struck and 139 side struck. The end flakes range in length/breadth from 134 × 109 mm to 14 × 13 mm, with an average of 44 × 32 mm; 288 are divergent, 109 convergent and eight approximately parallel sided. Lava flakes, in general, are larger than the quartzite; the average measurements for the two series being 50 × 41 mm and 38 × 29 mm. Lava predominates in the side flakes, with eighty-one examples against fifty-eight quartzite. Again, the average size for the lava series is greater, 40 × 57 mm, compared to 29 × 37 mm for the quartzite series. Quartzite is by far the most common material in both the broken flakes and the core fragments, amounting to 85 per cent and 96 per cent respectively.

Two cores of Naibor Soit quartzite consist of thick slabs from which flakes have been detached round the circumference. In the first specimen the largest negative scar is of a side flake measuring 72 × 82 mm. The core itself measures 156 × 121 × 80 mm and has been flaked from only one of the flat surfaces. The second specimen measures 132 × 119 × 61 mm and has been flaked on the circumference from two directions. Three flakes have been removed, the largest measuring 58 × 60 mm. The third quartzite core is conical, with a flat base, and is in fact a particularly large and unusual flake, since the base consists of a flake scar with a positive bulb of percussion. One flake has been detached from this surface and several others from a ridge running from the base to the top of the specimen. Maximum measurements are 132 mm (length of base) by 98 mm (height) by 112 mm (width of base). The most interesting core consists of a boulder of basalt or trachyandesite weighing just under 4 kg from which eight large flakes have been struck bifacially. The largest of these is again a side flake of which the scar now measures 98 × 113 mm but originally

UPPER BED IV

ANALYSIS OF THE INDUSTRIES FROM WK UPPER CHANNEL

	Quartzite	Fine-grained quartzite	Gneiss, feldspar, etc.	Phonolite	Nephelinite	Trachyte	Basalt/trachyandesite	Lava, indeterminate	Totals
Heavy-duty tools									
Bifaces, picks, choppers, polyhedrons, discoids, spheroids, subspheroids, scrapers	125	2	2	14	—	1	154	5	303
Light-duty tools									
Scrapers, burins, laterally trimmed flakes, *outils écaillés*, punches	96	16	1	4	1	1	7	—	126
Utilised material, heavy-duty									
Pitted anvils/hammerstones, anvils, cobbles and blocks	28	—	4	4 WT	2	2	140	1	181
Utilised material, light-duty									
Flakes, etc.	199	—	1	37	—	—	11	—	248
Debitage									
Whole flakes	179	14	—	29	4	7	170	2	405
Broken flakes	3714	131	1G	81	12	22	387	10	4358
Core fragments	5108	41	1F	28	—	16	83	2	5279
Cores	3	—	1P	—	—	—	1	—	5
	9452	204	11	197	19	49	953	20	10905

G = granite
F = feldspar
P = purple quartzite
WT = welded tuff

was longer since it has been truncated by the removal of another flake from the opposite face of the core. This specimen is the only core found at the site which could have provided the large lava flakes from which the handaxes and cleavers were made. If the quarry site could be discovered it is likely that many such cores would be found. Measurements are 183 × 142 × 130 mm.

The WK East sites and PDK Trenches I–III

These sites lie on either side of a long spur that slopes gently down into the gorge from the southern rim. WK East A and C are on the western edge of the spur and PDK Trenches I–III on the opposite edge only 25 m distant from WK East A. A small trench (WK East B) was also excavated further down the west side of the spur where artefacts and fauna were seen to be eroding, but the material proved to be in disturbed context, probably derived from the *in situ* horizon found at the other three sites and re-worked during the cutting of the gorge. This material is not described here.

The stratigraphic position of all three sites appears to be very similar. At each there was a channel deposit of sand with clay lenses and patches of conglomerate that became coarser at the base and filled a channel cut into Tuff IVB. The extent to which the tuff had been excavated varied. At WK East it has been entirely removed

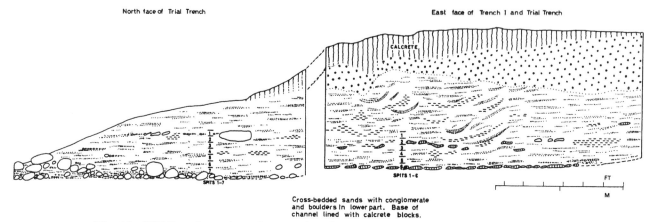

Fig. 5.5 WK East A: sections along north and east faces of Trench I and the Trial Trench

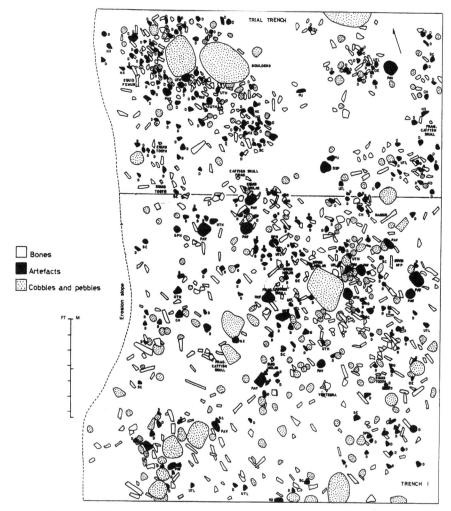

Fig. 5.6 WK East A: plan of finds in the upper part of the channel filling (Spits 1 to 3)
(for abbreviations used in Figs. 5.6–5.11, see p. xiv)

UPPER BED IV

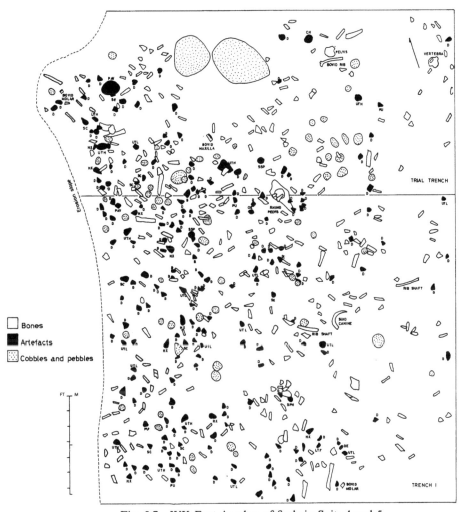

Fig. 5.7 WK East A: plan of finds in Spits 4 and 5

and the base of the channel with limestone blocks rested on the clay substratum. At WK East C, it had been reduced in thickness and now measures no more than 0.3 m in depth in contrast to areas where it has not been eroded and measures approximately 1.2 m. At PDK Trenches I–III it had barely been reduced and retained the greater part of its original thickness. Artefacts, faunal remains, cobbles and pebbles were found densely concentrated within the lower part of the channel filling; in general, the specimens were unrolled although a proportion were slightly abraded.

In view of the proximity of the three excavated sites and the close similarity of the stratigraphy exposed in the trenches excavated, there seems little doubt that all three are within one channel or a related system of contemporary channels, which appear to have continued to the southeast at least as far as WK.

WK East A

A trial trench 4 m long (east–west) and 1.8 m wide was first excavated, followed by a second parallel trench 3 m wide. The section exposed in the eastern face reached a maximum depth of just under 2 m. At the base, resting on the grey clay normally found underlying Tuff IVB, was a series of rather flat calcrete blocks forming the base of the channel. Above the calcrete was 60 to 70 cm of reddish bedded sands with a basal layer of large cobbles and small boulders in the southern

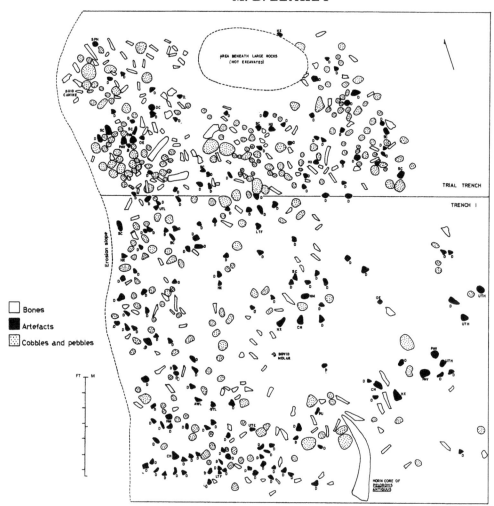

Fig. 5.8 WK East A: plan of finds in the upper part of Spit 6

part of the trenches (Fig. 5.5). Artefacts and faunal remains, cobbles and pebbles were found throughout the 60 to 70 cm of sands. This level was overlain by approximately 50 to 70 cm of cross-bedded sands, also reddish coloured, in which a number of small channels could be seen, sometimes no more than 70 cm wide and 20 cm deep. Above this there was a sandy clay of uneven thickness overlain by calcrete. To the north the upper part of the sandy clay appeared to have been eroded, perhaps during the cutting of the gorge.

For the purpose of excavation the lower 60 to 70 cm were divided into 10 cm spits and finds in each spit plotted (Figs 5.6 to 5.11). Spit 6, however, proved so exceptionally rich in artefacts and bones that it was necessary to plot the finds in three different levels. A curious feature in the higher spits was two large boulders of vesicular lava, 1.2 × 1 m and 0.8 × 0.7 m, that projected upwards into the overlying cross-bedded sands. Dr Hay considered these boulders too large to have been transported by the force of water indicated by deposits in the channel fill. The tops of both boulders appeared to have been artificially pitted and damaged and it is possible that they were brought in as large anvils or even as seats. The latter interpretation was put forward tentatively for the isolated cranium of *Giraffa jumae* found at the Acheulean site of EFHR in Bed II (Leakey 1971).

Apart from one horn core of *Pelorovis antiquus*

UPPER BED IV

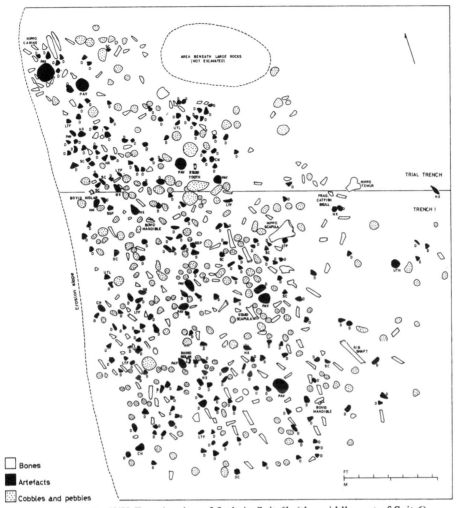

Fig. 5.9 WK East A: plan of finds in Spit 6b (the middle part of Spit 6)

the faunal remains were fragmentary and many small chips of bone occurred in addition to the larger numbered pieces shown in the plans.

The vertical positions of artefacts, fauna and cobbles in this part of the channel do not follow the same pattern as at WK East C nor as in the channel at HEB East although the horizontal distribution has features in common. Both artefacts and cobbles reach a peak in spit 6 whereas faunal specimens were most abundant in spit 4. All three categories decrease in the lowest spit (Fig. 5.12).

The industry
This consists of 17,783 specimens, comprising 564 tools, 759 utilised pieces and 16,460 debitage.

Tools

Bifaces (74 specimens) The series consists of fifty-one complete, two unfinished and twenty-one broken examples. Phonolite is the most common material with thirty specimens, twenty-one are of basalt or trachyandesite, fourteen of quartzite and nine of trachyte. Cleavers are not represented, and in all the handaxes where the tips are preserved they are sharply pointed with no tendency towards chisel ends. The retouch is coarse and even the phonolite specimens show irregular, deeply indented scars and some step flaking. This is in contrast to phonolite bifaces from other sites, notably WK, where the elaborate trimming contrasts markedly with the bold,

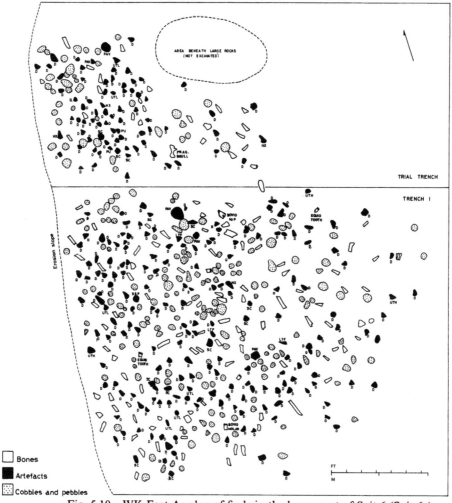

Fig. 5.10 WK East A: plan of finds in the lower part of Spit 6 (Spit 6c)

minimal flaking of the bifaces made from coarse-grained lavas. In general, the cross sections of the handaxes are thick relative to length and breadth. Some examples are made on side and end flakes in which the bulbs are quite shallow, although the primary surface has been flaked to a greater extent than in the thick flakes with protuberant bulbs characteristic of the WK series. Three of the quartzite specimens are made on pieces of tabular material and retain parts of the natural cleavage planes on the upper and lower faces; one example has the tip skewed out of the centre line. The phonolite series includes a number of small specimens that must be classed as bifaces on account of their morphology, but are diminutive in size. Eight are under 80 mm in length and none

in the series exceeds 112 mm. They are generally asymmetrical and do not conform to a regular pattern. A number of the larger phonolite specimens have been broken longitudinally, seemingly along fault planes in the rock. Eight of the basalt/trachyandesite specimens retain areas of cortical surfaces on the dorsal faces. Most of this series are crudely flaked and asymmetrical, but there are two elongated examples recalling *limandes* that are more regularly flaked and more symmetrical than the rest. The trachyte series is noticeably more heavily abraded than any other, and markedly variable in size. It includes two specimens with square, heavy butts, two ovates and one with a skewed tip similar to the quartzite specimens mentioned above. The average lengths

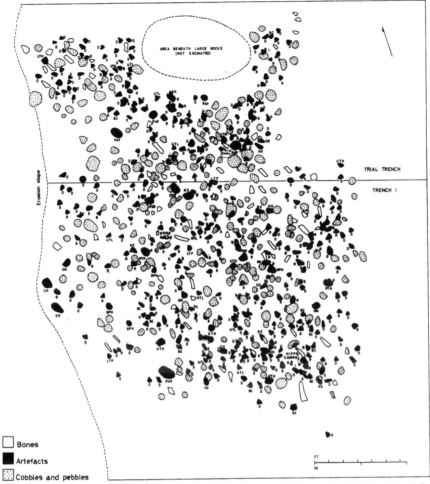

Fig. 5.11 WK East A: plan of finds in basal part of channel fill (Spits 6d and 7)

for the series in each of the four materials is: phonolite 81.5 mm, quartzite 89 mm, trachyte 112 mm and trachyandesite/basalt 108 mm. (This series is analysed by Drs Roe and Callow.)

Choppers (17 specimens) This series is variable in size, form, trimming and nature of the working edges. There appears to have been no selection of preferred shape in the stones from which the tools were made. The working edges have been prepared on random pieces of material such as whole and split cobbles or cuboid blocks. Five examples are of quartzite, one of trachyte and eleven of trachyandesite or basalt. All the choppers are rolled to some extent, in particular one of the end choppers made from trachyte. In this specimen and in two side choppers, one of which is unifacial, the working edges are unusually straight and even; the remainder are jagged, as usually seen in choppers. The largest side chopper, made on a flat, sub-triangular cobble, shows extensive shallow pecking on one face as well as on the butt, which is nearly vertical, but the indentations are too diffuse to rank as pits. The fourteen side choppers range in length/breadth/thickness from $122 \times 135 \times 166$ mm to $54 \times 68 \times 54$ mm. The measurements of the three end choppers are $106 \times 72 \times 53$ mm, $102 \times 73 \times 51$ mm and $72 \times 49 \times 28$ mm.

Polyhedrons (3 specimens) Two are of trachyandesite or basalt and one of phonolite. They are slightly rolled. The largest specimen ($94 \times 76 \times 60$ mm) has been extensively flaked over the

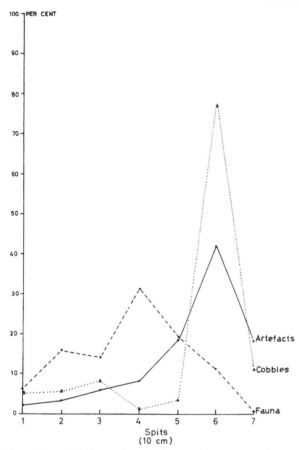

Fig. 5.12 WK East A: graph to show proportionate occurrences of artefacts, faunal remains and cobbles in the channel

Fig. 5.13 WK East A: sundry small tools

entire surface with numbers of small, intersecting scars. It resembles a faceted spheroid, although it is too asymmetrical to be classed even as a subspheroid. The two smaller specimens are similar in size (46 × 46 × 52 mm and 46 × 41 × 55 mm) and have been less extensively flaked, but exhibit a number of cutting edges which appear to have been used.

Discoids (12 specimens) These tools also vary greatly in size and form. They range from a massive, thick specimen measuring 98 × 77 × 62 mm to a miniature example measuring only 27 × 27 × 14 mm. Six are of trachyandesite or basalt, three of quartzite and one each of fine-grained quartzite, phonolite and trachyte. Two of the quartzite specimens are very similar in size and form, both are particularly thick in cross section, with a high keel on either face. The edges are heavily chipped and blunted. The most highly finished specimen is of fine-grained quartzite. It is not quite symmetrical and resembles a cordiform handaxe lacking the tip. The miniature discoid is regularly flaked on both faces round the entire circumference and also shows traces of wear. The trachyte specimen is included doubtfully in the series, on account of its marked asymmetry, but seems likely to have served the same purpose as standard discoids. The average length/breadth/thickness measurements are 72 × 57 × 35 mm.

Spheroids/subspheroids (20 specimens) There are no truly spherical specimens in this series, all are angular and rather asymmetrical. The majority, however, have been battered to a greater or lesser degree and are, therefore, placed in this category, although it is possible that some may be battered cores. The largest example (104 × 94

× 85 mm) is made of fine-grained quartzite. The remainder are all of the coarse quartzite from Naibor Soit. The smallest specimen measures 26 × 25 × 24 mm and is the most nearly symmetrical. The average length/breadth/thickness measurements for the twenty specimens are 51 × 47 × 42 mm.

Scrapers, heavy-duty (5 specimens) One consists of a slab of quartzite 28 mm thick in which one edge is curved and has been extensively chipped and blunted. The tool is 81 mm long (from working edge to butt) and 86 mm wide. One face is slightly pitted. Three specimens are on lava end flakes, two have relatively wide and steeply trimmed working edges at the tips and the third is a side scraper. In the larger example (76 × 78 × 33 mm) the working edge is uneven, with small projections at the intersections of the trimming scars. The working edge in the smaller specimen (69 × 59 × 22 mm) is more regularly trimmed and not so widely curved. The side scraper has an uneven working edge and measures 52 × 81 mm. The fifth specimen consists of a large, thick quartzite flake (80 × 54 mm) with a chipped notch on one lateral edge measuring 35 × 2.5 mm.

Scrapers, light-duty (223 specimens; Fig. 5.13) These may be subdivided into four categories, side, end, nosed and hollow or notched scrapers. A few might be classed as convergent scrapers, but they are so rare that it does not seem justifiable to set up a separate category. As usual in the Developed Oldowan, side scrapers are the most common type.

Side (105 specimens). All except seven are made of quartzite. A proportion are made on pieces of tabular material, but more often, the tools are on flakes or other fragments. Both size and form are variable. The working edges may be either straight or curved, sometimes with slight projections. Trimming also varies from steep to shallow. Four of the seven lava specimens are larger than any made from quartzite. They consist of scrapers on side flakes which have been trimmed along the edge opposite the butt. These specimens average 48 mm in length and 82 mm in width. The remainder of the series varies in length/breadth from 39 × 59 mm to 13 × 20 mm, with an average of 20 × 54 mm.

End (38 specimens). Five examples are of phonolite, seven of fine-grained quartzite and the balance of quartzite. They are made on whole and broken flakes and on angular fragments. In five specimens the working ends appear to have been broken off during use. Like the side scrapers, there is no standardisation of form, although none of the end scrapers is as large as the four lava side scrapers mentioned above. Only one example is on a piece of angular quartzite. The working edges are seldom symmetrical and may be nearly straight, convex or oblique. One example, measuring 26 × 18 mm, has two scraping edges, at right angles to one another, and is a combined end and side scraper. Length/breadth measurements range from 43 × 30 mm to 14 × 12 mm, with an average of 28 × 22 mm.

Nosed (16 specimens). Each of these tools has been trimmed to a projecting scraper edge, forming a 'nose'. In some examples, the trimming extends round the greater part of the circumference, as in the scrapers from the Developed Oldowan B site at BK, which were described as perimetal scrapers (Leakey 1971). In the present series, however, the trimming of the circumference is not such a constant feature as the projecting 'nose'. Trimming is generally steep and the edges tend to be more even than in either the side or end scrapers. Length/breadth measurements range from 36 × 38 mm to 15 × 14 mm, with an average of 23 × 23 mm.

Hollow or notched (64 specimens). In these tools, the notches seem to be the result of deliberate trimming rather than haphazard use, such as might be caused by pressure against a hard edge. There are, in addition, a number of other specimens in which the notches appear to be merely the result of use; these are described with the utilised material. The number of trimming scars as well as the size of the notches has been taken into account when separating the retouched from the merely utilised specimens. However, repeated use of the same notch could well produce numerous scars and it is probable that some specimens have been incorrectly classified. The notches are variable in length and depth and are situated in a haphazard manner, but in elongate specimens they are more often on a lateral edge than at the tip. Some specimens exhibit two notches, either on different edges or on the same edge but separated by an untrimmed area. Three of the four phonolite specimens and one of fine-grained quartzite are larger than any of Naibor Soit quartzite, from which the greater part of the series is made. The tools range in length and breadth from 61 × 48 mm to 18 × 17 mm, with an average of 30 × 24 mm. The notches range in width and depth from 25 × 3 mm to 10 × 1.5 mm, with an average of 15.5 × 2.1 mm.

Burins (6 specimens) All six specimens are of quartzite and appear to have had burin spalls intentionally removed from one lateral edge. However, the use of bipolar technique for flaking quartz could have resulted in the accidental removal of similar spalls. One example is an angle burin with a single spall struck from a trimmed notch at one end of a broken flake, 29 mm long and 15 mm wide. The second example also has one spall struck from a utilised edge at the end of a broken flake (30 × 20 mm). The third specimen has single spalls struck from either end of a broken flake (32 × 25 mm). These three tools show no trace of battering or crushing at the ends from which the spalls have been struck, but in the remaining three, single spalls have been detached from bruised extremities, so that it is possible they are the result of bipolar flaking. These three specimens measure 25 × 14 mm, 29 × 15 mm and 23 × 19 mm.

Awls (35 specimens) These small, sharply pointed tools are similar in all respects to the awls from Developed Oldowan sites in Upper Bed II. Except for one example of phonolite and one of vein quartz, all are of quartzite. The points are usually formed by a retouched notch on either side, generally trimmed from one direction only, but in four examples the trimming is from opposite directions. Although the points are sharp, they are generally quite short, not more than a few millimetres long. They are situated either at the tips of small flakes or on a lateral edge. The largest is the phonolite specimen which measures 54 × 39 mm, but a number of specimens do not exceed 20 mm in length. The average length/breadth measurements are 24 × 23 mm.

Laterally trimmed flakes (39 specimens) These flakes are all made of quartzite and an unusually high proportion are from the fine-grained variety (sixteen examples). They exhibit trimming on one or both lateral edges. The trimming is usually on one face only, but occasionally is bifacial. Some examples are bluntly pointed, but in others the trimming does not extend as far as the tip. In addition to the complete specimens there are four tip ends and one butt end. In general, the flakes tend to have broad and heavy butts, but this is not a constant feature and the series is quite variable. Trimming is usually shallow, but in a few examples it is relatively steep. Four complete and one broken specimen are bifacially flaked and resemble miniature bifaces, but the trimming scars are shallower and the edges consequently straighter than in any tools classed as bifaces, even the diminutive specimen of the Developed Oldowan. The length/breadth measurements for the whole series range from a lava specimen 53 × 37 mm to one of quartzite 20 × 17 mm. The average length/breadth is 37 × 27 mm.

Outils écaillés, punches and pieces with scaled utilisation) This series consists of three groups of artefacts which are clearly related and appear to be directly associated with the use of bipolar flaking. They consist of *outils écaillés*, both single and double ended, rod-like specimens termed 'punches' which have been battered at one or both extremities, sometimes also showing incipient scaling, and fragments with scaled utilisation similar to that of *outils écaillés*. The most typical examples of each of these three groups are easily distinguishable, but there are also a number of borderline specimens which could be allocated to one or other category.

Outils écaillés (62 specimens). The entire series is of quartzite, the majority being of the coarse-

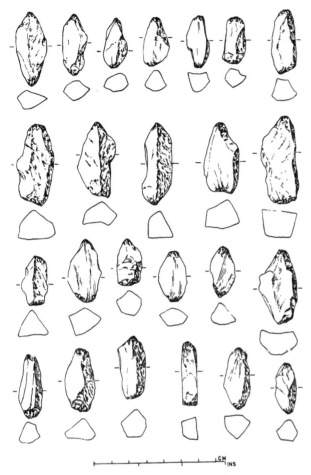

Fig. 5.14 WK East A: punches

grained variety from Naibor Soit. Thirty-four examples are double ended, twenty-six single ended and two probably broken. No specimen exceeds 47 × 26 mm and the average length/breadth for both the double and single-ended examples is 28 × 19 mm. A proportion of the working edges are typically concavo-convex, but the majority are nearly straight.

Punches (68 specimens; Fig. 5.14). Whereas a number of *outils écaillés* are made of fine-grained quartzite (sixteen out of sixty-two), only three punches are of this material, the rest being of Naibor Soit quartzite. The tools are elongate and narrow, usually battered at either end, but sometimes at one end only, and occasionally slightly scaled. The cross section is most often approximately rectangular, but may also be roughly triangular. Some are on parts of flakes, broken

previously, but they are essentially 'cores', that is residual pieces of material from which flakes have been removed. The range in length/breadth is from 56 × 26 mm to 21 × 8 mm.

Utilised material

Pitted anvils and hammerstones (68 specimens; Fig. 5.15) Although this series of pitted stones is smaller than that from the Acheulean site of WK, there is a clearer division between the weights and sizes of the four categories. The average weights of types (a) and (c), that is specimens with single and twin pits (see below), is less than half that of categories (b) and (d), with pits on opposite sides and multiple pits respectively. This suggests that the majority of specimens in the two heavier categories were used as anvils, whilst the lighter specimens served as hammerstones. This conclusion can only be applied in general terms since the stones in each of the four categories are by no means of uniform size except in the case of category (c), with twin pits, in which the heaviest specimen weighs 775 g and the average weight is 535 g. This group is also comfortable in shape, all the stones consisting of oblong cobbles in which the long axis of the pits is transverse or oblique to the length of the stones.

The following classification into four categories is based on well-marked pits. In all four groups the stones often show additional shallow pecked marks which have not been counted.

(a) *Single pits* (20 specimens). Sixteen examples consist of basalt or trachyandesite cobbles, ten of which are vesicular, two are of trachyte and a single specimen each is of phonolite and quartzite. All the pits are on relatively broad, flat surfaces. They are roughly circular, oblong or irregular in shape and do not include any unusually wide or deep examples. Three specimens consist of cobbles with battered utilisation at the extremities and in two examples the cobbles have been crudely flaked on the edges. The quartzite specimen consists of a cuboid block, flaked on opposite edges, with a shallow depression in the centre of a weathered flat surface. The range in weight and size for this series is from 1,330 g (148 × 104

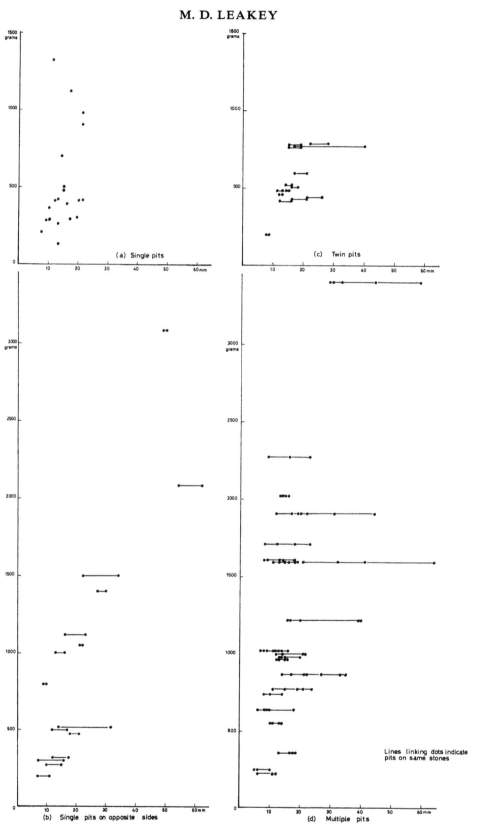

Fig. 5.15 WK East A: pitted anvils and hammerstones; diagrams to show size of pits (mean diameters) in relation to weight of stones

× 76 mm) to 130 g (88 × 56 × 22 mm), whilst the pits range in length, breadth and depth from 30 × 26 × 6 mm to 13 × 9 × 1 mm. (Average weights of stones and mean diameters of the pits in each category are shown as a table on p. 100.)

(b) *Pits on opposite faces* (19 specimens). Fourteen examples are of basalt or trachyandesite, four of which are vesicular, two of trachyte and one each of quartzite, gneiss and nephelinite. The five largest specimens, ranging in size from 192 × 156 × 96 mm to 116 × 114 × 93 mm, appear to have been artificially shaped. One example is circular, with smooth cortical surface remaining on the upper and lower faces, each of which bears a single, relatively deep circular pit, whilst the entire circumference is roughly pecked and no longer retains the original cortex. The remaining four examples are oblong, with rounded ends and relatively flat upper and lower surfaces, bearing single well-defined pits situated approximately in the centre. Both ends of these stones appear to have been roughly shaped by pecking and battering. In similar specimens of vesicular lava it is difficult to determine whether the pecked appearance of the surface is due to artificial agency or to natural hollows in the stone. One specimen in this series consists of a very heavily rolled oblong fragment of gneiss. The upper and lower faces are flat and exhibit single pits. There is also a broad groove on one edge, whilst two flakes have been struck from the opposite edge. The quartzite specimen is the smallest in the group, measuring 83 × 77 × 44 mm. It is approximately circular in shape, with nearly vertical sides and flat upper and lower faces, each of which bears a rather shallow pit. This series ranges in weight and size from 3,150 g (184 × 141 × 102 mm) to 200 g (67 × 59 × 35 mm) while the pits range in length, breadth and depth from 75 × 67 × 20 mm to 11 × 11 × 1 mm.

(c) *Twin pits* (7 specimens). This group is the most consistent in weight and form, as previously noted. Six specimens are oblong cobbles and one is a large end flake with twin pits on the dorsal surface; all are of basalt or trachyandesite. Four specimens bear only one set of twin pits, two have a pair on either face, and in two examples, in which the cobbles are triangular in cross section, there are three sets, one on each of the three flat surfaces. The pits are generally oblong and transverse to the long axis of the stones. In one example, the surface of the cobble is stained red while the interior of the pits and the battered ends lack red patina, indicating that the patina was prior to the pitting. It is possible that this stone was obtained from Bed III, where red staining is prevalent. Five of this series exhibit battering on the extremities in addition to the pits. The range in weight and size is from 775 g (130 × 80 × 63 mm) to 430 g (112 × 83 × 40 mm) while the pits range in length, breadth and depth from 46 × 43 × 5 mm to 20 × 13 × 4 mm.

(d) *Multiple pits* (20 specimens). The stones in this series are variable in form as well as in weight and size. Sixteen are of basalt or trachyandesite, six of which are vesicular, two of welded tuff and one each of quartzite and trachyte. The maximum number of pits on a single stone is twelve. These are on an oval cobble of coarsely vesicular basalt measuring 174 × 121 × 94 mm. In view of the nature of the material, it is possible that a number of other hollows, regarded as natural, may in fact be additional pits. This specimen contains one of the largest and deepest pits recorded; it measures 92 × 78 × 23 mm. The opposite face bears one relatively large central pit (38 × 34 × 10 mm) surrounded by a series of smaller pits, sometimes overlapping.

Two additional specimens are unusual. One of these is a hexagonal block of quartzite, measuring 73 × 73 × 62 mm, which bears a small but relatively deep pit on five of the six faces, all of which are relatively flat. The second specimen is of basalt and measures 105 × 90 × 74 mm. It bears a total of six deeply indented and four shallower pits as well as two grooves leading outwards from one of the large pits. In two of the large pits there are pairs of more deeply indented hollows, approaching twin pits, but not sufficiently distinct from one another to be classed as such. Four examples exhibit battering of the extremities and in two of the smaller specimens flakes have been detached from one edge. The series ranges in weight and size from 3,410 g (191 × 171 × 98 mm) to 225 g (72 × 67 × 44 mm) while the pits vary in length, breadth and depth from 92 × 78 × 23 mm to 9 × 9 × 1 mm.

	Average weight of specimens	Average mean diameter of pits
(a) single pits	453 g	14.0 mm
(b) pits on opposite sides	1040 g	20.5 mm
(c) twin pits	488 g	19.7 mm
(d) multiple pits	1250 g	16.0 mm

Hammerstones (9 specimens) These consist of cobbles with battering or bruising on one or both extremities. Seven examples have also had crude flakes removed. The scarcity of such hammerstones in this industry may be considered as confirmation of the prevalent use of the bipolar technique indicated by the pitted hammerstones and anvils. The length/breadth/thickness measurements range from 104 × 64 × 43 mm to 58 × 53 × 28 mm, with an average of 82 × 61 × 43 mm.

Cobbles (19 specimens) All, except one specimen of welded tuff and one of nephelinite, are of basalt or trachyandesite. The stones consist of water-worn cobbles, sometimes broken, that show evidence of utilisation, generally in the form of crude flakes and fractures. Two examples also show somewhat doubtful pitting. Length/breadth/thickness measurements range from 102 × 74 × 70 mm to 62 × 50 × 23 mm, with an average of 78 × 62 × 35 mm.

Blocks (24 specimens) Some of these specimens may, perhaps, be cores that have been used subsequently. All have had flakes removed and the edges are both chipped and battered. Two examples are of basalt or trachyandesite, one of quartzo-feldspathic gneiss and the balance of quartzite. One is broken. Sizes range from 166 × 86 × 57 mm to 33 × 32 × 27 mm, with an average of 78 × 60 × 45 mm.

Light-duty flakes, etc. (639 specimens) In addition to the flakes and other fragments with scaled utilisation, related to the *outils écaillés* and punches, there are a large number of specimens exhibiting chipping and blunting on the edges. Although it is impossible to subdivide them precisely, they can be classified into five groups, with chipping on straight, convex, concave or notched edges and with miscellaneous chipping on more than one edge. The wear is generally on one face only, but in a few specimens it is bifacial.

Fragments with scaled utilisation (38 specimens). These specimens probably represent early stages of wear in *outils écaillés* and punches, since the type of wear is similar but not so far advanced. Almost equal numbers are single ended (twenty) and double ended (eighteen). The single-ended examples usually have a flat, transversely broken edge forming the butts. The length/breadth measurements range from 42 × 30 mm to 27 × 13 mm, with an average of 29 × 19 mm.

Straight edges (148 specimens). These consist of broken flakes and angular fragments, as well as occasional whole flakes, in which there is chipping along one more-or-less straight edge. With the exception of one basalt specimen, all are of quartzite. The specimens do not conform to any pattern and the edges are seldom exactly straight. Utilisation is generally on one face only, but in a few examples chips have been removed from both faces. Length/breadth measurements range from 77 × 47 mm to 15 × 12 mm, with an average of 30 × 20 mm (in this and in the remaining three groups the length has been measured from the utilised edge to the opposite end, regardless of the shape of the specimens).

Notched edges (142 specimens). With the exception of four lava flakes the entire series is of quartzite. In common with the other categories of utilised material this group consists of small fragments of various shapes. One lava flake measures 55 × 38 mm, but the remainder are much smaller, with an average length/breadth of 19 × 22 mm. The notches are generally narrower and less deep than in the hollow scrapers. Many appear to have been caused by the edge being pushed against a hard object. The chipping within the notches is from one direction only and, in the

UPPER BED IV

ANALYSIS OF THE INDUSTRIES FROM WK EAST A

	Numbers	Percentages
Tools	564	3.2
Utilised material	759	4.2
Debitage	16460	92.5
	17783	99.9
Tools		
Bifaces	74	13.1
Choppers	17	3.0
Polyhedrons	3	0.5
Discoids	12	2.1
Spheroids/subspheroids	20	3.6
Scrapers, heavy-duty	5	0.9
Scrapers, light-duty	223	39.5
Burins	6	1.1
Awls	35	6.2
Laterally trimmed flakes	39	7.0
Outils écaillés	62	11.0
Punches	68	12.0
	564	100.0
Utilised material		
Pitted anvils/hammerstones	68	8.9
Hammerstones	9	1.2
Cobbles and blocks	43	5.7
Light-duty flakes and other fragments	639	84.2
	759	100.0
Debitage		
Whole flakes	360	2.2
Broken flakes	2970	18.0
Core fragments	13119	79.7
Cores	11	0.0
	16460	99.9

case of flakes or parts of flakes, it is usually from the primary flake surface. The average width and depth of the notches is 8.8 × 1.4 mm. (Measurements have been taken on half the series.)

Convex edges (170 specimens). This is the largest group of light-duty utilised material. All the specimens are of quartzite, with the exception of one broken flake of nephelinite. The average size is similar to that of the straight-edged and notched series. A few examples are on small pieces of tabular quartzite, but the majority are on broken flakes and angular fragments. The utilised edges are more irregular than in scrapers and are less intensively chipped. The length/breadth measurements range from 33 × 31 mm to 17 × 11 mm, with an average of 19 × 21 mm (measurements were taken on 50 per cent of the total).

Miscellaneous (141 specimens). The evidence of use is less marked in these specimens than in the preceding groups. Except for one broken phonolite flake and two chips of vein quartz, the whole series is of quartzite. The specimens consist mostly of broken flakes and angular fragments with a few complete flakes. The length/breadth measurements range from 42 × 36 mm to 19 × 12 mm, with an average of 23 × 18 mm.

Debitage (16,460 specimens)

The debitage was recovered in gradually increasing quantities from spit 1, which yielded 2.6 per cent of the total, to spit 6 with the maximum yield of 41.5 per cent. At the very base of the channel the quantity had declined to 18.1 per cent, a figure comparable to that from spit 5 (19.5 per cent). Although the coarse-grained quartzite generally exceeds the fine-grained variety in the ratio of 19:1, the fine-grained material is more common in the broken flakes recovered from spits 6 and 7, where it amounts to 35 per cent of the 1,646 quartzite specimens. In the complete flakes from these levels it amounts to 12 per cent. Core fragments or angular waste are the most common form of debitage and amount to 80 per cent of the total.

Complete flakes (360 specimens) In the quartzite series, 135 are end struck and eighty-two side struck. The end flakes range in length/breadth from 104 × 55 mm to 15 × 12 mm, with an average of 27 × 20 mm. Bulbs of percussion are very slight or often absent and only a few specimens bear striking platforms. These are variable and include some with unusually wide angles. This variability appears to be due to the flakes having been struck in a haphazard manner from blocks of quartzite and not from cores with prepared striking platforms. One hundred and fifty-nine are divergent, thirty-eight convergent and twenty-six approximately parallel-sided. The lava flakes amount to 143 specimens in which basalt and trachyandesite are the most common materials, followed by phonolite. Only eleven specimens are of nephelinite and trachyte. End and side flakes occur in almost equal numbers; the former range in

ANALYSIS OF THE INDUSTRIES FROM WK EAST A

	Quartzite	Fine-grained quartzite	Gneiss, feldspar, granite, welded tuff, etc.	Phonolite	Trachyte	Nephelinite	Basalt or trachyandesite	Lava, indeterminate	Totals
Heavy-duty tools									
Bifaces, choppers, discoids, polyhedrons, spheroids, subspheroids, scrapers	41	4	—	32	11	—	43	—	131
Light-duty tools									
Scrapers, burins, awls, laterally trimmed flakes, *outils écaillés*, punches	306	105	—	19	—	—	3	—	433
Utilised material, heavy-duty									
Pitted anvils/hammerstones, hammerstones, cobbles and blocks	23	1	4GN 1WT	1	5	2	83	—	120
Utilised material, light-duty									
Flakes, etc.	556	73	2VQ	2	—	1	5	—	639
Debitage									
Whole flakes	188	29	—	53	8	4	78	—	360
Broken flakes	1962	597	—	138	27	4	242	—	2970
Core fragments	12874	181	—	21	5	2	36	—	13119
Cores	8	2	—	—	1	—	—	—	11
	15958	992	7	266	57	13	490	—	17783

GN = gneiss
WT = welded tuff
VQ = vein quartz

length/breadth from 101 × 68 mm to 20 × 14 mm, with an average of 48 × 36 mm. The side flakes range from 60 × 114 mm to 16 × 23 mm, with an average of 37 × 48 mm; thus in overall size the end and side flakes stand very close. One hundred and ten specimens are divergent, seventeen convergent and sixteen approximately parallel-sided. The striking platforms are variable, as in the quartzite series, and range from broad, wide-angled platforms to minimal points of impact. A small proportion are shattered.

Broken flakes (2,970 specimens) Quartzite is again the most common material and accounts for 86 per cent of the total. Other materials are represented as follows: phonolite 4.6 per cent, basalt and trachyandesite 8 per cent, nephelinite and trachyte 1 per cent each. Spit 6 again provided a far greater number of specimens than any other, but there was not the graduated increase to be seen in the complete flakes. The broken flakes have not been measured and do not call for comment.

Core fragments (13,119 specimens) These constitute by far the largest group of debitage. Many specimens are chipped along the edges, but this seems more likely to be the result of indiscriminate bashing with hammerstones and anvils, using the bipolar technique, than to any form of utilisation as tools. Quartzite accounts for 99.5 per cent of this series, all other materials amounting to only 0.5 per cent. Naibor Soit quartzite accounts for 98.6 per cent of the quartzite specimens.

Cores (11 specimens) In addition to the debitage described above as core fragments, most of which probably represent residual 'cores' from bipolar flaking, there are eleven specimens in which regular flake scars are evident and which conform more nearly to the accepted definition of cores. It is probable that the flakes were detached from these cores by means of hammerstones by directed blows.

With the exception of one heavily rolled trachyte specimen, all these cores are of quartzite. They range in size from a massive example weighing 2.475 kg to a small specimen weighing only 40 g. The large specimen consists of an unusually thick slab of quartzite om which the natural flat cleavage plane has served as a striking platform for flakes struck round the circumference. A second, bifacially flaked platform is present on the opposite side, abutting on the flat surface at either end. The specimen can thus be described as a 'tea cosy' type of core. A second example, measuring 114 × 79 × 86 mm, is similar in type with a natural cleavage plane forming the main striking platform. The remaining specimens are formless, except for two which resemble crude tortoise cores. The average length/breadth/thickness for the eleven specimens is 68 × 57 × 48 mm.

WK East C

This locality was in close proximity to WK East A and the main purpose of the excavation was to identify the deposits overlying the channel since they had been removed by erosion at WK East A. For this purpose a step trench 1.2 m wide was dug from the rim of the gorge down to Tuff IVB (Fig. 5.16). The trench was then extended 1.8 m to the north. The section exposed was as follows: at the base, approximately 30 cm of Tuff IVB underlain by grey clay. Above the tuff in the northern part of the trench there was 1.3 m of bedded sands, similar to those at WK East A but with less evidence of current bedding. The lowest 40 to

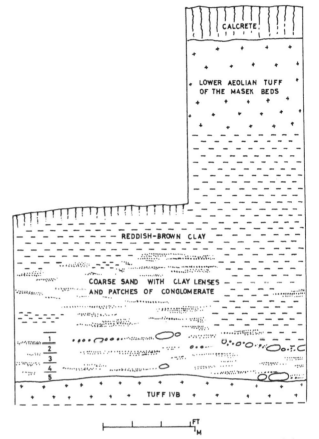

Fig. 5.16 WK East C: section along the east face of the excavation

50 cm likewise contained many artefacts and faunal remains but the basal layer was sparser and there were few large cobbles or boulders. In the southern part of the trench the sands were replaced by reddish clay with occasional sand lenses. The clay also extended into the northern part of the trench where it overlay the sands. One metre of the characteristic aeolian tuff of the lower Masek Beds occurred above the clay, overlain in turn by calcrete. The section exposed here demonstrated that the channel in this group of sites could confidently be placed in Upper Bed IV, above Tuff IVB, and that it was overlain by the Masek Beds.

The distribution of remains in the five plotted spits within the channel is shown on the graph in Fig. 5.17. A maximum depth of 50 cm or five spits is involved, whereas at WK East A specimens occurred throughout a depth of 70 cm. At Site C

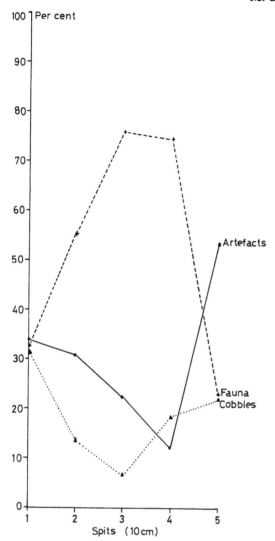

Fig. 5.17 WK East C: graph to show proportionate occurrences of artefacts, faunal remains and cobbles in the channel

artefacts and cobbles are most numerous at the base of the deposit, instead of 10 cm higher as at Site A. Fauna follows a similar pattern at the two sites, being most abundant in spits 3 and 4. In view of the similarity of distribution with the levels plotted at WK East A, no plans of WK East C are reproduced.

The industry
The industry consists of 4,983 specimens, comprising ninety-six tools, seventy-three utilised pieces and 4,814 debitage.

Tools

Bifaces (17 specimens) These comprise nine complete, three broken and five tip ends of handaxes. There are no cleavers. The complete specimens are variable in form and size and also in the degree of abrasion. The largest handaxe (167 × 91 × 30 mm) is made of trachyte. It is flat in cross section, slightly asymmetrical and step flaked round the circumference. Another trachyte specimen is heavily abraded and battered round the edges. Three small phonolite examples are asymmetrical and one has pronounced step flaking. They are weathered but not much abraded. The average size for these three specimens is 83 × 50 × 31 mm. There is no evidence that any of these handaxes was made on a flake, except possibly the flat trachyte specimen although no part of the primary flake surface remains. One of the basalt specimens appears to have been made on a cobble, trimmed round the circumference, whilst another is made on a split cobble of nephelinite. (The series is analysed by Drs Roe and Callow.)

Choppers (3 specimens) Two of these examples are of basalt. The largest, a side cobble, measures 103 × 126 × 79 mm and is in fresh condition. The working edge is relatively straight, but the trimming consists largely of step flakes. The butt is formed by cortex and has been considerably battered. The second example (59 × 66 × 44 mm) is heavily abraded and the working edge blunted. The third specimen is a bifacially flaked side chopper made of quartzite and is heavily abraded. It measures 51 × 55 × 35 mm and the butt has been battered.

Discoids (2 specimens) One of these is a flat-sectioned trachyandesite specimen, crudely flaked, with a jagged edge on the circumference. It is in sharp condition and measures 63 × 67 × 25 mm. The second specimen is a miniature quartzite discoid, regularly chipped over both faces and round the circumference. The edge is blunted by use (27 × 27 × 12 mm).

Spheroids/subspheroids (2 specimens) Both examples are of quartzite. The largest (68 × 64 × 47 mm) is symmetrical but faceted, although the projecting ridges have been partially rounded off by battering. The second example (48 × 50 × 38 mm) is rather less symmetrical, but battered over the greater part of the surface.

Scrapers, light-duty (24 specimens) The light-duty scrapers consist of five end, eleven side, seven hollow and one broken specimen. The end scrapers are roughly chipped with variously shaped working edges. They range in length/breadth from 46 × 35 mm to 24 × 20 mm, with an average of 37 × 29 mm. The side scrapers are also variable in form. A proportion have relatively straight working edges, others are curved or have a slight projection in the centre forming a small 'nose'. Measurements range from 39 × 61 mm to 21 × 25 mm, with an average of 23 × 36 mm. Two of the hollow scrapers are double, with a notch on opposite edges. The chipping of the notches is carried out from the primary flake surface, except in one double specimen where the chipping is from opposite faces. The length of the notched scrapers (from working edge to butt) ranges from 41 to 22 mm and the width from 53 to 30 mm. The notches vary in width and depth from 19 × 4 mm to 13 × 2 mm, with an average of 14 × 3 mm.

Awls (9 specimens) All these tools have short, carefully trimmed, rather thick points jutting out from one edge or from the end of the tool. The trimming is steep and with one exception is present on both sides of the points where it forms slight notches. The length (measured from the points to the opposite edge) ranges from 34 to 18 mm and the width from 36 to 19 mm, with an average of 30 × 25 mm.

Outils écaillés (13 specimens) These tools are all of quartzite and are relatively small, the largest specimen measuring 44 mm in length. Five are double ended and the balance single ended, although some evidence of battering or other utilisation is visible at the opposite end. Length/breadth measurements range from 44 × 20 mm to 30 × 16 mm, with an average of 27 × 17 mm.

Punches (26 specimens) These are the small, oblong quartzite tools described previously, triangular or rectangular in cross section. Twenty examples show crushing and scaling at both ends, one is broken in mid-shaft, and in five specimens only one end shows typical wear. In these, the butt ends are broad and formed by a flat, transverse surface. Length/breadth measurements range from 36 × 23 mm to 23 × 11 mm, with an average of 29 × 14 mm.

Utilised material

Pitted anvils and hammerstones (15 specimens) These consist of twelve complete and two broken specimens that have been used subsequent to the fracture, as well as a fragment with twin pits in which one pit has been broken in half. Thirteen examples are of basalt or trachyandesite, five of which are vesicular, and two are of an unidentified lava. The usual four categories are represented, i.e. (a) single pits, (b) pits on two sides, (c) twin pits and (d) multiple pits. Three examples with single pits have been extensively battered on the extremities of the cobbles and similar damage is also evident on one of the specimens with twin pits. One example, with a pit on either face, is discoidal and appears to have been artificially shaped and another, with multiple pits, also appears to have been artificially rounded. One of the specimens with twin pits is on a split cobble with pits on the cortical surface; the opposite side appears to be fractured rather than flaked, but small chips and flake scars on the circumference indicate additional utilisation.

(a) *Single pits* (4 specimens). These four specimens range in weight from 775 g to 120 g, with an average of 549 g. Length/breadth/thickness measurements range from 118 × 72 × 63 mm to 62 × 59 × 33 mm. The pits are unusually deep relative to their diameter, ranging from 32 × 24 × 6 mm to 19 × 16 × 4.5 mm.

(b) *Pits on two sides* (4 specimens). These vary in weight from 2,140 g to 225 g, with an average of 827 g. Length/breadth/thickness varies from $158 \times 135 \times 59$ mm to $78 \times 56 \times 36$ mm, with an average of $109 \times 82 \times 51$ mm. The pits are shallower than in type (a) and range from $36 \times 33 \times 4.5$ mm to $16 \times 15 \times 1.5$ mm.

(c) *Twin pits* (3 complete and 1 broken specimens). The three complete specimens are relatively light and weigh 275 g, 375 g and 500 g respectively, giving an average weight of 382 g. Measurements are $95 \times 73 \times 28$ mm, $94 \times 71 \times 51$ mm and $154 \times 120 \times 56$ mm. The size of the two pits on any one stone is variable and their dimensions do not seem to relate to the size of the stones. The pits range from $22 \times 25 \times 4$ mm to $15 \times 8 \times 1.5$ mm, with an average of $18 \times 12 \times 2.5$ mm.

(d) *Multiple pits* (3 specimens). These stones weigh 900 g, 1,075 g and 600 g, giving an average weight of 858 g, considerably greater than that for the single or twin pits. Length/breadth/thickness measurements are $125 \times 106 \times 54$ mm, $126 \times 108 \times 62$ mm and $92 \times 74 \times 73$ mm. The pits vary from $55 \times 60 \times 8$ mm to $23 \times 20 \times 2$ mm, with an average of $33 \times 32 \times 4$ mm (one specimen also has a large number of small, indeterminate pits which have not been counted).

The weights and dimensions of these fifteen pitted stones have been given but the results will not be plotted in view of the small size of the series. The average weights for the four categories may be noted, since they tally in proportion with the figures obtained for the more extensive collections from WK and WK East A:

(a) single pits   549 g
(b) pits on two sides   827 g
(c) twin pits   382 g
(d) multiple pits   858 g

Blocks (4 specimens) These consist of one cuboid fragment of tabular quartzite 46 mm thick and three irregularly shaped fragments, all of which show chipping and blunting on the edges. Sizes are $83 \times 63 \times 46$ mm, $85 \times 76 \times 51$ mm, $72 \times 61 \times 42$ mm and $65 \times 43 \times 22$ mm.

Light-duty flakes, etc. (54 specimens) The light-duty utilised material can be subdivided into five categories with straight, convex or concave edges, scaled utilisation related to *outils écaillés* and miscellaneous, with minimal chipping.

Straight edges (15 specimens). The largest of these specimens is a broken phonolite flake measuring 54×50 mm, but the majority, particularly those of quartzite, are smaller. Chipping along the edges is generally from one direction only. The average length/breadth measurements are 37×27 mm and the smallest specimen measures 21×15 mm.

Convex edges (3 specimens). These consist of one phonolite side flake and two quartzite flakes in which rounded edges show chipping. Measurements are 24×42 mm, 69×56 mm and 24×30 mm.

Concave edges (11 specimens). There is one broken phonolite flake, but the balance consists of broken flakes and angular fragments of quartzite bearing a chipped notch on one edge. The notches are not so pronounced as in the hollow scrapers. Overall measurements range from 39×42 mm to 18×27 mm with an average of 23×28 mm. The notches vary in width and depth from 17×2 mm to 6×1 mm, with an average of 11×1.5 mm.

Scaled edges (17 specimens). These quartzite fragments are generally oblong but broader and flatter than punches. In this respect they resemble *outils écaillés* but are not scaled to the same extent and never exhibit the typical concavo-convex edges. Utilisation is generally present at both ends but is sometimes at one end only. The length/breadth measurements range from 42×22 mm to 24×18 mm, with an average of 31×20 mm.

Miscellaneous (8 specimens). These consist of two broken phonolite flakes and sundry fragments of quartzite with evidence of utilisation on more than one edge. Sizes vary from 55×39 mm to 24×20 mm, with an average of 35×27 mm.

Debitage (4,814 specimens)

Complete flakes (113 specimens) Thirty-six complete flakes are of lava, mostly of basalt, trachyandesite and phonolite with only two specimens each of nephelinite and trachyte. Lava side flakes are more numerous than end flakes, with twenty-six examples against eleven. The few end flakes are mostly divergent. In the quartzite series, end flakes predominate with twenty-seven examples against eighteen side flakes. Length/breadth measurements are: lava side flakes, from 71 × 80 mm to 26 × 28 mm, with an average of 33 × 42 mm; end flakes from 48 × 38 mm to 29 × 26 mm, with an average of 30 × 21 mm. In the quartzite series, end flakes range from 51 × 34 mm to 16 × 12 mm, with an average of 30 × 21 mm. Side flakes range from 44 × 51 mm to 14 × 15 mm with an average of 23 × 28 mm. It will be seen that the lava flakes are larger in average size than the quartzite series.

Broken flakes (2,403 specimens) Sixty-seven per cent of the broken flakes are of quartzite while only 23 per cent of the complete flakes are of quartzite.

Core fragments (2,297 specimens) Quartzite is again the predominant material, to an even greater extent than in the broken flakes (99 per cent). The majority of these fragments are small, between 10 and 20 mm in length. There is little evidence of chipping on the edges, such as occurs on the larger fragments and blocks.

Core (1 specimen) There is a split boulder of basalt or trachyandesite weighing 5 kg from which three flakes have been detached along one edge, the fractured surface having been used as the striking platform. Although this core is very large and has the potential for a great many additional flakes being struck from it, there is a large crack in the rock which may have led to its being discarded (177 × 165 × 128 mm).

An unusually large, massive flake of trachyandesite, struck from a boulder was found at this site. It is in much sharper condition than the majority of the large tools and is probably extraneous. This flake would not be out of place in the WK industry, although it is more massive than the average from that site (164 × 139 × 75 mm).

ANALYSIS OF THE INDUSTRIES FROM WK EAST C

	Numbers	Percentages
Tools	96	1.9
Utilised material	73	1.5
Debitage	4814	96.6
	4983	100.0
Tools		
Bifaces	17	17.7
Choppers	3	3.1
Discoids	2	2.1
Spheroids/subspheroids	2	2.1
Scrapers, light-duty	24	25.0
Awls	9	9.4
Outils écaillés	13	13.5
Punches	26	27.0
	96	99.9
Utilised material		
Pitted anvils/hammerstones	15	20.5
Cobbles and blocks	4	5.5
Light-duty flakes etc.	54	73.9
	73	99.9
Debitage		
Whole flakes	113	2.3
Broken flakes	2403	49.9
Core fragments	2297	47.7
Core	1	0.02
	4814	99.92

PDK Trenches I–III

This was the first area where artefacts and fossils were observed eroding from the horizon above Tuff IVB and where excavations at this level were undertaken. As noted above, the site lies 25 m east of WK East A. The first trench to be excavated was begun in January 1970 and was parallel to the edge of the cliff. It measured 4.6 m north–south by 2.2 m east–west (Trench I), and was subsequently extended 7.3 m to the south (Trench II), giving a total length from north to south of just under 12 m. A further extension 3.65 m east–west and 3.4 m north–south (Trench III) was dug at right angles and to the west of the first trenches. It yielded an abundance of artefacts and faunal specimens. The section revealed was similar to that subsequently found at WK East A except for the capping calcrete which was missing.

ANALYSIS OF THE INDUSTRIES FROM WK EAST C

	Quartzite	Fine-grained quartzite	Gneiss, feldspar, granite, welded tuff, etc.	Phonolite	Trachyte	Nephelinite	Basalt/trachyandesite	Lava, indeterminate	Totals
Heavy-duty tools									
Bifaces, choppers, discoids, spheroids/subspheroids	8	—	—	5	3	1	7	—	24
Light-duty tools									
Scrapers, awls, *outils écaillés*, punches	61	6	—	3	—	—	2	—	72
Utilised material, heavy-duty									
Pitted anvils/hammerstones, blocks	4	—	—	—	—	—	13	2	19
Utilised material, light-duty									
Flakes, etc.	44	2	—	7	—	—	1	—	54
Debitage									
Whole flakes	71	6	—	11	2	2	21	—	113
Broken flakes	2136	104	1	64	4	6	88	—	2403
Core fragments	2244	31	—	3	—	1	18	—	2297
Core	—	—	—	—	—	—	1	—	1
	4568	149	1	93	9	10	151	2	4983

The deposits exposed measured 1.5 m in depth, the lowest 30 to 40 cm containing the artefacts, faunal material and cobbles. Here, as at WK East, there were two large boulders of vesicular lava measuring 1.4 × 1.1 m and 0.6 × 0.5 m respectively. The larger of the two had a relatively flat upper surface and both appeared to have been pitted artificially. The higher part of the section consisted of 80 cm of bedded reddish sands overlain by approximately 60 cm of sandy clay similar to that at WK East A.

A feature not noted elsewhere in this group of sites was a group of three roughly circular depressions in the surface of Tuff IVB that appeared to be pot-holes (Fig. 5.18, see pg. viii). The two deepest were approximately 23 cm deep, with a shallower depression linking the two larger cavities. They contained few remains except for a concentration of pebbles and some cobbles.

The channel deposit was excavated in two levels, not in 10 cm spits as in the WK East sites.

The proportions of artefacts, faunal remains and cobbles, therefore, cannot be exactly matched but comparison shows that cobbles are most numerous in the upper level, with 62 per cent against 44 per cent in the lower, whereas both artefacts and fauna are scarcer than in the lower level. They amount to 22 per cent and 14 per cent in the upper level with 28 per cent and 27 per cent in the lower level.

The industry

This consists of 1,746 specimens, comprising ninety-seven tools, sixty-seven utilised pieces and 1,582 debitage.

Tools

Bifaces (15 complete, 7 broken specimens) In addition to the bifaces found *in situ* which are described here, four handaxes were recovered from the scree below the site. These are not included since it is possible they were derived from the same artefact-bearing horizon at the base of Bed IV that was excavated at PDK IV. With the exception of one small ovate made from a fine-grained material, the series is crude and lacks refined retouch. The small ovate is symmetrical, evenly flaked over both faces as well as round the circumference. It lacks the step flaking seen on other specimens and is elliptical in cross section with a sinuous but not jagged edge. One handaxe is made on part of an oblong cobble of basalt or trachyandesite flaked steeply along one edge from both the upper and lower faces. The opposite lateral edge consists of cortical surface. One end is sharply pointed and shows additional flaking. This specimen resembles a crude pick. The remaining specimens are relatively thick in cross section with crude flaking in which the scars are large and often deeply indented. The quartzite bifaces are in sharp condition but the remainder are weathered and abraded. (The series is analysed by Drs Roe and Callow.)

Choppers (15 specimens) These comprise eleven side, three end and one double-edged chopper. Seven of the side choppers are made on cobbles of basalt or trachyandesite, two on nephelinite cobbles and two on blocks of quartzite. The three end choppers and the double-edged specimen are on basalt or trachyandesite cobbles. The side choppers are all bifacially flaked along one edge and generally have jagged working edges showing variable degrees of blunting and wear. The two quartzite choppers are sharp and in fresh condition, but all those made from lava are weathered and abraded. In one of the larger specimens the angle of the working edge is unusually obtuse and approximates 105°. Two examples show battering on the butts. The working edge in one of the quartzite specimens is less jagged than in the remainder of the series and forms a relatively even arc, extending partially down either side of the tool and resembling the butt end of a handaxe. The range in length/breadth/thickness in the side choppers is from $92 \times 98 \times 74$ mm to $52 \times 83 \times 37$ mm, with an average of $67 \times 86 \times 51$ mm. The three end choppers are made on oblong cobbles in which one extremity has been minimally flaked on both faces. In two examples the working edges are straight and transverse, while in the third, now considerably damaged, the end appears to have been pointed. All three specimens are weathered and abraded. In two examples the butts have been battered. Length/breadth/thickness measurements are $86 \times 58 \times 50$ mm, $67 \times 45 \times 35$ mm and $70 \times 50 \times 24$ mm. The double-edged specimen is a double side chopper. Both edges are steeply flaked and both ends are battered. It measures 52 mm from edge to edge, is 63 mm wide and 37 mm thick.

Discoid (1 specimen) This is made on part of a slightly vesicular basalt or trachyandesite cobble. Both upper and lower faces retain part of the original cortical surface. The greater part of the circumference has been bifacially flaked, but an area 37 mm long has been battered to a blunt, rounded edge. Measurements $63 \times 55 \times 34$ mm.

Spheroids (5 specimens) Two specimens are of basalt or trachyandesite, one of quartzite, one of nephelinite and one of an unidentified fine-grained material. The two examples of basalt/trachyandesite are the largest and are made on cobbles although they have been so extensively flaked and battered that only small areas of the cortex remain. The quartzite specimen is the most symmetrical and has been battered over the entire surface. The remaining two examples show both battering and faceting. Measurements range from $127 \times 119 \times 85$ mm (weight 1,800 g) to $57 \times 45 \times 42$ mm (weight 200 g) with an average of $90 \times 81 \times 70$ mm (weight 890 g).

Scrapers, heavy-duty (2 specimens) These consist of a flat cobble of basalt or trachyandesite trimmed at one end to a regularly curved scraping edge 55 mm long. One specimen is weathered and measures $80 \times 71 \times 34$ mm. The second specimen is made on an oblong slab of quartzite 40 mm thick, 133 mm long and 58 mm wide. One side has been roughly flaked to a scalloped edge approximately 86 mm long.

Scrapers, light-duty (21 specimens) The majority of these tools are made from lava. Sixteen can be subdivided summarily into side and end scrapers on the basis of length or width in relation to the working edges, but they are essentially formless, made on irregularly shaped flakes and fragments. The working edges are likewise irregular, with projections and small indentations. One broken example cannot be allocated to either category. The twelve examples classified as side scrapers range in length/breadth from 47 × 54 mm to 19 × 24 mm, with an average of 28 × 37 mm. The four end scrapers range from 57 × 41 mm to 29 × 20 mm with an average of 38 × 29 mm. The remaining specimens are hollow scrapers with well-defined notches. They range in size from 37 × 50 mm to 26 × 35 mm with an average of 32 × 40 mm. The notches are evenly curved and appear to have been deliberately flaked. The length/breadth measurements for the four notches are 8 × 3.5 mm, 22 × 3.5 mm, 17 × 3.5 mm and 21 × 1 mm.

Burin (1 specimen) There is one double angle burin with a trimmed notch at the tip from which a burin spall has been struck from either side. The tool is made from fine-grained quartzite and measures 32 × 17 mm.

Awls (3 specimens) These consist of one quartzite and two phonolite flakes with sharp points at the tips. The points are formed by a retouched notch on one or both sides. Length/breadth measurements: 36 × 29 mm, 33 × 19 mm and 35 × 30 mm.

Laterally trimmed flakes (4 specimens) These specimens are all made of phonolite. The butts are broad and the tips narrow, although not pointed. Retouch is present on one or both lateral edges, but the edges are not symmetrical. Length/breadth measurements: 53 × 35 mm, 47 × 26 mm, 45 × 32 mm and 40 × 35 mm.

Outils écaillés (3 specimens) All three specimens are of quartzite and are double ended. They show the scaled utilisation typical of these tools. Length/breadth measurements: 43 × 32 mm, 28 × 18 mm and 27 × 20 mm.

Punches (20 specimens) These small, rod-like specimens are all of quartzite and are thick-set in cross section and, as usual, are rectangular or triangular. One or both ends are pointed and show crushing and scaling. Length/breadth measurements range from 43 × 17 mm to 23 × 12 mm, with an average of 30 × 16 mm.

Utilised material

Pitted anvils and hammerstones (14 specimens) These consist of twelve cobbles of basalt or trachyandesite and two battered pieces of granular, coarse-grained quartzite. The series includes the following:

(a) *Single pits* (2 specimens). One of these consists of a fragment of quartzite measuring 64 × 48 × 35 mm and weighing 160 g. Both ends and one lateral edge have been battered smooth. The single pit is not central and the opposite lateral edge may be broken. The pit is wide and shallow, measuring 27 × 18 × 1.5 mm. The second specimen is an oval basalt pebble measuring 62 × 54 × 40 mm, weight 225 g. The greater part of the circumference shows battered utilisation and the single pit measures 21 × 13 × 2 mm.

(b) *Twin pits*. The single specimen in this category consists of a basalt cobble with three sets of twin pits, each on a different face. The cobble has been extensively battered at both extremities. Measurements: 91 × 72 × 65 mm, weight 575 g. The pits measure 24 × 16 × 2 mm and 21 × 14 × 1.5 mm, 21 × 15 × 1.5 mm and 17 × 13 × 15 mm, 24 × 11 × 0.5 mm and 23 × 11 × 1 mm. (Two sets of twin pits are also present on one of the specimens with multiple pits described below.)

(c) *Pits on two sides* (7 specimens). These comprise six basalt or trachyandesite cobbles and one block of quartzite. The largest lava example (125 × 75 × 52 mm) is triangular in shape. It shows hammerstone type of utilisation at one end, whilst at the opposite end flakes have been detached, probably as a result of use rather than deliberate flaking. The quartzite specimens and

UPPER BED IV

ANALYSIS OF THE INDUSTRIES FROM PDK TRENCHES I–III

	Numbers	Percentages
Tools	97	5.5
Utilised material	67	3.8
Debitage	1582	90.7
	1746	100.0
Tools		
Bifaces	22	22.7
Choppers	15	15.5
Discoid	1	1.0
Spheroids/subspheroids	5	5.2
Scrapers, heavy-duty	2	2.1
Scrapers, light-duty	21	21.6
Burin	1	1.0
Awls	3	3.1
Laterally trimmed flakes	4	4.1
Outils écaillés	3	3.1
Punches	20	20.6
	97	100.0
Utilised material		
Pitted anvils/hammerstones	14	20.9
Hammerstone	1	1.5
Cobbles and blocks	24	35.8
Light duty flakes, etc.	28	41.8
	67	100.0
Debitage		
Whole flakes	80	5.0
Broken flakes	514	32.5
Core fragments	988	62.5
	1582	100.0

another of basalt exhibit hammerstone type of utilisation round the edges. Weights, overall dimensions and measurements of the pits (when sufficiently well defined to be measurable) are as follows: range in weight from 500 g to 175 g with an average of 329 g. Length/breadth/thickness measurements range from 123 × 74 × 53 mm to 68 × 46 × 33 mm. The pits range in length/breadth/depth from 34 × 32 × 3 mm to 12 × 11 × 1 mm.

(d) *Multiple pits* (4 specimens). These stones are larger and heavier than those in the preceding categories. All consist of basalt or trachyandesite cobbles. One example with crude flaking along one edge shows two sets of oblong twin pits on different faces and a single oblong pit on a third face. Another example has a total of eleven small, shallow pits distributed over the surface. The largest specimen, measuring 182 × 89 × 72 mm, is triangular in shape and has been battered and roughly flaked at one extremity; another shows hammerstone utilisation along one side. The range in weight for the four specimens is from 1,475 g to 625 g, with an average of 1,010 g. Length/breadth/thickness measurements range from 182 × 89 × 72 mm to 93 × 92 × 64 mm, with an average of 122 × 90 × 66 mm. The pits vary in length/breadth/depth from 35 × 18 × 3 mm to 10 × 10 × 1 mm, with an average of 21 × 12 × 1.3 mm.

Hammerstones (1 specimen) The only specimen showing hammerstone utilisation not combined with pitting consists of a sub-rectangular block of quartzite, possibly broken. It measures 66 × 62 × 36 mm. There are also two small fragments possibly derived from hammerstones, on the basis of the type of utilisation exhibited on one face.

Cobbles and blocks (24 specimens) Sixteen lava cobbles and one small enough to be classed as a pebble show rough cobbles and battering, probably resulting from heavy utilisation, although one specimen may, perhaps, be an unfinished chopper. Length/breadth/thickness range from 116 × 114 × 73 mm to 37 × 33 × 17 mm, with an average of 77 × 62 × 40 mm. There are also seven utilised pieces consisting of amorphous lumps of quartzite with some degree of flaking or battering. They are mostly of very coarse-grained quartzite which may have proved unsuitable for flaking. Measurements range from 93 × 80 × 63 mm to 58 × 41 × 46 mm.

Light-duty flakes, etc. (28 specimens) Four small cores or core fragments and twenty-four whole and broken flakes show chipping along one or more edges. The chipping is generally irregular and appears to be the result of utilisation rather than retouch. Lava and quartzite are represented in equal proportions. Length/breadth measurements range from 63 × 49 mm to 18 × 11 mm, with an average of 36 × 28 mm.

Debitage (1,582 specimens)

Whole flakes (80 specimens) Fifty-five flakes are of lava. They are mostly of basalt/trachyandesite

ANALYSIS OF THE INDUSTRIES FROM PDK TRENCHES I–III

	Quartzite	Fine-grained quartzite	Gneiss, feldspar, etc.	Phonolite	Nephelinite	Trachyte	Basalt/trachyandesite	Lava, indeterminate	Totals
Heavy-duty tools Bifaces, choppers, discoids, spheroids, heavy-duty scraper	5	1	—	6	6	3	21	3	45
Light-duty tools Scrapers, awls, burin, *outils écaillés*, punches, laterally trimmed flakes	30	3	—	13	—	—	6	—	52
Utilised material, heavy-duty Pitted anvils/hammerstones, hammerstones, blocks	10	—	—	1	—	4	24	—	39
Utilised material, light-duty Flakes, etc.	9	5	—	9	—	—	5	—	28
Debitage									
Whole flakes	23	2	—	16	1	5	33	—	80
Broken flakes	314	12	—	84	1	8	95	—	514
Core fragments	915	18	—	25	1	5	24	—	988
	1306	41	—	154	9	25	208	3	1746

and phonolite, but also include a single flake of trachyte and five of nephelinite. Twenty-four of the lava flakes are side struck and thirty-one end struck. In the quartzite series (twenty-five specimens) end flakes are the most numerous, with nineteen examples. Divergent flakes outnumber other types in both the lava and quartzite series. The lava side flakes range in length/breadth from 56 × 57 mm to 8 × 19 mm, with an average of 32 × 40 mm. Lava end flakes range from 64 × 57 mm to 21 × 14 mm, with an average of 37 × 30 mm. As usual, the quartzite flakes are rather smaller, with a range for the side flakes from 50 × 54 mm to 17 × 22 mm and an average of 28 × 33 mm. The end flakes range from 41 × 35 mm to 19 × 17 mm, with an average of 29 × 24 mm.

Broken flakes (514 specimens) These are parts of irregularly shaped flakes in which quartzite, basalt/trachyandesite and phonolite are the most common materials.

Core fragments (988 specimens) All except seventy-three of these pieces consist of quartzite fragments. They appear to be mainly derived from use of the bipolar method of flaking. A random sample of fifty specimens gives average length/breadth/thickness measurements of 23 × 17 × 12 mm.

Sundries (The following are not included in the analysis of the artefacts.)

Large flake One unusually large, flat flake of ?trachyandesite was found *in situ*. It appears to be entirely out of context, both on account of its size and its technology. It may possibly have been obtained from an Acheulean site and brought in as raw material from which to detach the smaller

ANALYSIS OF THE INDUSTRIES FROM HEB WEST
LEVEL 1

	Numbers	Percentages
Tools	17	4.0
Utilised Material	22	5.0
Debitage	391	91.0
	430	100.0
Tools		
Biface	1	5.9
Chopper	1	5.9
Polyhedrons	2	11.8
Discoids	2	11.8
Spheroids/subspheroids	3	17.6
Scraper, heavy-duty	1	5.9
Scrapers, light-duty	5	29.4
Outils écaillés	2	11.8
	17	100.0
Utilised material		
Pitted anvils/hammerstones	2	9.1
Cobbles	5	22.7
Light-duty flakes, etc.	15	68.2
	22	100.0
Debitage		
Whole flakes	38	9.7
Broken flakes	297	76.0
Core fragments	53	13.5
Cores	3	0.8
	391	100.0

flakes usual in this industry. Measurements 174 × 178 × 34 mm.

Unmodified blocks Four relatively large blocks of quartzite show no evidence of flaking or utilisation and were, presumably, brought to the site as raw material but not used. Average measurements 100 × 91 × 53 mm. Average weight 1,006 g.

HEB West Level 1

This consists of a thin, irregular layer of claystone immediately overlying the grey siltstone. It occurs in all the trenches dug at HEB West but was not present at HEB nor in the intervening area. The artefacts and faunal remains were dispersed in the area excavated and not so concentrated as in the lower archaeological levels. A small number of artefacts and fossil bones which were in the sand a few inches above the claystone of Level 1 have been included in the series in order to augment the material available for study. (For description of the site see pp. 52–5.)

The industry
This consists of 430 specimens, comprising seventeen tools, twenty-two utilised pieces and 391 debitage.

Tools

Biface (1 specimen) A small example, made on a trachyandesite end flake. The primary surface has been trimmed only at the tip. Trimming on the dorsal aspect is confined to the left lateral edge (114 × 60 × 31 mm).

Chopper (1 specimen) A small quartzite side chopper. It is crudely flaked and the edges are slightly abraded (50 × 62 × 39 mm).

Polyhedrons (2 specimens) A small specimen made on fine-grained quartzite is in sharp condition although the working edges are chipped (44 × 39 × 33 mm). The second example, made of lava, is very abraded and weathered (52 × 45 × 44 mm).

Discoids (2 specimens) A large lava specimen measuring 115 × 105 × 60 mm is very abraded and in similar condition to the lava polyhedron. It is possible that these two tools are derived. A somewhat oblong quartzite discoid is in fresh condition. It measures 62 × 47 × 34 mm and has been flaked all round the circumference as well as over the upper and lower faces.

Spheroids/subspheroids (3 specimens) All three examples are asymmetrical although some of the projecting parts have been battered. They measure 34 × 32 × 37 mm, 42 × 35 × 26 mm and 34 × 34 × 30 mm.

Scraper, heavy-duty (1 specimen) This is a hollow scraper and consists of a lava flake measuring 90 × 51 mm with a well-defined notch on one lateral edge. The notch is 15 mm wide and

ANALYSIS OF THE INDUSTRIES FROM HEB WEST LEVEL 1

	Quartzite	Fine-grained quartzite	Gneiss or granite	Phonolite	Nephelinite	Trachyte	Basalt or trachyandesite	Lava, indeterminate	Totals
Heavy-duty tools	5	1	—	—	1	—	3	—	10
Light-duty tools	7	—	—	—	—	—	—	—	7
Utilised material, heavy-duty									
Pitted anvils/hammerstones, cobbles	—	—	—	—	—	—	7	—	7
Utilised material, light-duty									
Flakes, etc.	15	—	—	—	—	—	—	—	15
Debitage									
Whole flakes	35	—	—	2	—	—	—	1	38
Broken flakes	282	3	—	5	—	—	7	—	297
Core fragments	52	—	—	1	—	—	—	—	53
Cores	3	—	—	—	—	—	—	—	3
	399	4	—	8	1	—	17	1	430

4.5 mm deep and has been flaked from the primary flake surface only. This tool was found standing vertically in the deposit, but the edge that was uppermost shows no signs of unusual utilisation.

Scrapers, light-duty (5 specimens)   There are five quartzite examples, of which one is discoidal and four are side scrapers. They are very similar in size and the average length/breadth measurements are 33 × 28 mm.

Outils écaillés (2 specimens)   There are two double-ended quartzite specimens. They measure 28 × 22 mm and 20 × 18 mm respectively.

Utilised material

Pitted anvils/hammerstones (2 specimens)   Two basalt or trachyandesite cobbles exhibit pitted hollows. The largest specimen is oblong and measures 204 × 75 mm. It is flattened on opposite faces, both of which bear shallow, irregular pits. The edges of the cobble have also been extensively battered by use. The second specimen, measuring 108 × 84 × 79 mm, has a single hollow on one face. It is 33 mm in maximum diameter and 5 mm deep.

Cobbles (5 specimens)   Three basalt or trachyandesite cobbles have been battered at both ends. They are similar in size and the average measurements are 81 × 60 × 51 mm. Two further cobbles have been broken and also show rudimentary flaking.

Light-duty flakes, etc. (15 specimens)   Fifteen quartzite flakes show evidence of wear on the edges. They vary in length/breadth from 83 × 50 mm to 24 × 17 mm.

Debitage (391 specimens)

Whole flakes amount to thirty-eight specimens. End-struck flakes are the most numerous, with twenty-four examples. Most of the flakes are divergent and there are only two in which the greatest width is at the butt end. The length/breadth measurements for the end flakes range from 68 × 66 mm to 15 × 13 mm, with an average of 45 × 31 mm, and for the side flakes from 30 × 37 mm to 14 × 22 mm, with an average of 22 × 28 mm. Among the 297 broken flakes, quartzite is by far the most common material. Sundry angular fragments amount to fifty-three; with the exception of a single piece of Engelosen phonolite they are of quartzite. Three irregular blocks of quartzite have had a number of flakes detached and are probably cores; they vary in size from 84 × 75 × 62 mm to 53 × 53 × 33 mm.

The industry from Level 1 is too scanty for valid assessment but it does not appear to differ to any appreciable extent from the industries found in the lower levels except in the scarcity of bifaces and this is probably due to the small quantity of material recovered.

6

THE MASEK BEDS AND SITES IN UNCERTAIN STRATIGRAPHIC POSITIONS

M. D. LEAKEY

FLK Masek Beds

The site

Excavations at this site were undertaken following the discovery of part of a hominid mandible (O.H.23) by E. Kandindi during October 1968. The hominid fossil was *in situ* in the Masek Beds about 0.5 m above a level that yielded an extensive Acheulean industry when excavations were undertaken.

The site is on the west side of the Main Gorge, at the southern end of a cliff that overlooks the gully where the cranium of *Australopithecus boisei* was found in Bed I (FLK). In the northern section of the cliff, the deposits exposed are characteristic of Bed IV, but to the south they have been cut through by two channels, both of which can be shown to be within the Masek Beds. The level containing the mandible and the Acheulean industry was in the lower part of the earlier of the two channels that had cut down into Bed IV. As exposed in the cliff face and later in the trenches, the channel filling consisted of approximately 5 m of earthy, horizontally bedded deposits with a coarse rubble layer at the base containing artefacts and bone debris. This rubble also contained blocks of calcrete and of the lower aeolian tuff of the Masek Beds, as well as pieces of Tuff IVA from Lower Bed IV. To the north, this channel was cut through by the second channel that can be equated with the upper part of the Masek Beds (Norkilili Member). This second channel measures 46 m across from north to south where the banks are exposed in the cliff and the southern bank is almost vertical (Fig. 6.1). The filling consisted of conglomerate and coarse to fine-grained sandstone, some of which was strongly current bedded. A few heavily rolled artefacts and fragments of fossil bones occurred, their condition being in marked contrast to those from the rubble at the base of the earlier channel, most of which are quite unabraded.

The stratigraphic position of this Acheulean industry is well documented; it postdates the lower aeolian tuff of the Masek Beds, since blocks of this tuff occurred in the base of the channel with the artefacts; equally, it antedates the upper part of the Masek Beds since the tool-bearing channel is cut through by a channel of that period. No other sites are known within the Masek Beds and the industry is the most recent known at Olduvai that is in a sound geological context. Sites HK and TK Fish Gully are certainly more recent but cannot be fitted into the stratigraphic sequence. Until the discovery of the mandible, the possibility that an occupation site might exist in this area had not been envisaged, since there were virtually no artefacts or fossil bones visible on the slope below the cliff.

The detailed sequence of deposits exposed in the cliff face and in the trenches was as follows (from top to bottom):

(a) 30 to 40 cm of calcrete
(b) 4 m of brown, earthy deposit, horizontally bedded, barren of remains
(c) approximately 30 cm of silty clay, also barren
(d) the bone and artefact bearing horizon, consisting of a ferruginous quartz sand and rubble with unabraded blocks of the lower

THE MASEK BEDS

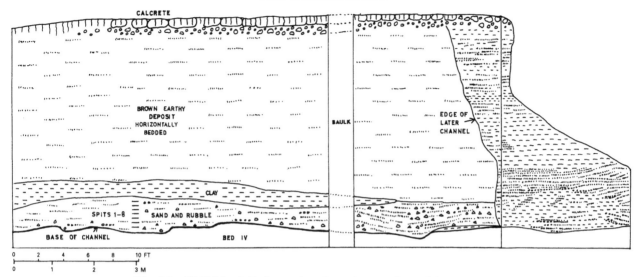

Fig. 6.1 FLK Masek Beds: section along the west face of the excavations

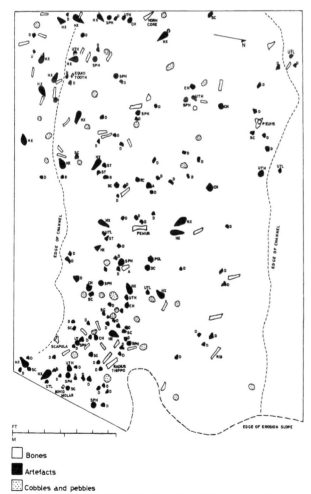

Fig. 6.2 FLK Masek Beds: plan of finds in the lower part of the channel (for abbreviations, see p. xiv)

Masek aeolian tuff, calcrete, Tuff IVA and pebbles of Naabi ignimbrite, the latter derived from the north and indicating drainage from that direction. The sand and rubble reached a maximum thickness of 80 cm in the deepest part of the channel

(e) the eroded surface of Bed IV.

The lower part of the channel, where it had cut into Bed IV, was orientated in a general east–west direction, sloping downwards to the east, with a difference in height of approximately 15 to 20 cm within the length exposed in the trenches (7.5 m). The base of the channel was uneven and consisted partly of a smooth clay surface and partly of a coarse grey sand which was easily distinguishable from the ferruginous sandy filling of the channel. There were several circular depressions in the stream bed, from 55 to 30 cm in diameter, that appeared to be pot-holes, as well as a number of minor channels or runnels. The basal part of the main channel measured from 3.5 to 4 m in width and did not exceed 40 cm in depth.

The implementiferous deposit was excavated in 10 cm spits. The upper levels contained few artefacts or fossil bones and the greatest number of remains occurred in the main stream bed 10 to 20 cm above the base of the channel (Fig. 6.2). This level also contained more large handaxes

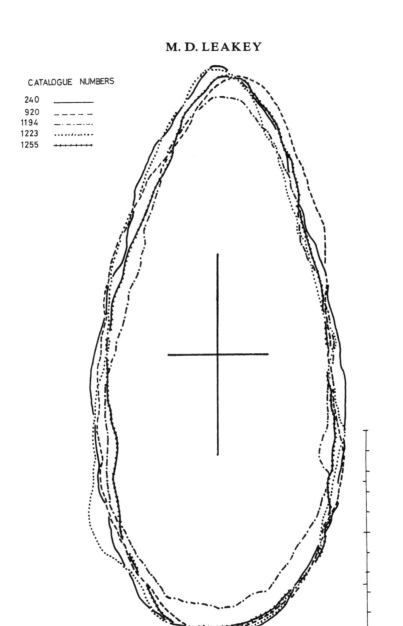

Fig. 6.3 FLK Masek Beds: superimposed outlines of five handaxes showing similarity in size and form

and bigger cobbles than either the higher or lowest levels. There was no apparent evidence of water-alignment and most of the artefacts, particularly the large bifaces made from quartzite, were in unusually fresh condition. Unmodified lava cobbles or manuports were relatively scarce. Those that occurred are mostly of lavas from Lemagrut, and were almost certainly introduced by humans since contemporary drainage was from the opposite direction, to the northwest.

The first excavations consisted of a Trial Trench 6 m long and 1.5 m wide, aligned northwest–southeast. This trench was located 6 m to the south of the spot where the fragment of hominid mandible had been found. It yielded two large quartzite bifaces, a number of flakes and many bone fragments, some of which were abraded. A trench 4 m wide was next cut back into the cliff, to the north, followed by two adjacent parallel trenches in the intervening area, where the overburden had already been partly removed by erosion. These trenches measured 5

THE MASEK BEDS

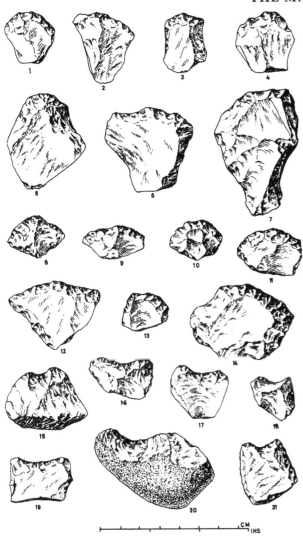

Fig. 6.4 Scrapers from FLK, Masek Beds

and 7 m in length respectively and were aligned northeast–southwest, at right angles to the trial trench. The total area finally excavated amounted to 180 sq m. In places, up to 4 m of overburden had to be removed before the artefact-bearing level was reached.

The industry

The industry consists of 2,465 specimens, of which 193 are tools, forty-seven utilised pieces and 2,225 debitage. Quartzite from the nearby inselberg of Naibor Soit is by far the most common raw material and accounts for 94.8 per cent of the total number of artefacts. Only 3.7 per cent are of lava, including Engelosen phonolite, trachyandesite and basalt from Lemagrut, with only a very few specimens of trachyte or nephelinite.

Tools

Bifaces (22 whole, 9 broken and 7 probably unfinished specimens) The series consists solely of handaxes and does not include any cleavers. With the exception of three lava specimens all are made from quartzite (Pls. 20 and 21). The five largest specimens (all about 27 cm in length) show a remarkable similarity in form as well as being closely comparable in size (Fig. 6.3). They are elongate, with delicately trimmed tips. The cutting edges extend round the entire circumference and are particularly even. In spite of the material being coarse grained, and intractable, these tools have been elaborately trimmed over both faces. The close similarity in technique, size and form suggests the possibility that they may have been the work of a single craftsman who either had an extremely accurate visual appreciation of the tools he wished to make, or else employed some form of measuring device. Four out of the five were found close together in the channel filling and the fifth approximately 10 ft distant. The length/breadth/thickness measurements for these five specimens and the minimum scar counts are as follows:

		L B T	Scars
FLK	240	289 × 132 × 72 mm	44
	928	268 × 124 × 83 mm	29
	1194	249 × 116 × 72 mm	25
	1223	277 × 129 × 69 mm	30
	1255	270 × 117 × 67 mm	32

Of the remaining specimens, thirteen are bifacially trimmed over the greater part of both surfaces and four are on flakes in which the primary surface shows only a minimum of trimming; one of these, made on a large end flake, has been steeply trimmed on the dorsal face to form a high, longitudinal keel, while the lower face is flat. Only a small number of specimens exhibit the elaborate trimming seen on the five large handaxes described above; two examples are sharply pointed, but in the majority the tips are slightly rounded. The cutting edges vary from nearly straight to uneven and jagged, depending on the nature of the trimming scars. The size of the handaxes is also very variable. The broken bifaces consist of one broad ovate lacking part of the butt, three butt ends, three tip ends and two specimens lacking the tips. The seven examples classed as unfinished are very crude and appear to represent rough-outs. (The bifaces are analysed by Drs Roe and Callow in chapters 8 and 9.)

Choppers (11 specimens) All the choppers were recovered from the deeper part of the channel and did not occur in the upper levels. Two examples are made on quartzite blocks and the balance on lava cobbles. These are mostly of basalt or trachyandesite but also include one specimen of phonolite. There are six side and five end choppers. In three examples the working edges are flaked from a single negative scar and in the remainder the flaking is bifacial. One quartzite end chopper is made on a slab of tabular material and exhibits a rounded, splayed cutting edge that is the widest part of the tool. The series is crude in appearance and the specimens do not conform closely to one another either in size or in style of flaking. The butt of one example has been used as a hammerstone. The range in size is from $104 \times 100 \times 64$ mm to $36 \times 48 \times 27$ mm, with an average of $70 \times 81 \times 41$ mm.

Discoids (3 specimens) These three tools are made from quartzite. They are thick in cross section and markedly biconvex. Two examples are symmetrical, with a cutting edge all round the circumference, but the third is somewhat oblong and the cutting edge extends round only part of the circumference. It is possibly unfinished, although the edge is chipped and blunted and appears to have been used. Length/breadth/thickness measurements are $66 \times 63 \times 41$ mm, $69 \times 56 \times 41$ mm and $42 \times 37 \times 30$ mm.

Spheroids/subspheroids (34 specimens) There is only one well-finished stone ball; it is made of granite and has been broken on one side, but the edges of the broken surface are worn in such a way as to suggest that the tool continued to be used after it was broken. It measures $76 \times 71 \times 54$ mm and weighs 500 g. The remainder of the series consists of subspherical, angular pieces of quartzite only a few of which are symmetrical. The range in size is from $73 \times 64 \times 58$ mm to $28 \times 24 \times 26$ mm, with an average of $44 \times 39 \times 34$ mm.

Scrapers, heavy-duty (12 specimens) These are all made of irregular blocks or pieces of tabular quartzite. They comprise nine side scrapers and one double-edged, one nosed and one steeply trimmed core-scraper. The scraping edges are either straight or convex and usually uneven. Trimming is from one direction only and is generally irregular. Length/breadth/thickness varies from $80 \times 110 \times 56$ mm to $50 \times 66 \times 57$ mm. (Fig. 6.4).

Scrapers, light duty (77 specimens) These may be divided into side, end and hollow or notched scrapers.

Side (35 specimens). These scrapers are mostly made on broken flakes or irregular fragments; only a few are on complete flakes. The scraping edges correspond with the long axis of the tools, and tend to be slightly irregular with minor projections and are rarely even. Three well-made specimens, all about 15 mm in length, would not be out of place in a microlithic industry. The largest specimen in the series measures 37×54 mm and the average length/breadth for the whole series is 21×32 mm.

End (25 specimens). The forms of the scraping edges are very variable. They are sometimes broad and mainly straight, but more often rounded, either forming a broad working edge extending across the width of the implement or else a relatively narrow, 'nosed' edge. The trimming is generally steep and some examples are more evenly trimmed than any of the side scrapers. Five examples are made on end flakes that retain the bulbar extremities, but most of the series are made on broken flakes and other fragments. The largest specimen measures

THE MASEK BEDS

ANALYSIS OF THE INDUSTRIES FROM FLK MASEK BEDS

	Numbers	Percentages
Tools	193	7.8
Utilised material	47	1.9
Debitage	2225	90.3
	2465	100.0
Tools		
Bifaces	38	19.7
Choppers	11	5.7
Discoids	3	1.5
Spheroids/subspheroids	34	17.6
Scrapers, heavy-duty	12	6.2
Scrapers, light-duty	77	39.9
Laterally trimmed flakes	10	5.2
Outils écaillés	3	1.5
Punches	5	2.6
	193	99.9
Utilised material		
Pitted anvil	1	2.1
Anvils	4	8.5
Hammerstones	2	4.3
Cobbles	6	12.8
Light-duty flakes, etc.	34	72.3
	47	100.0
Debitage		
Whole flakes	276	12.4
Broken flakes	1320	59.2
Core fragments	622	28.0
Cores	7	0.3
	2225	99.9

61 × 44 mm and the smallest of the series is 33 × 17 mm. The average length/breadth for the series is 33 × 27 mm.

Hollow (17 specimens). These are all made from quartzite. They consist mostly of broken flakes and other fragments and include only two complete flakes. The notches vary considerably in size and are trimmed from one direction only. In specimens made on flakes, the trimming is generally from the primary surface. The largest specimen measures 57 × 45 mm and the smallest 24 × 20 mm. The average length/breadth for the series is 36 × 26 mm. The notches vary in width and depth from 18 × 5 mm to 10 × 2 mm, and the average figure for the series is 25 × 2.9 mm.

Laterally trimmed flakes (10 specimens) These consist of flakes that have been trimmed along both lateral edges to a point. The flaking is generally on the dorsal surface only. The specimens have thick, relatively broad striking platforms. Although all these flakes are trimmed to points, they vary considerably: some are sharp and others bluntly rounded. The largest specimen measures 68 × 46 mm and the smallest 30 × 31 mm, with an average of 46 × 34 mm.

Outils écaillés (3 specimens) All three examples are made from quartzite. One is double ended and the remaining two have been utilised on one end only. The length/breadth measurements are 29 × 17 mm, 24 × 27 mm and 27 × 21 mm.

Punches (5 specimens) All five examples are made from quartzite. They are oblong and thickset. One example is perhaps on part of a flake that had been broken longitudinally. They vary in size from 38 × 20 mm to 32 × 15 mm, the average being 34 × 19 mm.

Utilised material

Pitted anvil (1 specimen) There is one large oblong cobble of vesicular basalt, with relatively flat upper and lower faces on both of which there is a pecked depression. These are wide and shallow, measuring 70 × 62 × 10 mm and 63 × 50 × 8 mm respectively. The interiors of the hollows are rough and irregular, with no trace of abrasion. The cobble measures 170 × 142 × 84 mm.

Anvils (4 specimens) Four examples consist of cuboid blocks of quartzite in which edges that are approximately 90° have been battered. They vary in size from 112 × 108 × 70 mm to 54 × 44 × 39 mm, with an average of 88 × 70 × 52 mm.

Hammerstones (2 specimens) Two oblong cobbles are battered and bruised at the extremities, indicating utilisation as hammerstones. They measure 97 × 81 × 57 mm and 84 × 47 × 36 mm.

Cobbles (6 specimens) These cobbles are battered and sometimes cracked but do not show the bruising typical of the hammerstones. They are very close to one another in size and the average is 77 × 67 × 52 mm.

	Quartzite	Fine-grained quartzite	Gneiss, feldspar, etc.	Phonolite	Nephelinite	Trachyte	Basalt/trachyandesite	Lava, indeterminate	Totals
Heavy-duty tools									
Bifaces, choppers, discoids, spheroids/subspheroids, scrapers	84	—	1	1	—	—	12	—	98
Light-duty tools									
Scrapers, laterally trimmed flakes, *outils écaillés*, punches	84	4	—	2	—	—	5	—	95
Utilised material, heavy-duty									
Pitted anvils/hammerstones, anvils, hammerstones, cobbles	4	—	—	—	—	—	9	—	13
Utilised material, light-duty									
Flakes, etc.	33	—	—	1	—	—	—	—	34
Debitage									
Whole flakes	233	9	—	9	—	2	23	—	276
Broken flakes	1237	29	—	23	2	3	26	—	1320
Core fragments	605	9	—	4	1	—	1	2	622
Cores	6	—	—	—	—	1	—	—	7
	2286	51	1	40	3	6	76	2	2465

Light-duty flakes, etc. (34 specimens) These consist of broken flakes and other fragments in which one or more edges have been chipped or blunted, apparently by utilisation. They vary in size from 51 × 50 mm to 26 × 13 mm, with an average of 31 × 23 mm.

Debitage

There are 2,225 specimens that appear to be waste material. They comprise 276 whole flakes, 1,320 broken flakes, 622 angular fragments and seven cores. By far the most common material is the quartzite from Naibor Soit inselberg which accounts for 93.4 per cent of the total. Other rocks represented are fine-grained quartzite (2.2 per cent), Engelosen phonolite (2 per cent), nephelinite (0.1 per cent), trachyte (0.2 per cent) and basalt/trachyandesite (2 per cent).

Whole flakes The proportion of end-struck flakes is almost identical in the quartzite and lava series, with 72.5 per cent and 73.5 per cent respectively. The lava flakes vary in length between 79 mm and 21 mm, with an average length of 40 mm. In the quartzite series, however, there are some massive flakes up to 110 mm long that may have been detached while blocking out handaxes.

Broken flakes Quartzite is by far the most common material, with 1,266 specimens as against fifty-four in lava.

Core fragments This category of debitage is unusually scarce in comparison with the broken flakes and amounts to only 622 specimens, of which 605 are quartzite.

Cores These consist of six irregular blocks of quartzite and one of trachyte from which flakes have been detached. With the exception of the trachyte specimen in which there are three striking platforms, the flaking has been carried out from a single platform. Sizes range from 123 × 73 × 68 mm to 46 × 42 × 37 mm, with an average of 80 × 50 × 45 mm.

HK (Hopwood's Korongo)

During the 1931 expedition to Olduvai an extensive collection of handaxes and some cleavers was made from an excavation at HK, a site on the north side of the gorge, about 1.5 km above the confluence of the two branches. While the exact stratigraphic position of the site was considered uncertain, it was assumed to be within Bed IV and to be at a higher level than the sites at JK, WK, etc., where industries referred to Stage 9 of the Chelles Acheul sequence had been found. The industry from HK, therefore, was described by L. S. B. Leakey (Leakey 1951) as Stage 10 of the sequence, or as Acheulean 5. However, when a small trench was dug at the site in 1969 in order to obtain a sample of the industry, it became evident that the Acheulean tools were not in their original context but in a disturbed deposit of relatively recent date. About 450 handaxes and cleavers were recovered during the 1931 excavations as well as a further ninety specimens from the slope below the site. The excavated series was described as lying on a fine-grained sand, overlain by a silty clay. Waste material occurred with the bifaces, as well as faunal remains that included the greater part of the skeleton of a hippopotamus. Spheroids were noted to be unusually scarce. The site was interpreted as being either a camping place or else a kill site where the tools had been discarded after cutting the meat from the carcass of the hippopotamus.

The site is at the top of a small hill that stands about 13 m above the floor of the gorge; it is separated from the main exposures in the north wall of the gorge by a small erosion gully. The artefact level is visible for a distance of approximately 20 m on the south side of the hill, where there is a steep erosion slope. The deposits on the north and west sides of the hill are obscured by accumulations of recent volcanic ash.

The trench dug in 1969 was situated at the northeast end of the site and the total area excavated amounted to 54 sq m. The artefacts were found in the lower part of 2.5 m of rubble mixed with patches of silt and sand (presumably corresponding to the silty clay that was noted in 1931) and lay on a fine-grained sand, composed largely of quartz. Small derived blocks of the upper aeolian tuff of the Masek Beds and of tuff from the Ndutu Beds were present in the same deposit as the artefacts as well as in the upper part of the level, indicating that it is of more recent date than either of these tuffs. Massive deposits of a cream-coloured tuff of Upper Ndutu age also occurred on the southwest side of the hill.

In order to establish the relationship of the artefact-bearing level with the Ndutu tuff and further verify the stratigraphic position of the tools, the upper part of the erosion slope was cleaned and cut back to a vertical face for a distance of 22 m. It became evident that the Upper Ndutu tuff lay directly on deposits of Bed IV, banked against an eroded slope at an angle of 27° (Fig. 6.5). Further to the northeast the Bed IV deposits were also eroded into an uneven surface and were overlain by 60 cm of earthy rubble containing fragments of the Lower Ndutu tuff as well as blocks of calcrete and the upper aeolian tuff of the Masek Beds (the Norkilili Member).

Thus, from the evidence obtained during excavation and from the section exposed on the south side of the HK hill it is clear that although the artefacts are Acheulean and may have been derived from Bed IV or the Masek Beds the deposit in which they now occur is of recent origin, since it postdates not only the upper part of the Masek Beds but also the Ndutu tuff. The fact that most of the tools are in sharp condition, seldom showing more than slight abrasion of the edges, indicates that they have not travelled far from their original locality. This is further supported by the hippopotamus skeleton found in 1931, that appeared to be of a single animal. The only indication as to the possible direction from which the tools may have been derived is in the

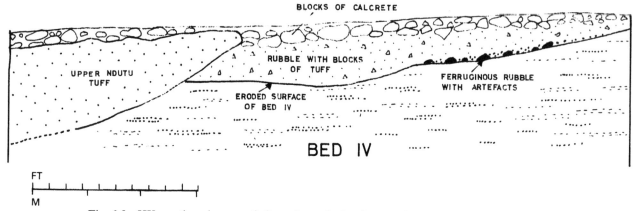

Fig. 6.5 HK: section along north face of trench showing stratigraphic position of artefacts

slope of the surface on which they lay. This dips downwards to the northeast, suggesting an origin to the southwest, possibly in the area now cut through by the gorge. (A careful search of the deposits in the vicinity has failed to reveal any comparable material *in situ*.)

Although the original excavation yielded the hippopotamus skeleton as well as a number of other bones, relatively few faunal remains were found in the trenches dug during 1969; these total about 700 specimens, of which 650 are small, unidentifiable fragments of bone.

The industry

The sample of the industry recovered during 1969 consists of 349 specimens, of which eighty-five are tools, twenty-one utilised pieces and 243 debitage. Bifaces constitute a high proportion of the tools and amount to 54.8 per cent of the total.

Bifaces (38 complete, 13 broken, 3 probably unfinished specimens) With the exception of one cleaver, a broken handaxe of basalt and one of nephelinite the series is of Naibor Soit quartzite. Small, elaborately trimmed ovates are characteristic. Most have been flaked over both faces and are symmetrically biconvex but in eight examples made on end flakes the primary surface has only received minimal flaking. Four others are made on thin slabs of quartzite with small areas of the natural cleavage surfaces remaining on the upper and lower faces. The cutting edges are generally straight and have been carefully trimmed. The only two cleavers in the collection are very similar in size and morphology. The butts are rounded and the maximum width is median, both sides converging towards the cleaver edges. In a few of the more elaborately trimmed small ovates there is a suggestion of an S-twist, but it is usually not very pronounced and sometimes on one edge only. A few specimens are abraded, but most are in mint condition, even though the edges have been chipped. The broken handaxes consist of ten specimens in which the tips are missing, one that is broken lengthwise, one tip end and one of phonolite in which almost the entire surface has exfoliated. (The bifaces are analysed by Drs Roe and Callow in chapters 8 and 9.)

Choppers (5 specimens) All five are side choppers. They are made on blocks of quartzite with bifacially flaked, jagged working edges. The butts are formed by untrimmed flat surfaces that are at right angles to the working edges (a feature found commonly in the quartzite choppers from Upper Bed II). These tools range in size from 106 × 134 × 78 mm to 50 × 70 × 36 mm, with an average for the five specimens of 69 × 89 × 50 mm.

Discoids (6 specimens) Four examples are of quartzite and two of basalt. Three of the quartzite specimens are thick in cross section and markedly biconvex, but there are also two examples, made on a flake and on a piece of tabular quartzite, that are flat on the upper and lower faces and relatively

THE MASEK BEDS

ANALYSIS OF THE INDUSTRIES FROM HK

	Numbers	Percentages
Tools	85	24.3
Utilised material	21	6.1
Debitage	243	69.6
	349	100.0
Tools		
Bifaces	54	63.5
Choppers	5	5.9
Discoids	6	7.0
Spheroids/subspheroid	1	1.2
Scrapers, heavy-duty	10	11.8
Scrapers, light-duty	9	10.5
	85	99.9
Utilised material		
Anvils	2	9.5
Hammerstones	2	9.5
Light-duty flakes	17	80.9
	21	99.9
Debitage		
Whole flakes	42	17.3
Broken flakes	150	61.7
Core fragments	47	19.3
Cores	4	1.6
	243	99.9

thin. The sixth specimen, made on a basalt flake, is steeply trimmed on the circumference from a flat lower face that is presumably the primary surface. The range in size is from $107 \times 98 \times 70$ mm to $70 \times 65 \times 41$ mm, with an average of $89 \times 78 \times 44$ mm.

Spheroid (1 specimen)   This is of quartzite and measures $53 \times 50 \times 45$ mm. The projecting ridges have been battered and rounded off, but not sufficiently to make a symmetrical spheroid.

Scrapers, heavy-duty (10 specimens)   These tools are all of quartzite and are made either on side flakes or on pieces of tabular quartzite. They include seven well-made side scrapers of a type that is unusual at Olduvai in which the bilateral width exceeds the length from working edge to butt. The working edges are evenly trimmed from the lower faces and extend in a gentle curve across the whole width of the tools. The seven specimens are very similar in size; they range from $74 \times 152 \times 32$ mm to $64 \times 106 \times 36$ mm, with an average of $70 \times 126 \times 33$ mm. Two of the remaining heavy-duty scrapers are made on large flakes and the third on an irregular block in which one edge has been steeply trimmed to a curved working edge. The average size for these three specimens is $96 \times 86 \times 44$ mm.

Scrapers, light-duty (9 specimens)   These comprise three side, two end, three hollow and one core scrapers. With the exception of one end scraper made on a basalt flake they are of quartzite. The basalt specimen is symmetrical with an evenly rounded and regularly trimmed working edge at the tip of the flake. In the remaining specimens the working edges are rather uneven. The range in size for the whole series is from $69 \times 51 \times 13$ mm to $31 \times 35 \times 11$ mm.

Utilised material

Anvils (2 specimens)   These are both of quartzite and consist of a discoidal slab and a high-backed, flat-based example. Battered utilisation is present on the edges of the flat surfaces. The specimens measure $89 \times 82 \times 36$ mm and $71 \times 68 \times 56$ mm.

Hammerstones (2 specimens)   A single subspherical cobble of trachyandesite shows battering on several parts of the surface ($76 \times 70 \times 59$ mm). The second example is a basalt cobble measuring $66 \times 62 \times 53$ mm that has been considerably battered and from which a number of flakes have been detached, probably as a result of utilisation.

Flakes and other fragments (17 specimens)   These vary from large fragments measuring 135×105 mm to small pieces 24×20 mm, but in each example one or more edges has been chipped or blunted. There is a single flake of fine-grained pink quartzite in which one edge is serrated: it has a series of ten small indentations with intervening projections along a length of 71 mm. (This series is too heterogeneous for average measurements to be meaningful.)

Debitage (243 specimens)

The debitage comprises forty-two whole flakes, 150 broken flakes, forty-seven angular fragments and four cores. One complete and two broken

	Quartzite	Fine-grained quartzite	Gneiss, feldspar, etc.	Phonolite	Nephelinite	Trachyte	Basalt/trachyandesite	Lava, indeterminate	Totals
Heavy-duty tools									
Bifaces, choppers, discoids, spheroid/subspheroid, scrapers	71	—	—	—	—	1	4	—	76
Light-duty tools									
Scrapers	8	—	—	—	—	—	1	—	9
Utilised material, heavy-duty									
Anvils	2	—	—	—	—	—	—	—	2
Hammerstones	—	—	—	—	—	—	2	—	2
Utilised material, light-duty									
Flakes, etc.	16	1	—	—	—	—	—	—	17
Debitage									
Whole flakes	39	2	—	—	—	—	1	—	42
Broken flakes	145	1	—	2	—	—	2	—	150
Core fragments	47	—	—	—	—	—	—	—	47
Cores	3	—	—	—	—	—	1	—	4
	331	4	—	2	—	1	11	—	349

flakes are of basalt and two broken flakes of phonolite, the remainder being of quartzite.

Whole flakes The greater part of these are small flakes characteristic of sites where quartzite was the principal raw material, but there are also seven large flakes in which the average length is 116 mm, which may be rough-outs for handaxes, or even have served the same purpose as handaxes, since they are oval in shape and the edges are generally chipped and blunted. End and side flakes are almost equally represented, with nineteen of the former and twenty-three of the latter. Omitting the seven large flakes, the end flakes range in size from 73×72 mm to 21×13 mm, with an average of 42×32 mm, and the side flakes from 53×56 mm to 15×24 mm, with an average of 30×37 mm.

Broken flakes These amount to 150 specimens, of which 145 are quartzite.

Core fragments There is a total of only 47 specimens all of which are of quartzite. This small number is in contrast to the large number at sites where pitted anvils and hammerstones indicate that the bipolar technique was in general use.

Cores There are two large blocks of quartzite measuring respectively $122 \times 110 \times 107$ mm and $150 \times 125 \times 86$ mm that exhibit negative flake scars and appear to be cores. A smaller fragment ($79 \times 48 \times 36$ mm) of fine-grained quartzite as well as a piece of basalt ($95 \times 90 \times 44$ mm) can also be classed as cores.

TK Fish Gully (Teal's Korongo)

This site lies on the west side of the gully, opposite the site in Bed II that was excavated in 1963. It was first recorded in 1932 when it yielded handaxes, catfish remains and a well-preserved skull of *Megalotragus kattwinkeli*. During 1962 the late

THE MASEK BEDS

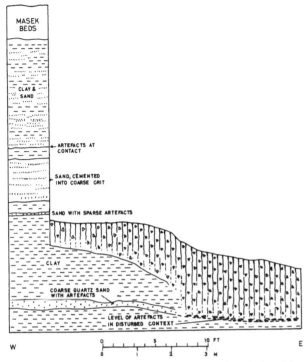

Fig. 6.6 TK Fish Gully: section to show the relationships of artefacts *in situ* to those in disturbed context

Dr John Waechter carried out excavations at this site and recovered a series of large elaborately trimmed Acheulean handaxes, mostly made from quartzite, as well as abundant remains of catfish. Both artefacts and catfish bones were densely concentrated within a small area and many had also eroded on to the slope below. In 1963 the nature and stratigraphy of Beds III and IV were little understood and the material was assumed to be *in situ*, in its original context. In 1970, however, when investigation of Bed IV was in progress, it was noted that only Holocene rubble and hillwash overlay the horizon from which the artefacts had been excavated. The rubble contained pieces of the aeolian tuffs of the Masek Beds and was clearly unconformable to the undisturbed strata of Bed IV upon which it rested. A trench measuring 3.6 × 2.5 m adjoining the area excavated by Dr Waechter, and to the west of it, was then dug in an endeavour to ascertain the original stratigraphic position of the artefacts. A layer of undisturbed coarse quartz sand 10 to 30 cm thick containing bifaces and other tools as well as catfish remains and a number of indeterminate fragments of mammalian bones was found at approximately the same level as the artefacts previously excavated (Fig. 6.6). It lay at a depth of 7.5 m below the base of the Masek Beds and was both overlain and underlain by clay horizons. The exact stratigraphic position of this horizon was not determined other than that it was within Upper Bed IV. Among the tools from the quartz sand are eleven handaxes, five of which are of quartzite, but only one of these exhibits the refined trimming seen on most of those recovered by Dr Waechter. Four of the six lava specimens, however, are symmetrical and comparable in form and technology. The fact that a proportionately greater number of lava handaxes than quartzite occurred in the *in situ* material may be coincidental and it is possible that the original concentration of handaxes was derived from this horizon. The density of the occurrence may have been brought about when the *in situ* deposits were exposed and the sand, together with the lighter elements of the industry, was mostly removed, leaving the heavier handaxes as a residual lag. In many respects this site resembles HK, where bifacial tools, also elaborately trimmed, were found under similar conditions but with even less evidence as to their original position in the sequence.

In the course of his excavations at TK Fish Gully Dr Waechter dug three other trenches, at different stratigraphic levels. Unfortunately notes on these excavations have not come to light among his papers. It is, therefore, not possible to determine whether all the artefacts belong to the assemblage under review or whether some are from the trenches dug above or below this level. The large, unusually finely trimmed specimens are known to have come from the level tested in 1970, but it is likely that the smaller lava bifaces may be from a higher level. The writer was not present during the concluding stages of this excavation, but the late Mrs S. C. Savage remarked on the differences to be seen in the bifaces from the higher level, particularly their smaller size and the more prevalent use of lava.

The artefacts recovered by Dr Waechter consist of the following: sixty-eight complete handaxes

ANALYSIS OF THE INDUSTRIES FROM TK FISH GULLY: THE *IN SITU* MATERIAL

	Numbers	Percentages
Tools	24	3.8
Utilised material	13	2.1
Debitage	586	94.1
	623	100.0
Tools		
Bifaces	11	45.8
Discoid	1	4.2
Spheroids/subspheroids	4	16.6
Scrapers, light-duty	7	29.2
Laterally trimmed flake	1	4.2
	24	100.0
Utilised material		
Anvils	2	15.4
Light-duty flakes, etc.	11	84.6
	13	100.0
Debitage		
Whole flakes	71	12.1
Broken flakes	224	38.2
Core fragments	291	49.7
	586	100.0

and one cleaver, of which fifty-three are made from quartzite, seven from phonolite, eight from basalt or trachyandesite and one of nephelinite. There are also six broken bifaces, one discoid, four spheroids/subspheroids, a few flakes and some rough, unmodified blocks of quartzite. With the exception of two light-duty scrapers there are no small tools. The quartzite handaxes are among some of the largest and most elaborately trimmed known from Olduvai. Although more varied in form, they compare in technology with those from the Masek Beds at FLK and some of the larger specimens from HK.

The *in situ* artefacts

The industry consists of 623 specimens, comprising twenty-four tools, thirteen utilised pieces and 586 debitage.

Tools

Bifaces (9 complete specimens, 1 with the tip missing and 1 rough-out) Four examples and the rough-out are of quartzite, two of phonolite

ANALYSIS OF THE INDUSTRIES FROM TK FISH GULLY: THE *IN SITU* MATERIAL

	Quartzite	Fine-grained quartzite	Gneiss, etc.	Phonolite	Nephelinite	Trachyte	Basalt/trachyandesite	Totals
Tools								
Bifaces	5	—	—	2	—	—	4	11
Discoid	1	—	—	—	—	—	—	1
Spheroids/subspheroids	3	—	—	—	—	—	1	4
Light-duty scrapers	7	—	—	—	—	—	—	7
Laterally trimmed flake	—	—	—	—	—	—	1	1
Utilised material								
Anvils	2	—	—	—	—	—	—	2
Light-duty flakes	11	—	—	—	—	—	—	11
Debitage								
Whole flakes	63	—	—	2	—	—	6	71
Broken flakes	212	—	—	3	—	—	9	224
Core fragments	287	—	—	1	—	—	3	291
	591	—	—	8	—	—	24	623

and four of basalt or trachyandesite. One of the phonolite specimens is made on a flake. It is unusually thin, flat on both faces and cordiform in shape (119 × 91 × 19 mm). The second phonolite handaxe is an elongated ovate with a blunt tip, and measures 174 × 91 × 34 mm. One of the other four lava specimens is triangular in shape and sharply pointed. It is made on a cobble with a relatively thick and heavy butt formed by weathered cortex (93 × 71 × 48 mm). The remaining three specimens are elongated ovates, one of which shows unusually elaborate trimming with a minimum of thirty-six trimming scars. It is symmetrical, with a sharp, slender point and a straight, even edge on the whole circumference (181 × 91 × 49 mm). The last two specimens show less refinement of trimming. They measure 183 × 87 × 51 mm and 136 × 87 × 51 mm. One of these and the small triangular biface are heavily abraded or weathered. The five quartzite specimens vary considerably in size and workmanship, from a miniature example measuring 76 × 41 × 24 mm to a rather crude specimen made on a piece of tabular material, possibly unfinished, measuring 154 × 75 × 63 mm. Another example measures 152 × 84 × 49 mm. The specimen with a broken tip now measures 121 × 85 × 50 mm but was probably 134 mm long when complete. The rough-out consists of an oblong slab of quartzite with a minimum of crude flaking along one lateral edge (167 × 106 × 36 mm). (This series has been analysed by Drs Roe and Callow.)

Other tools consist of the following: one quartzite discoid radially flaked on both faces (34 × 32 × 24 mm); one basalt or trachyandesite subspheroid and three of quartzite, all of which are angular. The lava specimen is the largest and measures 76 × 66 × 65 mm while the quartzite specimens measure 49 × 46 × 39 mm, 45 × 40 × 36 mm and 39 × 36 × 29 mm. There are seven light-duty quartzite scrapers, of which five are convex edged but made on irregular-shaped fragments; in four examples the working edges are blunted by use. Length/breadth measurements are 29 × 41 mm, 16 × 28 mm, 36 × 41 mm (side scrapers) and 29 × 19 mm, 29 × 24 mm (end scrapers). Two examples are notched and the interiors of the notches have been blunted. They are situated on the lateral edges of oblong fragments measuring 30 × 20 mm and 25 × 15 mm. The notches measure 9 × 2.5 mm and 9 × 2 mm. There is one laterally trimmed flake of trachyandesite measuring 92 × 55 mm. It has been crudely flaked along both lateral edges on both faces. There is a projecting shoulder on one side and the tip is blunt; it is either weathered or abraded.

The utilised material consists of the following: two angular blocks of quartzite showing anvil type of battering on the ridges (90 × 77 × 77 mm and 88 × 86 × 59 mm). Eleven flakes and other fragments showing various types of chipping on the edges, including slight scaling. The length/breadth measurements range from 50 × 42 mm to 24 × 17 mm, with an average of 38 × 27 mm.

The debitage includes seventy-one complete flakes of which sixty-three are quartzite, two phonolite and six basalt or trachyandesite. End-struck flakes number forty-nine and average 36 × 25 mm in length/breadth, side flakes amount to twenty-two and average 30 × 35 mm in length/breadth. There is a total of 224 broken flakes of which 212 are quartzite, three phonolite and nine basalt or trachyandesite. Core fragments amount to 291 specimens of which 298 are quartzite, one phonolite and three basalt or trachyandesite. In addition to the above artefacts there are four unmodified blocks of quartzite.

7

THE FAUNA

M. D. LEAKEY

The fauna from Beds III–IV and the Masek Beds is somewhat better known now than it was in 1965 when L. S. B. Leakey wrote a preliminary report on the Olduvai fauna in volume 1 of the Olduvai monographs. Excavation of sites in these beds has yielded material from securely established stratigraphic horizons, but owing to its generally fragmentary condition it has proved less informative than the fauna from Beds I and II.

The most notable contribution to the understanding of the Olduvai fauna has been Dr and Mrs Gentry's study of the Bovidae published in two British Museum Bulletins (Gentry and Gentry 1978). A revision of the Suidae by Drs J. M. Harris and T. D. White (Harris and White 1979) has served to eliminate much of the confusion in respect to nomenclature that existed in the Plio-Pleistocene Suidae, including those from Olduvai. The paper by Dr J.-J. Jaeger (1976) on the Muridae of Beds I and II sheds considerable light on the palaeoenvironment, but rodents from Beds III and IV are generally scarce except for the site of HEB in Lower Bed IV. Other micro-mammalian and avian fossils as well as pollen are also rare in the upper part of the Olduvai sequence, so that the palaeoenvironment cannot be reconstructed with the same confidence that was possible for Beds I and II.

Bovidae are the most informative group of large mammals. This is due to a number of factors, but principally because they are the most common taxon at all the excavated sites and because horn cores are one of the most durable parts of any mammal.

A faunal list has been compiled for Beds III, IV, the Masek Beds and later sites (Table 7.1). This has been prepared from identifications kindly made by Dr J. M. Harris, Chief Curator, Division of Earth Sciences, Los Angeles County Natural History Museum, and from the lists of Bovidae published by Dr and Mrs Gentry. The list has been divided into five columns, as follows:

(1) Bed III
(2) Base of Bed IV
(3) Lower Bed IV
(4) Upper Bed IV
(5) Masek Beds

The faunal remains from the two excavated sites HK and TK Fish Gully, that proved to be in disturbed deposits, are listed neither under the sites nor in the comparative tables since their stratigraphic positions are unknown. The data regarding fauna from other sites in Beds III and IV, either isolated finds or surface material, have not altered appreciably since the publication of Volume 1 of the Olduvai monographs (Leakey 1965). The principal disparities lie in nomenclature, and some of the names used then are now obsolete. All the known species of Bovidae have been listed by Dr and Mrs Gentry. For convenience, the Beds III–IV faunal lists from their paper are incorporated here as well as the list from Volume 1 (Table 7.2). When necessary, names have been changed to agree with current usage.

It will be seen that in the lists of taxa from excavated sites identification to specific level has been possible in a fair number of cases. But there are also many entries where specific identification has been impossible owing to the fragmentary condition of the specimens. It is evident that the greatest number of individual mammals and different species are generally from the most extensive excavations. Their absence from smaller excavations is probably of no significance. On the other hand, some taxa such as catfish and

THE FAUNA

Table 7.1 *List of fauna from sites excavated in Beds III, IV and the Masek Beds*

	Bed III	Base of Bed IV			Lower IV	Upper IV			Masek Beds
	JK	WK Lower Channel	WK Hippo Cliff	PDK Trench IV	HEB, HEB East, HEB West	WK	WK East, A and C	PDK Trenches I–III	FLK
Mollusca									
Unio sp.	—	—	—	—	×	×	—	—	×
cf. *Limicolaria*	—	—	—	—	—	×	—	—	×
Clariidae									
Clarias sp.	×	×	×	×	×	×	×	×	×
Amphibia									
sp. indeterminate	×	—	—	—	×	—	×	—	—
Chelonia									
sp. indeterminate	×	×	—	—	×	×	×	×	×
Reptilia									
Crocodylus sp.	×	×	×	—	×	×	×	—	×
Aves									
spp. indeterminate	×	×	—	—	×	×	×	—	×
Struthio sp.	×	×	—	×	×	×	×	—	×
Primates									
Colobinae	×	—	—	—	—	—	—	—	—
Cercopithecoidea kimeui	—	—	—	—	×	×	—	—	—
Papionini	—	—	—	—	—	—	—	×	—
Theropithecus oswaldi				(not found at any excavated sites)					
Leporidae									
Lagomorpha sp. indeterminate	×	—	—	—	—	×	—	—	—
Rodentia									
Otomys sp.	—	—	—	—	×	—	—	—	—
Dasymus incomptus	—	—	—	—	×	—	—	—	—
Mastomys natalensis	—	—	—	—	—	—	—	—	×
Mastomys sp.	—	—	—	—	×	—	—	—	—
Arvicanthus niloticus	—	—	—	—	×	—	—	—	—
Aethomys cf. *lavocati*	—	—	—	—	×	—	—	—	—
Zelotomys leakeyi	—	—	—	—	×	—	—	—	—
Thalomys sp.	—	—	—	—	×	—	—	—	—
Saccostomus sp.	—	—	—	—	×	—	—	—	—
Tatera sp.	—	—	—	—	×	—	—	—	×
Steatomys sp.	—	—	—	—	×	—	—	—	—
Heterocephalus sp.	—	—	—	—	×	—	—	—	—
Jaculus sp.	—	—	—	—	—	—	—	—	×
Muridae sp. indeterminate	—	—	—	—	—	×	—	—	—
Hystrix sp. indeterminate	—	—	—	—	—	×	—	—	—
Carnivora									
Viverridae sp.	—	—	—	—	×	×	—	—	—
Hyaenidae sp.	×	—	—	—	—	—	—	—	—
Panthera leo	—	—	—	—	—	—	—	—	—
Lutrinae sp.	×	—	—	—	—	—	—	—	—
Aonyx sp.	×	—	—	—	×	—	—	—	—
Canidae sp.	×	—	—	—	—	—	—	—	—

(*continued on next page*)

Table 7.1 (cont.)

	Bed III	Base of Bed IV			Lower IV	Upper IV			Masek Beds
	JK	WK Lower Channel	WK Hippo Cliff	PDK Trench IV	HEB, HEB East, HEB West	WK	WK East, A and C	PDK Trenches I–III	FLK
Elephantidae									
Elephas recki	×	—	—	—	×	×	×	×	—
Equidae									
Equus oldowayensis	×	×	×	×	×	×	×	×	×
Hipparion cf. *ethiopicum*	×	—	—	—	×	×	×	×	×
Rhinocerotidae									
Ceratotherium simum	—	—	×	—	×	×	×	—	—
Diceros bicornis	?×	×	—	—	×	×	×	—	—
Suidae									
Kolpochoerus limnetes	×	×	—	—	×	×	×	—	×
K. major	—	—	—	—	—	×	×	—	—
Metridiochoerus modestus	—	×	—	—	×	×	—	—	—
M. compactus	×	—	—	—	×	×	×	×	×
Phacochoerus antiquus	—	—	—	—	—	×	×	—	×
Hippopotamidae									
Hippopotamus gorgops	×	×	×	×	×	×	×	×	×
Giraffidae									
Giraffa jumae	—	×	—	—	×	×	×	?×	—
G. stillei	—	—	—	—	—	×	—	—	—
Sivatherium maurusium	×	×	—	—	×	×	×	—	—
Bovidae									
Tragelaphus strepsiceros grandis	—	—	—	—	—	×	—	×	×
Tragelaphus sp. indeterminate	×	—	—	—	×	×	×	—	—
Taurotragus cf. *arkelli*	×	×	—	—	—	×	—	—	—
Pelorovis oldowayensis	×	—	—	—	—	—	—	—	—
P. antiquus	—	—	—	—	—	—	×	—	—
Syncerus aceolatus	×	—	—	—	—	—	×	—	—
Kobus ellipsiprymnus	×	—	—	—	—	—	—	—	—
K. kob	×	—	—	—	—	—	—	—	—
Reduncini sp. indeterminate	×	—	—	—	—	×	×	—	×
Hippotragus gigas	×	—	—	—	—	—	—	—	—
Megalotragus cf. *kattwinkeli*	×	×	—	—	—	—	—	—	—
Connochaetes taurinus olduvaiensis	(not represented at excavated sites)								
Parmularius rugosus	×	—	—	—	—	—	—	—	—
Damaliscus niro	×	—	—	—	—	—	—	—	—
D. agelaius	×	—	—	—	—	—	—	—	—
Rabaticeras arambourgi	×	—	—	?×	—	—	—	?×	—
Alcelaphini sp. 1	×	—	—	—	—	×	×	—	—
Alcelaphini sp. 2	×	—	—	—	—	×	×	—	—
Alcelaphini sp. 3	—	—	—	—	—	—	—	—	×
Neotragini sp. indeterminate	—	—	—	—	—	—	—	—	×
Cephalophini sp. indeterminate	—	—	—	—	—	—	—	—	×
Antidorcas sp.	×	—	—	—	×	—	—	—	—
Antilopini sp.	—	—	—	—	—	×	×	—	—
Gazellini sp.	×	×	—	—	—	×	—	—	×

THE FAUNA

Table 7.2 *List of all known fauna from Beds III, IV and the Masek Beds*

	Bed III	Bed IV	Masek Beds
Mollusca			
Unio sp.	×	×	×
cf. Limicolaria	—	×	×
Clariidae			
Clarias sp.	×	×	×
Amphibia			
sp. indeterminate	×	×	—
Chelonia			
sp. indeterminate	×	×	×
Reptilia			
Crocodylus sp.	×	×	×
Aves			
spp. indeterminate	×	×	×
Struthio sp.	×	×	×
Primates			
Colobinae	×	×	—
Cercopithecoidea kimeui	—	×	—
Papionini	—	×	—
Theropithecus oswaldi	×	×	—
Leporidae			
Lagomorpha sp. indeterminate	×	×	—
Rodentia			
Otomys sp.	—	×	—
Dasymus incomputus	—	×	—
Mastomys natalensis	—	—	×
Mastomys sp.	—	×	—
Arvicanthus niloticus	—	×	×
Aethomys cf. lavocati	—	×	—
Zelotomys leakeyi	—	×	—
Thalomys sp.	—	×	—
Saccostomus sp.	—	×	—
Tatera sp.	—	×	×
Steatomys sp.	—	×	—
Heterocephalus sp.	—	×	—
Jaculus sp.	—	—	×
Muridae sp. indeterminate	—	×	—
Hystrix sp. indeterminate	—	×	—
Carnivora			
Viverridae sp.	—	×	—
Hyaenidae sp.	×	×	—
Panthera leo	—	×	—
Lutrinae	×	—	—
Aonyx sp.	—	×	—
Canidae sp.	×	×	—

Table 7.2 (*cont.*)

	Bed III	Bed IV	Masek Beds
Elephantidae			
Elephas recki	×	×	—
Equidae			
Equus oldowayensis	×	×	×
Hipparion cf. ethiopicum	×	×	×
Rhinocerotidae			
Ceratotherium simum	×	×	×
Diceros bicornis	?×	×	—
Suidae			
Kolpochoerus limnetes	×	×	×
K. major	×	×	—
Metridiochoerus modestus	×	×	—
M. hopwoodi	×	×	—
M. compactus	×	×	×
Phacochoerus antiquus	—	×	×
Hippopotamidae			
Hippopotamus gorgops	×	×	×
Giraffidae			
Giraffa jumae	—	×	—
G. stillei	—	×	—
Sivatherium maurusium	×	×	—
Bovidae			
Tragelaphus strepsiceros grandis	×	×	—
T. aff. spekei or scriptus	—	×	—
Taurotragus arkelli	—	×	—
Pelorovis oldowayensis	×	—	—
P. antiquus	—	×	—
Synercus aceolatus	×	×	—
Kobus ellipsiprymnus	×	—	—
K. kob	×	—	—
Redunca sp.	×	—	×
Thaleroceros radiciformis	—	?×	—
Hippotragus gigas	×	—	—
Megalotragus kattwinkeli	×	×	—
Connochaetes taurinus olduvaiensis	×	×	—
Parmularius rugosus	×	×	—
Damaliscus niro	×	×	—
D. agelaius	×	×	—
Rabaticeras arambourgi	×	×	—
Alcelaphini sp. 1	—	×	—
Alcelaphini sp. 2	—	—	×
Neotragini sp.	—	—	×
Cephalophini sp.	—	—	×
Antidorcas recki	×	×	—
Antidorcas sp.	—	×	—
Antilopini sp.	—	×	—
Gazellini sp.	—	—	×

hippopotami are ubiquitous and occur at nearly all sites, no matter how small an area was excavated.

The fauna described here has been analysed in a different manner from that of Beds I and II where most of the faunal material was recovered from primary contexts. The remains found at any given site, therefore, could be assumed to represent the undisturbed debris left *in situ* by the hominid population or by predators and scavengers. Numbers of individuals were not counted, but recognisable parts belonging to different taxa were totalled to give approximate figures of the greater or lesser abundance of certain groups. In nearly every case Bovidae proved to be the most numerous. As has been shown, the sites described in this volume are almost invariably in channels where both the faunal and archaeological material had perhaps been disturbed by water action. Lists of the taxa represented as well as lists of the parts of different animals from each site have been made and the probable numbers of individuals have been estimated from these data (Tables 7.3, 7.4). It should be noted that at every site in Beds III and IV there were very large numbers of small, indeterminate bone fragments which could not be assigned to skeletal parts, more numerous than similar fragments from sites in Beds I and II.

Table 7.3 *Minimum numbers of individual mammals from sites excavated in Beds III, IV and the Masek Beds*

	Bed III		Base of Bed IV		Lower Bed IV		Upper Bed IV		Masek Beds	
	Nos	%	Nos	%	Nos	%	Nos	%	Nos	%
Primates	2	3.0	1	3.1	3	6.0	4	3.5	1	4.2
Rodentia	6	9.3	—	—	?	—	7	6.1	4+	16.6
Carnivora	4	6.1	1	3.1	3+	6.0	4	3.5	2+	8.3
Elephantidae	7	10.8	1	3.1	3	6.0	6+	5.3	—	—
Equidae	10	15.4	6+	18.7	7	14.0	19+	16.7	3	12.5
Rhinocerotidae	1	1.5	2	6.3	3	6.0	6+	5.3	1	4.2
Suidae	4+	6.2	2	6.3	5+	10.0	19+	16.7	3	12.5
Hippopotamidae	8+	12.3	7	21.9	6+	12.0	10	8.7	1+	4.2
Giraffidae	2	3.0	2	6.3	4	8.0	7	6.1	1	4.2
Bovidae	21+	32.3	10	31.2	16+	32.0	32+	28.1	8+	33.3
Totals	65+	99.9	32+	100.0	50+	100.0	114+	100.0	24+	100.0

Note: Rodentia from Lower Bed IV are omitted since the collection from HEB is not available.

Table 7.4 *Minimum numbers and percentage of mammals from sites excavated in Beds III, IV and the Masek Beds*

	Nos.	Per cent
Primates	11	3.8
Rodentia	17+	6.0
Carnivora	14+	5.0
Elephantidae	17+	6.0
Equidae	45+	15.8
Rhinocerotidae	13+	4.6
Suidae	33+	11.5
Hippopotamidae	32+	11.2
Giraffidae	16	5.6
Bovidae	87+	30.5
	285+	100.0

Bed III

JK

(For description of the site see pp. 15–18.)

In the overall faunal table for this site (Table 7.5), the four levels recognised have been pooled in view of the disturbed condition of the deposits, but separate identifications and estimated numbers of individual animals are given for each level. Owing to the extensive excavations carried out by Dr M. R. Kleindienst and by the writer this site has yielded the second largest number of identifiable faunal specimens from Beds III–IV.

There is no obvious association of parts of any one animal at JK. Moreover, there is reason to believe that skeletal parts were widely dispersed in the channel, the most striking example being the mandible and maxilla of an elephant which were found 20 m apart and in different horizons. There is, of course, no absolute proof that these two specimens belonged to the same animal, but judged on the degree of wear on the teeth and general condition of the bones it seems a reasonable assumption. Thirty-eight taxa have been identified at JK and it is estimated that a minimum of sixty-five individual larger animals are represented, but it is possible that this figure may be too high if parts of the same animal were transported and redeposited in different levels.

Bovidae are the most common taxon and amount to twenty-one individuals, 32.3 per cent of the estimated number of mammals. They include nineteen genera and species. Among these is the latest known example of *Pelorovis oldowayensis*, which appears to have been replaced by

THE FAUNA
Table 7.5 *Fauna from JK*

JK CLAY ABOVE THE PINK SILTSTONE
Clariidae: Cranial fragments 9
Struthionidae: Eggshell fragments 8

MAMMALIA
Hippopotamidae: Fragment of femur
Bovidae: Cranium with horn cores, fragment left side of mandible, 4 teeth, part of scapula
Taxa indeterminate: Cranial fragments 2, 1 broken rib shaft, 1 broken phalanx, 1 complete and 1 broken vertebra, 1 fragment acetabulum, 10 fragments limb bone shafts. Three individuals probably represented

JK PINK SILTSTONE
Clariidae: Cranial fragments 23, bones 18

	Cercopithecidae	Rodentia	Carnivora	Elephantidae	Equidae	Rhinocerotidae	Suidae	Hippopotamidae	Giraffidae	Bovidae	Taxa indeterminate	Totals
Skull parts, horn cores	—	—	—	—	—	—	—	—	—	—	7	7
Maxillae/mandibles	—	—	—	—	—	—	—	—	—	1	4	5
Teeth, isolated	—	—	1	—	—	—	—	10	—	15	34	60
Axial: vertebrae, ribs	—	—	—	—	—	—	—	—	—	—	11	11
Scapulae	—	—	—	—	—	—	—	—	—	—	12	12
Pelves	—	—	—	—	—	—	—	—	—	—	2	2
Fore limb, long bones	—	—	—	—	—	—	—	—	—	—	1	1
Hind limb, long bones	—	—	—	—	—	—	—	—	—	—	1	1
Podials, patellae, etc.	—	—	—	—	—	—	—	—	—	2	—	2
Estimated number of individuals	—	—	1	—	—	—	—	1	—	2	—	4
Indeterminate bone fragments												580

JK FINE-GRAINED FERRUGINOUS SAND
Clariidae: Cranial fragments 137, bones, various 111
Chelonia: Broken scutes 2, limb bone 1
Crocodylidae: Teeth 80, broken scute 1

MAMMALIA

	Cercopithecidae	Rodentia (inc. Lagomorpha)	Carnivora	Elephantidae	Equidae	Rhinocerotidae	Suidae	Hippopotamidae	Giraffidae	Bovidae	Taxa indeterminate	Totals
Skull parts, horn cores	—	—	—	—	—	—	—	—	—	6	25	31
Maxillae/mandibles	—	—	—	2	3	—	—	—	—	4	9	18
Teeth, isolated	4	5	1	3	6	—	3	11	—	37	72	142
Axial: vertebrae, ribs	—	—	—	1	2	—	1	23	1	10	70	108
Scapulae	—	—	—	—	1	—	—	—	—	6	15	22
Pelves	—	—	—	—	—	—	—	—	—	—	9	9
Fore limb, long bones	—	2	1	1	1	—	1	1	—	10	16	33
Hind limb, long bones	—	3	—	—	—	—	—	2	—	4	10	19
Podials, patellae, etc.	1	—	—	—	2	—	2	12	—	52	12	81
Estimated number of individuals	1	4	1	5	5/6	—	2+	4+	1	3+		36+
Indeterminate bone fragments												10,455

(continued on next page)

Table 7.5 (cont.)

JK COARSE GREY SAND
Clariidae: Cranial fragments 130, various bones 104
Chelonia: Broken scute 1
Crocodylidae: Teeth 104, cranial fragment 1, scute 1, centrum of vertebra 1

MAMMALIA

	Cercopithecidae	Rodentia	Carnivora	Elephantidae	Equidae	Rhinocerotidae	Suidae	Hippopotamidae	Giraffidae	Bovidae	Taxa indeterminate	Totals
Skull parts, horn cores	—	—	—	1	—	—	—	—	—	2	9	12
Maxillae/mandibles	1	2	—	—	1	—	1	—	—	—	—	5
Teeth, isolated	4	5	2	6	10	4	3	7	1	31	58	131
Axial: vertebrae, ribs	—	2	—	5	—	—	—	—	—	3	18	28
Scapulae	—	—	—	—	—	—	—	—	—	1	1	2
Pelves	—	—	—	1	—	—	—	—	—	—	1	2
Fore limb, long bones	—	—	—	—	—	—	—	1	—	1	2	4
Hind limb, long bones	—	—	—	—	—	—	—	1	—	1	3	5
Podials, patellae, etc.	1	—	—	1	—	1	—	1	—	8	8	20
Estimated number of individuals	1	2	2	2	4	1	2+	3+	1	6	—	24+
Indeterminate bone fragments												3475

P. antiquus during Bed IV times. Both *Kobus ellipsiprymnus* and *K. kob* are present at this site but have not been identified positively from any other site in Beds III or IV except at EF-HR where a surface specimen of *K. ellipsiprymnus* was found at the Beds III–IV contact.

The faunal material recovered by Dr Kleindienst has not been included in Table 7.5 which refers only to specimens from the four levels excavated by the writer.

Base of Bed IV

Thirty taxa are represented in the faunal material from the three sites at the base of Bed IV, namely WK Hippo Cliff, PDK Trench IV and WK Lower Channel. Clariidae, Equidae, *Hippopotamus cf. gorgops* and *Connochaetes* sp. are common to all three.

WK Hippo Cliff

(For description of the site see p. 36).

The principal preserved parts of the hippopotamus skeleton are as follows:

Cranium, lacks top of skull
Mandible, almost complete, but crushed
Right scapula, complete (length from glenoid hollow to lower margin, 50 cm)
Axis, complete
Atlas, parts of
5 thoracic vertebrae, complete
3 cervical vertebrae, complete
2 centra, thoracic vertebrae
1 neural spine
2 centra, indet.
Right femur, complete
Left tibia, complete
Left femur, proximal and distal ends (shaft missing)
Left humerus, shaft only
4 ribs, complete
3 parts rib shafts
2 proximal ends of metapodials

This is a particularly massive specimen of *H. gorgops*, but owing to crushing and distortion no accurate measurements can be taken. Other

THE FAUNA

Table 7.6 *Fauna from WK Lower Channel*

Clariidae:	Cranial fragments 213
Chelonia:	Scutes 15, limb bones 2
Crocodylidae:	Teeth 18, broken scute 1, vertebra 1
Aves indet.:	Bones 3
Struthionidae:	Fragments of eggshell 17

MAMMALIA

	Cercopithecidae	Rodentia	Carnivora	Elephantidae	Equidae	Rhinocerotidae	Suidae	Hippopotamidae	Giraffidae	Bovidae	Taxa indeterminate	Totals
Skull parts, horn cores	—	—	—	—	—	—	—	—	—	3	7	10
Maxillae/mandibles	—	—	—	—	2	—	—	—	—	1	3	6
Teeth, isolated	—	—	—	1	3	—	2	7	1	11	6	31
Axial: vertebrae, ribs	—	—	—	—	—	1	—	—	1	—	15	17
Scapulae	—	—	—	—	—	—	—	1	—	1	2	4
Pelves	—	—	—	—	—	—	—	—	—	—	4	4
Fore limb, long bones	—	—	—	—	—	—	—	—	1	—	4	5
Hind limb, long bones	—	—	—	—	—	—	1	2	—	1	3	7
Podials, patellae, etc.	—	—	—	1	4	—	—	3	—	8	—	16
Estimated number of individuals	—	—	—	1	4+	1	2	3	2	8		21+
Indeterminate bone fragments												6978

faunal remains associated with the hippopotamus skeleton were:

Clariidae:	Cranial fragments 3
Crocodylidae:	Tooth 1
Equidae:	Left astragalus
Rhinocerotidae:	Part of right side of mandible with molars, fibula 1
Hippopotamidae:	Very small deciduous molar 1, fragments of adult canines, 2, adult molars 2, adult premolars (representing a second adult and a juvenile individual) 2
Bovidae, indet.:	Molar 1, fragment of humerus, fragment of acetabulum
Indet.:	Humerus shaft 1, ? rib shaft 1, fragments of bone 4

PDK Trench IV

(For description of the site see pp. 39–40.)

Clariidae are represented by nineteen skull fragments and twenty-one sundry bones. There is also a single fragment of ostrich eggshell.

Five large mammals appear to be represented by the following specimens: Cercopithecidae, 1 upper incisor; Carnivora, 1 small phalanx; Equidae, 1 right metatarsal IV and a damaged lateral incisor; Hippopotamidae, part of a very broken maxilla; Bovidae, part of a horn core, a complete right tibia with articulating astragalus and calcaneum, and one cuneiform.

Five additional bones cannot be identified with certainty although the neural spine of a large vertebra and one unfused proximal end of a large phalanx are probably of hippopotamus. One sesamoid and a broken phalanx are indeterminate as are 480 small bone fragments.

WK Lower Channel

(For description of the site see pp. 42, 75–6.)

The exposure by erosion of a well-preserved horn core cf. *Megalotragus kattwinkeli* led to a small excavation (7.5 m) being excavated at this level. It proved rich in fauna and yielded seven-

Table 7.7 *Fauna from HEB East*

Mollusca:	Bivalves 10
Clariidae:	Cranial fragments 65, various bones 16
Chelonia:	Fragment of limb bone
Crocodylidae:	Tooth 1
Struthionidae:	Fragments of eggshell 11

MAMMALIA

	Cercopithecidae	Rodentia	Carnivora	Elephantidae	Equidae	Rhinocerotidae	Suidae	Hippopotamidae	Giraffidae	Bovidae	Taxa indeterminate	Totals
Skull parts, horn cores	—	—	—	—	—	—	—	—	—	3	1	4
Maxillae/mandibles	—	2	—	—	—	—	—	—	—	1	3	6
Teeth, isolated	—	—	—	2	4	—	4	3	—	5	3	21
Axial: vertebrae, ribs	—	—	—	—	—	—	—	—	—	2	7	9
Scapulae	—	—	—	—	—	—	—	—	—	—	—	—
Pelves	—	—	—	—	—	1	—	—	—	—	—	1
Fore limb, long bones	—	—	—	—	—	—	—	—	1	3	1	5
Hind limb, long bones	—	—	—	—	—	—	—	—	—	1	1	2
Podials, patellae, etc.	—	—	1	—	2	2	—	—	—	—	1	6
Estimated number of individuals	—	1	1	1	2	1	1	1	1	4		13
Indeterminate bone fragments												805

teen taxa and the remains of over twenty-one individual mammals (Table 7.6).

Lower Bed IV

Faunal remains from the four sites excavated in Lower Bed IV, namely HEB East, HEB, HEB West and WK Intermediate Channel, were as follows:

HEB East

(For description of the site see pp. 45–6.)

The small area excavated at this site yielded an estimated number of thirteen individual mammals (Table 7.7). Ten bivalve shells were recovered and a proportion were intact, with the two halves still attached to one another, indicating that they were a natural occurrence and not food debris. A well-preserved frontlet with both horn cores of *Antidorcas recki* was also found.

HEB and HEB West

(For description of the sites see pp. 45, 52–5.)

The fauna from these two adjacent sites has been pooled for analysis. The few specimens from Level 1, above the grey siltstone, have been omitted, since they are not from Lower Bed IV. Material from the channel in Trench IV has been included, however, since the channel represents a local disturbance with material apparently derived from the *in situ* levels at HEB and HEB West. The most comprehensive collection of rodents known from Lower Bed IV was collected from HEB West. It comprises eleven species, some of which indicate dense vegetation with permanent water and others dry savannah conditions with Acacia trees (Jaeger 1976).

A total of twenty-nine taxa are represented and

Table 7.8 *Fauna from HEB and HEB West*

Clariidae:	Cranial fragments 134, various bones 116
Amphibia:	Bones 2
Chelonia:	Scutes 2
Crocodylidae:	Teeth 18
Aves, indet.:	Bones, various 25
Struthionidae:	Eggshell fragments 16

MAMMALIA

	Cercopithecidae	Carnivora	Elephantidae	Equidae	Rhinocerotidae	Suidae	Hippopotamidae	Giraffidae	Bovidae	Taxa indeterminate	Totals
Skull parts, horn cores	—	—	—	—	—	—	—	—	1	4	5
Maxillae/mandibles	—	—	—	3	—	1	1	—	3	1	9
Teeth, isolated	3	3	3	11	1	13	11	—	20	18	83
Axial: vertebrae, ribs	—	—	—	—	—	—	11	4	—	22	37
Scapulae	—	—	—	1	—	—	—	—	—	4	5
Pelves	—	—	1	2	—	—	—	—	—	9	12
Fore limb, long bones	1	1	—	—	—	—	5	1	2	6	16
Hind limb, long bones	—	—	—	—	—	3	5	—	4	5	17
Podials, patellae, etc.	—	—	1	3	1	1	—	3	20	5	34
Estimated number of individuals	3	2+	1	4	2	3+	3+	2	9+	—	29+
Indeterminate bone fragments											3454

(Note: specimens of Rodentia are not available for listing.)

there is an estimated number of over twenty-nine individual mammals (Table 7.8).

WK Intermediate Channel

(For description of the site see pp. 123–4.) fauna see Table 7.9.

Upper Bed IV

No additional comments are required here on the faunal remains from the **WK Upper Channel** (see pp. 75–6 and Table 7.10). At **WK East A** (see pp. 87–91 and Table 7.11), a minimum of thirty-two individual mammals has been recorded and eighteen individuals were represented at **WK East C** (see pp. 87–9, 103–4 and Table 7.12). The total number of taxa at these two sites is thirty-one. At **PDK Trenches I–III** (see pp. 87–9, 107–8 and Table 7.13), nineteen taxa and an estimated number of about sixteen individual mammals are represented. At all these sites, most bones were in fragmentary condition.

The Masek Beds

FLK

(For description of the site see pp. 116–19.)
Both Cephalophini (duikers) and Neotragini (pygmy antelopes) occur at this site. Neither has been found at any site in Beds II, III or IV and the only possible occurrence is at FLK N I in Upper Bed I where a pair of horn cores has been tentatively assigned to the Neotragini.

Twenty-three taxa and over twenty-four in-

Table 7.9 *Fauna from Intermediate Channel*

| Clariidae: | Cranial fragments 20, various bones 9 |
| Crocodylidae: | Teeth 10 |

MAMMALIA

	Cercopithecidae	Rodentia	Carnivora	Elephantidae	Equidae	Rhinocerotidae	Suidae	Hippopotamidae	Giraffidae	Bovidae	Taxa indeterminate	Totals
Skull parts, horn cores	—	—	—	—	—	—	—	1	—	—	—	1
Maxillae/mandibles	—	—	—	—	—	—	1	—	—	—	—	1
Teeth, isolated	—	—	—	1	2	—	2	4	—	2	—	11
Axial: vertebrae, ribs	—	—	—	—	—	—	1	2	—	—	3	6
Scapulae	—	—	—	—	—	—	—	2	—	—	—	2
Pelves	—	—	—	—	—	—	—	—	—	—	3	3
Fore limb, long bones	—	—	—	—	—	—	—	—	—	—	—	—
Hind limb, long bones	—	—	—	—	—	—	—	2	—	—	—	2
Podials, patellae, etc.	—	—	—	—	—	—	—	2	1	2	—	5
Estimated number of individuals	—	—	—	1	1	—	1	2	1	3+	—	9+
Indeterminate bone fragments												1343

dividual animals are present (Table 7.14). They include four species of rodents of which *Arvicanthus niloticus* and two genera, *Mastomys* and *Tatera*, are common in Bed IV, but there is also a jerbil *Jaculus* which has not been found elsewhere at Olduvai and has not hitherto been recorded as far south as East Africa (Jaeger 1976).

Later than the Masek Beds

HK

(For description of the site see pp. 123–4.)

A few faunal remains were found with the abundant bifacial tools at this site during the 1969 excavations: some were exceedingly rolled and resembled pebbles, whilst others, partially encrusted with a gritty matrix, were in relatively fresh condition. Parts of a hippopotamus skeleton lay on the surface nearby and presumably belonged to the skeleton found in the 1930s. They were similarly coated with gritty matrix.

Clariidae:	Cranial fragments 10, bones, various 8
Crocodylidae:	Tooth 1
Aves indet.:	Fragment of tibia 1
Struthionidae:	Fragments of eggshell 13

MAMMALIA
Rhinocerotidae:	Distal end right humerus 1
Suidae:	Fragment of molar 1, phalanx 1
Hippopotamidae:	Fragment of canine 1
Bovidae:	Fragments of horn cores 3, lower molar 1, distal end of right humerus 1, proximal end left ulna, phalanx 1, scaphoid 1.

Estimated number of individuals: Rhinoceros 1, suid 1, hippopotamus 1, bovidae 2. Indeterminate bone fragments: 650.

TK Fish Gully

(For description of the site see pp. 126–7.)

Mammalian remains were very scarce at this site although cranial fragments and bones of Clariidae were particularly abundant both in the undisturbed deposit and in the disturbed context. The mammalian fauna consisted of one complete

THE FAUNA

Table 7.10 *Fauna from WK Upper Channel*

Mollusca:	Fragments of bivalves 6, fragment of gastropod 1
Clariidae:	Cranial fragments 609 (including 19 central parts of skulls), various bones 347
Chelonia:	Whole and broken scutes 33, limb bones 2
Crocodylidae:	Teeth 49, vertebrae 2, broken scute 1
Aves, indet.:	Limb bones 8
Struthionidae:	Eggshell fragments 39

MAMMALIA

	Cercopithecidae	Rodentia (inc. Lagomorpha)	Carnivora	Elephantidae	Equidae	Rhinocerotidae	Suidae	Hippopotamidae	Giraffidae	Bovidae	Taxa indeterminate	Totals
Skull parts, horn cores	—	—	—	1	—	—	—	1	—	6	30	38
Maxillae/mandibles	—	3	—	—	—	1	2	—	—	6	13	25
Teeth, isolated	1	4	7	6	72	25	39	17	4	49	83	307
Axial: vertebrae, ribs	—	—	—	—	—	—	1	4	—	—	75	80
Scapulae	—	—	—	—	1	—	—	1	2	1	12	17
Pelves	—	1	—	—	1	—	—	—	—	—	5	7
Fore limb, long bones	1	4	—	—	—	—	—	1	—	4	19	29
Hind limb, long bones	—	1	—	2	—	—	—	2	—	8	4	17
Podials, patellae, etc.	1	—	1	—	5	—	—	6	2	37	12	64
Estimated number of individuals	2	5	4	3+	7+	2+	9+	5	3	8		48+
Indeterminate bone fragments												7349

canine and one broken molar of a hippopotamus, one broken bovid molar and one navicular, as well as a few indeterminate bone fragments, some of which were heavily rolled. The cranium of *Megalotragus kattwinkeli* (found in 1932) came from higher in the sequence, at the head of the gully. A skull of a lion identified as *Panthera leo*, in crushed and fragmentary condition, was found a few feet above the artefact-bearing quartz sand on the sloping side of the gully, eroding from the *in situ* deposits.

Conclusions

The faunal assemblages from excavated sites in Bed IIIs and IV consistently show a similar overall pattern of genera and species. There are, however, some marked differences in the fauna of the Masek Beds, confirming the change in ecology postulated by Dr R. L. Hay on the basis of the geological evidence. However, the scarcity of micro-mammalian fauna in Beds III and IV is a severe handicap in assessing the ecological conditions since small mammals are a surer indication of habitat than large animals who are more capable of adapting to altered environmental conditions.

Remains of Clariidae, sometimes in considerable quantities, indicate the presence of fresh rather than alkaline water in Beds III and IV, since catfish do not tolerate high alkalinity. Similarly, the large bivalves found in Bed III at MNK and in Bed IV at HEB East, WK and TK Fish Gully confirm the presence of fresh water. These molluscs are, in fact, so intolerant of

Table 7.11 *Fauna from WK East A*

Clariidae:	Cranial fragments 259, various bones 100
Chelonia:	Scutes 3 (medium size)
Amphibia:	Limb bone 1
Crocodylidae:	Teeth 18
Struthionidae:	Fragments of eggshell 90

MAMMALIA

	Cercopithecidae	Rodentia	Carnivora	Elephantidae	Equidae	Rhinocerotidae	Suidae	Hippopotamidae	Giraffidae	Bovidae	Taxa indeterminate	Totals
Skull parts, horn cores	—	—	—	—	—	—	—	—	—	1	4	5
Maxillae/mandibles	—	—	—	—	—	—	—	—	—	2	1	3
Teeth, isolated	1	—	—	2	32	9	18	2	3	37	11	115
Axial: vertebrae, ribs	—	—	—	1	—	—	—	1	—	2	3	7
Scapulae	—	—	—	—	—	1	—	1	—	—	4	6
Pelves	—	1	—	—	—	—	—	—	—	—	—	1
Fore limb, long bones	—	—	—	—	—	—	—	—	—	—	2	2
Hind limb, long bones	—	—	—	—	1	1	—	1	1	—	2	6
Podials, patellae, etc.	—	1	—	—	—	—	—	1	11	16	2	31
Estimated number of individuals	1	2	—	2	5	2	5	2	2	11+		32+
Indeterminate bone fragments												9461

The single bovid horn core has been identified by A. W. Gentry as *Pelorovis antiquus*.

alkalinity that they are unable to live in Lake Turkana today although this body of water is only slightly alkaline and abounds in Clariidae. Whether the Clariidae found at excavated sites were *in situ* in their original riverine habitats or whether they were brought in as food by early man is impossible to determine but the latter seems to be more likely in view of the very fragmentary condition of the bones and the fact that they are usually associated with artefacts. In the case of bivalves, the unopened pairs that were found at HEB East indicate that at this site, if not at others, the shells were not food debris.

Among other acquatic taxa is *Hippopotamus gorgops*, whose remains were not only found at each of the excavated sites, but were also common as isolated finds in Beds III and IV. It is estimated that parts of over thirty individual hippopotami were recovered from the excavations, representing 11 per cent of the total number of mammals. Crocodiles occur at all except two sites (PDK Trenches I–III and PDK Trench IV) but they are usually represented by only a few teeth and it is likely that there were far fewer crocodiles than hippopotami.

The abundance of equid remains, with an estimated figure of over forty-five individual animals, amounting to 15 per cent of the mammalian total, indicates open, grassy surroundings. Both *Equus* and *Hipparion* are present. The former was found at every excavated site but the latter appears to be absent from the three sites excavated at the junction of Beds III and IV, namely WK Lower Channel, WK Hippo Cliff and PDK Trench IV. Bones and teeth estimated to represent thirteen rhinoceroses were recovered. They include both *Ceratotherium simum* and *Diceros bicornis* which occurred in approximately equal proportions and were found together at three localities in Bed IV: at HEB East, HEB and

THE FAUNA

Table 7.12 *Fauna from WK East C*

Clariidae:	Cranial fragments 75, various bones 50
Chelonia:	Scutes 4 (2 costal, 1 marginal and 1 neural)
Crocodylidae:	Teeth 15
Aves indet.:	Small left coracoid 1
Struthionidae:	Fragments of eggshell 7

MAMMALIA

	Cercopithecidae	Rodentia	Carnivora	Elephantidae	Equidae	Rhinocerotidae	Suidae	Hippopotamidae	Giraffidae	Bovidae	Taxa indeterminate	Totals
Skull parts, horn cores	—	—	—	—	—	—	—	—	—	1	3	4
Maxillae/mandibles	—	—	—	—	1	—	—	—	—	—	1	2
Teeth, isolated	1	—	—	—	16	5	6	2	1	24	6	61
Axial: vertebrae, ribs	—	—	—	—	—	—	—	—	—	2	5	7
Scapulae	—	—	—	—	—	—	—	—	—	—	1	1
Pelves	—	—	—	—	—	—	—	—	—	—	3	3
Fore limb, long bones	—	—	—	—	—	—	—	—	—	1	2	3
Hind limb, long bones	—	—	—	—	—	—	—	—	—	2	2	4
Podials, patellae, etc.	—	—	—	—	1	—	1	—	—	2	1	5
Estimated number of individuals	1	—	—	—	3+	1	3+	2	1	7+		18+
Indeterminate bone fragments												8010

Table 7.13 *Fauna from PDK Trenches I–III*

Clariidae:	Cranial fragments 4, various bones 6
Chelonia:	Broken scute 1

MAMMALIA

	Cercopithecidae	Rodentia	Carnivora	Elephantidae	Equidae	Rhinocerotidae	Suidae	Hippopotamidae	Giraffidae	Bovidae	Taxa indeterminate	Totals
Skull parts, horn cores	—	—	—	—	—	—	—	1	—	4	1	6
Maxillae/mandibles	—	—	—	—	2	—	—	—	—	—	—	2
Teeth, isolated	—	—	—	—	20	3	5	1	—	25	—	54
Axial: vertebrae, ribs	—	—	—	—	—	—	—	—	—	—	5	5
Scapulae	—	—	—	—	—	—	—	—	—	—	1	1
Pelves	—	—	—	—	—	—	—	—	—	—	—	—
Fore limb, long bones	—	—	—	—	—	2	—	—	2	—	—	4
Hind limb, long bones	—	—	—	1	—	—	—	—	—	4	5	10
Podials, patellae, etc.	—	—	—	—	1	—	—	7	3	7	3	21
Estimated number of individuals	—	—	—	1	4	1	2	1	1	6+		16+
Indeterminate bone fragments												1200

Table 7.14 *Fauna from FLK Masek*

Mollusca:	Gastropods 15, a few fragments of large bivalves
Clariidae:	Cranial fragments 519, including 10 central parts of skull, various bones 120
Chelonia:	Scutes 16
Crocodylidae:	Teeth 21, vertebra 1, broken scutes 2
Reptilia:	Vertebrae 3
Aves indet.:	Bones 2
Struthionidae:	Eggshell fragments 112

MAMMALIA

	Cercopithecidae	Rodentia	Carnivora	Elephantidae	Equidae	Rhinocerotidae	Suidae	Hippopotamidae	Giraffidae	Bovidae	Taxa indeterminate	Totals
Skull parts, horn cores	—	—	—	—	—	—	—	—	—	4	—	4
Maxillae/mandibles	—	6	1	—	—	—	—	—	—	4	—	11
Teeth, isolated	1	11	2	—	7	6	5	2	1	28	1	64
Axial: vertebrae, ribs	—	—	1	—	—	—	—	—	—	—	—	1
Scapulae	—	—	—	—	—	—	—	—	—	—	—	—
Pelves	—	2	—	—	—	—	—	—	—	—	—	2
Fore limb, long bones	1	—	—	—	—	—	—	—	—	2	—	3
Hind limb, long bones	—	6	—	—	1	1	—	—	—	6	—	14
Podials, patellae, etc.	—	1	3	—	6	—	—	—	—	41	—	51
Estimated number of individuals	1	4+	2+	—	3	1	3	1+	1	8+	—	24+
Indeterminate bone fragments												7142

HEB West, WK and WK East C. There is an estimated total of over thirty-three suids representing 11 per cent of the mammals from the excavated sites. Giraffidae are not common. None was found at JK and *G. jumae* only occurs at five sites. *G. stillei* occurs throughout the Olduvai sequence from Bed I to Bed IV but is only represented at the excavated sites by a single tooth from WK Upper Channel. Seventeen elephants are believed to be among the excavated faunal material; all belong to Stage 3, the advanced form of *E. recki* (Maglio 1970, 1973). The primate and carnivore remains are so scarce that very little can be said about them.

Bovidae, as usual, are by far the most common animals and a total of over eighty-seven individuals is estimated, or 30 per cent of the 285 mammals represented. Eighteen species have been identified as well as three undetermined Alcelaphini, the most numerous tribe among the bovids found at Olduvai. It is interesting to note that *Pelorovis oldowayensis*, which first appears in the lower part of Middle Bed II at HWK East, appears only to survive into Bed III, making its final known appearance at JK. Thereafter it is replaced by *Pelorovis antiquus*.

The habitat indicated by the Bovidae from Olduvai is best summed up in an extract from the paper by Dr and Mrs Gentry (Gentry and Gentry 1978).

The ecology of the Olduvai bovids may be compared with that of their close living relatives... As a whole, they indicate rather dry and open environment. The only tragelaphine which is even moderately common is the greater kudu which today has a widespread distribution in diverse habitats which have some degree of cover. It is commoner in Bed I than in the later beds. The eland, a gregarious and mainly browsing antelope, is very rare at Olduvai. Like the greater kudu and unlike other tragelaphines it occurs in a variety of habitats, even in fairly dry areas. The buffalo today is an unselective grazer often found among denser, ideally riverine vegetation, and it would not have occurred away from areas in which it could find places to wallow. *Hippotragus gigas* could be pictured as grazing in non-arid areas of long grass adjacent to woodland, if it was at all like the modern *H. equinus*. It is present but not common in Beds I–III. The Alcelaphini are the most characteristic antelopes of Olduvai, especially in middle Bed II and later horizons. They would have grazed on short grass in open country or in clearings with good visibility and have tolerated greater aridity than the reduncines or *Hippotragus*. *Antidorcas* is the commonest antilopine, notably in Bed I. As with living springbok an important component of its diet may have been semi-arid dicotyledonous plants.

Writing on the size of the Olduvai Bovidae Dr and Mrs Gentry continue:

In the past Olduvai Gorge has been noted as a site for giant-sized animals. One important factor contributing to this reputation was the old misidentification of *Pelorovis oldowayensis* as a relative of the sheep. Once *Pelorovis* had been recognised as a bovine its large size became less startling, for its nearest living relative was the slightly smaller buffalo *Syncerus caffer* and not the very much smaller sheep. Nevertheless the Olduvai bovidae as a whole continue to give the impression of being larger animals than those living today, even if one cannot call them giant. This impression is derived from a number of examples at different taxonomic levels of comparison. Two of the extinct lineages, *Pelorovis* and *Megalotragus*, are larger than the surviving members of the same tribes and *Thaleroceros radiciformis* could also be larger than the living reduncines, although this is not certain. *Hippotragus gigas* and *Beatragus antiquus* are larger than the living members of the same genera, and *Tragelaphus strepsiceros grandis* and the Bed III *Kobus kob* than the living members of the same species. Also at species level of comparison, *Damaliscus niro* is larger than its conspecific descendants at the later South African sites.

The fauna from the Masek Beds contains the majority of the commoner elements found in Beds III and IV but the Bovidae are poorly represented. They consist only of *Redunca* sp., Alcelaphini, Neotragini, Cephalophini, Antilopini and Gazellini, none of the last five being identifiable to more than tribal level. However, the occurrence of Neotragini and of Cephalophini is of interest. Neither of these tribes is present in Beds III or IV and there is only a single pair of horn cores from Bed I and FLK North that has been attributed to Neotragini. Cephalophini are otherwise unknown in the Olduvai sequence. The presence of these two small antelopes suggests less open country with more bush and cover than was prevalent during earlier times when large numbers of Alcelaphini flourished in the area.

8

A METRICAL ANALYSIS OF SELECTED SETS OF HANDAXES AND CLEAVERS FROM OLDUVAI GORGE

D. A. ROE

Introduction

A study of this kind necessarily depends on, and indeed largely consists of, a mass of extremely dull figures. I cheerfully accept that very few readers will wish to study these in detail. At the same time I cannot omit them, for several reasons, of which two will suffice. First, the tables of figures and supporting diagrams comprise the actual evidence on which all the general conclusions must rest, even to the extent that they can tell us how much confidence to place in those conclusions. No specialist report can reasonably consist of its conclusions alone, so the metrical and statistical diagrams and tables must be made available. Secondly, though some readers will find it hard to believe, future researchers may wish to carry out similar exercises to this one using other bodies of broadly comparable Lower Palaeolithic artefacts from Africa or elsewhere. If so, they will need the Olduvai results in full to carry out comparative studies, and they are also likely to need them if they wish to check or add to the Olduvai results themselves, using other statistical techniques.

With these considerations in mind, I have set out to state in simple terms in the relatively short text of this chapter what I did in studying the Olduvai artefacts, and what pattern of likes and unlikes thereby emerges amongst the sets of bifaces studied, which were drawn from several different levels of the great Olduvai sequence, within Bed II, Bed IV, the Masek Beds and two hill-wash deposits whose date is relatively late but not known exactly. Many people will take the figures for granted and read only the text; some will probably read only the conclusions. Perhaps a few will take the time and trouble to follow up the supporting quantitative statements and if so they will avoid making the mistake of quoting as absolute facts conclusions which are no more than measurable probabilities.

Scope of the research

This study was first suggested by Dr Mary Leakey as long ago as 1969. At that time, I had recently published the results of my doctoral research at Cambridge, which included newly devised metrical analyses of British Acheulean handaxes (1964, 1968), and Professor J. Desmond Clark had invited me to carry out a similar operation on the Large Cutting Tools from the stratified Acheulean and 'Sangoan' floors from Kalambo Falls, Zambia. The main collections of Kalambo Falls artefacts were at Livingstone, Zambia, and the journey there involved a stop at Nairobi, making possible, through the kindness of the late Dr Louis Leakey, a short visit to Olduvai, where I met Mary Leakey for the first time. She asked me to undertake a comparable metrical analysis of various sets of Olduvai bifaces, mainly from Beds II and IV, and I agreed in principle to do so as soon as time and funds were forthcoming.

In 1970, I returned to Olduvai and Nairobi and

collected a major part of the necessary data for a programme of metrical analysis. However, in subsequent years, as the whole research operation at Olduvai proceeded, the scope of the proposed metrical analyses altered and expanded and it became desirable to include other sets of Olduvai bifaces, for which the data had not yet been gathered, to test various hypotheses. I made a further visit to East Africa in 1974, to study collections at Nairobi, Olduvai and Dar-es-Salaam. Later still, a major sorting and cataloguing exercise by Dr Mary Leakey's staff on all the excavated and collected stone artefacts from Olduvai brought to light a substantial number of specimens which had not previously been available; a few specimens were added in this way even as late as August 1980.

What had begun as a relatively simple project of metrical analysis had over ten years grown into something much larger, by this time involving work by Dr Paul Callow as well as myself. The main changes stemmed from Dr Leakey's suggestion that a study of the number of flake scars on each implement might show an interesting pattern of change through time at Olduvai, reflecting for example a simple technology of biface manufacture in Bed II, where the first handaxe industries occur, and a more sophisticated one in Bed IV. Consideration of this idea suggested to me and to Dr Callow that rather more factors were involved than just the quantity of flake scars. How did different rock types flake? Were the size and disposition of the flake scars, as opposed to the mere number of them, also important? Should one not, when counting the number of flake scars, also take into account the overall size and shape and indeed if possible the true surface area of each implement? Some of these questions are considered by Dr Callow in chapter 9 of this volume.

I had assumed that the study of the Kalambo Falls handaxes, referred to above, would have been published in the mid-1970s, and that the comparable account of the Olduvai material would follow not too long afterwards, perhaps after a pause to digest the reactions of colleagues to the methods of study and the conclusions reached. The Kalambo study was duly completed and submitted on schedule in 1973, intended for publication in Volume 3 of the *Kalambo Falls* monograph series edited by Professor J. D. Clark. Regrettably however that volume has not yet appeared and no publication date has been set for it. The delays to the production of the volume have evidently been many and various and wholly outside any control of mine, and it is possible that the metrical analyses of the Kalambo Falls Large Cutting Tools (Roe n.d.) will eventually appear elsewhere separately. At the time of writing I do not know whether that report will be available before this account is published in the *Olduvai Gorge* series, but in any case I will describe the analytical methods used in as brief a manner here as conveniently possible, and will spell them out more fully in the Kalambo Falls report. The same methods were used in a brief study of certain handaxes and cleavers from Swaziland, carried out by the writer for Dr D. Price-Williams in 1977 (Price-Williams and Lindsey 1978), but that report is unlikely to have been widely read. Various scholars, most of them having a special interest in the African Acheulean, have studied privately the typescript of the Kalambo Falls report, and reaction so far has seemed generally favourable.

The present text was completed and submitted in June 1981, with a certain amount of subsequent revision, mainly in 1983–4, as the rest of the volume took shape.

Acknowledgements

It is a pleasure to record with gratitude the support of the Wenner-Gren Foundation for Anthropological Research Inc. for my original visit to Nairobi *en route* for Livingstone in 1969 (Grant no. 2614-1834) and of the Boise Fund of Oxford University for my visit to East Africa in 1970. I am also grateful to all those who have in one way or another helped, advised or encouraged my work on the Olduvai material over ten years, notably Louis, Mary and Richard Leakey, the late Glynn Isaac, Desmond Clark and John

Gowlett; also, members of the staff of the Antiquities Service and the National Museum in Tanzania, notably Mr A. A. Mturi and Dr F. Masao. To Dr Paul Callow special thanks are due for devising and achieving at Cambridge the computing involved in this operation: generously investing a great deal of his own precious time, he has made painless for me the processes of converting the raw data, i.e. the measurements taken directly from the artefacts or from special outline drawings of them, into frequencies, means, standard deviations and so forth, of carrying out the statistical comparisons, and even of drawing the shape diagrams themselves. The computing expertise is all his, but any errors or shortcomings so far as interpretation is concerned must count as my own.

The samples studied

General considerations

The object of this study can be stated quite simply: to quantify and compare the morphology of handaxes and cleavers from selected sites at various levels in the Olduvai sequence and to draw conclusions from the results. It is important in this kind of work to use as 'pure' samples as possible: that is to say, the implements studied from any one archaeological occurrence should belong together within close limits of time and space so far as the archaeologist can tell. Ideally, they would be the output of a single group of people during one episode of occupation, and they should have been recovered under controlled archaeological circumstances from a site of known extent. Few Lower Palaeolithic sites anywhere in the world meet such a specification and the nature of both the original deposition and the more recent erosion at Olduvai makes many of the extant sets of bifaces unsuitable for inclusion, though it was felt to be worth including certain samples from rather dubious contexts to make specific comparisons or assessments. After all, the metrical analyses will produce objective morphological assessments, whether a sample is pure or mixed, and there is no harm in being objective about every sample, whatever its quality. It remains the responsibility of the analyst, and to some extent of the reader, to remember the origin of individual samples and not to 'overinterpret' the results.

Another matter in which the analyst is usually at the mercy of his material is sample size. The relatively simple and standard methods of statistical comparison used here, and others like them, were never designed with Lower Palaeolithic stone tools in mind. With groups of handaxes, it is hardly ever a question of having several large populations available from which to take proper random samples of equal size; instead, one must use every piece one can lay hands on, still ending up with samples of very unequal sizes, many of them pitifully inadequate. It would be nice to have at least a hundred implements in each of the Olduvai samples, but a glance at Table 8.1 will show that some samples are in single figures and that only two out of eighteen reach a really adequate size, those from WK and HK. If one is dealing with a group of nine handaxes and working in percentages, each single implement has a value of $11 \cdot 1\%$, which is ridiculous. But there is no way of increasing the sample sizes in the present case, so one must either abandon the exercise or press on, noting the extra degree of uncertainty that attends the results. In my own view, the exercise remains worthwhile and the situation encourages me to keep the measurements and statistics at a simple level. It is also worth remembering that we really do not know the relationship of our 'samples' to the 'populations' from which they derive. How many other bifaces originally lay beside these before erosion revealed and partially destroyed each site? How many others may still remain in place beyond the limits of the excavation trenches? In any case, what did the original inhabitants remove, and what did they leave behind, when they abandoned the site? All these considerations are imponderable, and they warn us once more against overinterpretation.

Finally, it must be stressed that this is *not* a total study of a series of Lower Palaeolithic industries: it is an account of the morphology of the handaxes and cleavers from them. It does not seek to assess the role of the bifaces in the whole tool-kits to which they belong, nor is it concerned with details of their technology and function. Research under all these headings is extremely important, none more so than the study of function by analysis of microwear traces, if someone should some day take that on for the fresher Olduvai artefacts. I am wholly in favour of all these approaches, and well aware of the very fruitful results they can achieve, but they lie outside the present brief and I merely offer the comment that those who undertake them in due course may find the results of this purely morphological study of two particular tool classes of some relevance.

The Olduvai bifaces

I myself had no part in the primary classification of the stone artefacts from the various sites; the handaxes and cleavers had already been sorted out by Mary Leakey when I first saw them and I did no more than glance through the other material to ensure that the samples were as complete as possible. There was very little disagreement between Dr Leakey and myself on points of classification, with the exception of a few cleavers here and there, notably from the Bed II site, EF-HR. In this case I recognised a total of six cleavers, instead of one cleaver (later two) listed from EF-HR (Leakey 1971). Classifications must be expected to differ in this kind of way between different analysts. African cleavers are typically made from large flakes, often side struck, with a variable number of primary scars on the dorsal face. Secondary work to regularise the shape of the cleaver, after the detaching of the big flake itself from its shaped parent block, may be extensive but it may also be minimal or even absent, if the flake comes away in a shape suitable for immediate use, as a skilled worker can often contrive. It follows that there is an inevitable merging of at least three categories here:

(a) complete flake cleavers which have little or no secondary work because it was not needed;

(b) blanks for cleavers, intended for secondary shaping but for some reason never made into finished implements;
(c) large side-struck flakes which may by chance approximate to the shape of cleavers but are in fact no more than waste.

Ultimately, these three categories may be indistinguishable from each other. It is conceivable that one day the study of utilisation traces might sometimes allow positive identifications in category (a) to be made, but even then there remains the possibility of completed cleavers which were never used. All things considered, individual observers must be allowed to differ; it is perhaps unlikely that the 'borderline cases' will often make up a major component of an industry.

At Olduvai, my brief was to study the handaxes and cleavers from the selected sites, grouped together by Dr Leakey as 'bifaces' (English pronunciation, not French). At Kalambo Falls, I had studied for Desmond Clark the 'Large Cutting Tools', as that term was used by M. R. Kleindienst (1961, 1962), which included handaxes and cleavers but also core axes and knives. There are no core axes of the 'Sangoan' kind at Olduvai. Dr Leakey did not regard the 'knives' as a special tool class distinct from handaxes, and it is certainly my own impression that there are few if any knives at Olduvai that resemble the carefully made ones found – admittedly in small numbers – at Kalambo.

At Olduvai there are two groups of industries that contain bifaces, both represented in the present study. They have been called respectively 'Developed Oldowan' and 'Acheulean'. Different workers have interpreted in different ways the evident broad coexistence of these two traditions or facies over a very substantial period of time (Leakey 1971, 1975a, 1978; Stiles 1977, 1979; Davis 1980): they are both present in the upper part of Bed II and also in Bed IV; the situation in Bed III is less clear because much of the archaeological material is derived and in fact no Bed III sample is involved here. The presence of these two kinds of industry together over a period of perhaps $\frac{3}{4}$ million years is crucial to any

Table 8.1 *The biface samples studied*

Site	Stratigraphic position	Designation	Total bifaces	% value of each implement	Handaxes No.	%	Cleavers No.	%
HK	Hill wash later than Masek	Acheulean	387	0.3	350	90.4	37	9.6
TK FG	Hill wash later than Masek	Acheulean	71	1.4	71	100.0	–	–
FLK MASEK	Masek	Acheulean	23	4.3	21	91.3	2	8.7
PDK Trenches I–III	Upper Bed IV	Dev. Old. C	18	5.6	18	100.0	–	–
WK East A	Upper Bed IV	Dev. Old. C	42	2.4	42	100.0	–	–
WK East C	Upper Bed IV	Dev. Old. C	9	11.1	9	100.0	–	–
WK	Upper Bed IV	Acheulean	148	0.7	86	58.1	62	41.9
HEB West 2a	Lower Bed IV	Acheulean	35	2.9	28	80.0	7	20.0
HEB West 2b	Lower Bed IV	Acheulean	43	2.3	28	65.1	15	34.9
HEB West 3	Lower Bed IV	Acheulean	60	1.7	50	83.3	10	16.7
HEB East	Lower Bed IV	Acheulean	44	2.3	42	95.5	2	4.5
PDK Trench IV	Beds III–IV junction	Acheulean	18	5.6	11	61.1	7	38.9
BK	Upper Bed II	Dev. Old. B	67	1.5	66	98.5	1	1.5
TK Upper	Upper Bed II	Dev. Old. B	15	6.7	15	100.0	–	–
TK Lower	Upper Bed II	Acheulean	9	11.1	9	100.0	–	–
SHK	Upper Bed II	Dev. Old. B	47	2.1	46	97.9	1	2.1
EF–HR	Middle Bed II	Acheulean	37	2.7	31	83.8	6	16.2
MLK	Middle Bed II	Acheulean	29	3.4	28	96.6	1	3.4

understanding of the prehistory of Olduvai and perhaps of all Africa, whatever interpretation of their coexistence one may favour. It is therefore a matter of considerable interest to see what can be said about the morphology of the bifaces in the industries concerned. How consistent are they within each of the two traditions, and how great are the differences when samples of Developed Oldowan and Acheulean bifaces are compared?

It was on the basis of everything that has been said in this section so far, both the general considerations and the local ones, that the list of samples to be studied was finally drawn up. Eighteen sets of bifaces are involved, and the details are given in Table 8.1. A certain degree of stratigraphic information bearing on the relative ages of these samples exists, even if it is not quite possible to place them all in exact order of age or assign a precise chronometric date to each. For the purposes of the table, the industries are designated Developed Oldowan or Acheulean, just as they had been classified by Dr Leakey before this study began. The detailed discussions of their positions in the Olduvai stratigraphy will be found elsewhere in this volume, or in other volumes in the Olduvai series (Leakey 1971; Hay 1976), and in Table 8.1 they are merely assigned to the appropriate Bed. Brief forms of the site names are also used. The sites are listed in this same order in all subsequent Tables. The figures in Table 8.1 should not be regarded as the actual sample sizes for each site for every one of the metrical or statistical analyses carried out. The reason for this is that not every implement yielded a full set of data. Broken implements were never discarded by me unless there was no measurement at all that could be taken from them; such unusable fragments are not included in the totals in Table 8.1 at all. A handaxe that has lost, for example, its tip may be unable to yield measurements of weight, length, etc. but it may still be usable for thickness and breadth and therefore for at least one important ratio. Where damage is only minor, the missing part can usually be 'reconstructed' for the sake of measurement. It has also to be remembered that more than one person was involved in the collection of the metrical data. The majority was done by myself, but it has already been explained that some implements and certain assemblages (SHK, TK Upper and TK Lower) were added later, after my own visits to East Africa, and at the time when the data in respect of these were collected it was not always possible for the implements to be weighed accurately by those concerned, or for some of the profile measurements to be taken. For all these reasons, it will be found that the sample sizes for some sites vary greatly over the whole range of analyses. Each table accordingly states what are the sample sizes for the particular measurement or ratio concerned, and the percentage values for individual implements are also quoted each time. The reader is asked to accept that in every case I have used the maximum amount of data and information available. In Table 8.1, it is the full number of implements from each site that is given.

Consideration of Table 8.1 will show that by no means all the sites yielded viable samples of cleavers. Only at WK and HK were substantial numbers present, sixty-two and thirty-seven respectively, followed by two of the HEB sites with fifteen and ten. Three sites yielded only a single cleaver, which is useless for purposes of statistical analysis. The remaining cleaver samples are ridiculously small, but have been treated statistically for the sake of objective comparison; there are thus nine sets of cleavers for analysis from the eighteen sites. All the Olduvai sites produced viable samples of handaxes for study, i.e. eighteen sets of handaxes, and I also produced a complete set of analyses for the eighteen whole sets of bifaces, that is to say for the eighteen sets of handaxes and cleavers treated together.

Methods of study

A longer account of the metrical analyses is given in my Kalambo Falls report already referred to (Roe n.d.). It was originally envisaged by Desmond Clark and Mary Leakey that I should apply to the Kalambo and Olduvai implements exactly the same metrical methods that I had devised in studying the British Lower and Middle Palaeolithic industries, of which various accounts have appeared (1964, 1968, 1981: 152–62) but, although the general principles were the same, in the event a number of special adjustments were made which reflect certain basic differences between handaxe industries in Britain and East Africa.

The most important difference is that for the African sites separate studies were made of each main tool class, as well as a study of all the implements together, whereas in Britain a single study of the combined assemblage sufficed. Cleavers are rather rare in Britain and, since they are usually made bifacially, very much in the mode of handaxes, the categories 'handaxe' and 'cleaver' grade imperceptibly into each other. A very few flake cleavers do exist in Britain (cf. for example Sainty 1927, fig. 13; Cranshaw 1983) but, generally speaking, English flint in nodule or cobble form is more amenable to the production of cleavers by bifacial flaking. In Africa the raw material is often some hard rock occurring in large blocks – lava boulders, for example – and African cleavers, as we have already seen, are commonly made from large flakes removed from such blocks by a simple but regular knapping procedure; although there is some overlap with handaxes (many or even most of which may also be made from large flake blanks in African industries), the two classes remain far more distinct than they do in Britain and can usefully be studied individually. At the same time, however, it is well worth producing figures for the sets of handaxes and cleavers combined, precisely to catch those differences between individual African industries which reflect the strength or weakness of the cleaver element. Finally, there are African cleaver types which are wholly unknown in Britain – one thinks especially of some of the 'guillotine' forms, with their deliberately oblique cleaver edges – and the writer's original shape diagrams were designed to deal with plan-forms of an approximately symmetrical nature. The extreme guillotine forms of cleaver could give gravely misleading results if lumped together on a single diagram with generally symmetrical handaxes, and it was therefore necessary to devise a completely new cleaver-shape diagram. How this operates is explained below (pp. 154–8).

Generally speaking, a cleaver is defined by its possession of a characteristic transverse or

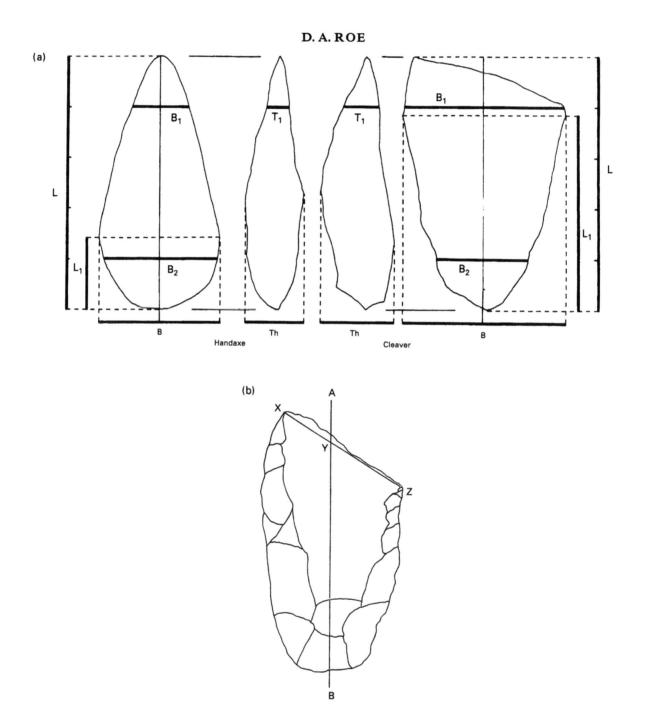

Fig. 8.1 Measurements taken from the bifaces: (a) basic measurements for handaxes and cleavers; (b) special measurements for cleavers: cleaver edge angle and cleaver edge length (AB is the long axis of the implement, XZ is a straight line joining the two points of intersection between the cleaver edge and a side edge of the implement, and intersecting the long axis line at Y. The length of XZ is the cleaver edge length. The angle AYX is the cleaver edge angle.)

oblique cutting edge at the tip end, having distinct points of junction with the implement's sides (which may be blunt or have worked edges of their own). Various workers have defined handaxes and cleavers, including myself in the Kalambo Falls report. There can be some overlap between cleavers and square-ended handaxes, and I do not myself recognise the 'ultra-convergent cleavers' of Kleindienst and others. The only point of metrical definition that need be repeated here is that if an implement is to qualify as a cleaver, the length of the distinctive transverse or oblique edge or 'bit' must be greater than half the implement's breadth. If not, the implement counts as a square-ended handaxe. (From this it follows that values for the ratio cleaver edge length/breadth (CEL/B) referred to below and used in the cleaver shape diagrams will always be greater than 0·500.)

Fig. 8.1 shows the measurements actually taken from the implements. They are as previously used by me on the British material, plus two new measurements taken from cleavers (Fig. 8.1(b)):

(a) Cleaver edge length (CEL), being the length (in mm to the nearest mm) of the straight line between the two extremities of the cleaver edge or 'bit';
(b) Cleaver edge angle, measured as the angle of intersection (not exceeding 90°) between the straight line measured as cleaver edge length and the long axis of the cleaver as aligned for measurement.

The other linear measurements are common to all the bifaces, and are taken parallel or perpendicular to the long axis as appropriate. All are measured in mm to the nearest mm except T_1, which is measured to the nearest half-millimetre. The measurements are:

L   Length
B   Breadth
Th Thickness (h)
T_1 thickness at a point one fifth of the length distant from the tip end
B_1 breadth at a point one fifth of the length distant from the tip end
B_2 breadth at a point one fifth of the length distant from the butt end
L_1 distance from the butt end to the point along the implement's long axis at which the position of maximum breadth occurs in the plan view. *Note:* if the implement is somewhat asymmetrical, there will be a 'zone' rather than a single point of maximum breadth. In this case, L_1 is measured from the butt end to the midpoint of this zone of maximum breadth, along the long axis. However, if the width of the zone of maximum breadth is found to be greater than one third of the implement's length, then the implement is regarded as too asymmetrical for use in this test, and no L_1 measurement is taken. Such pieces are in fact very rare.

In addition to the linear measurements already listed, and the cleaver edge angle measurement for cleavers, each implement is weighed, in grams to the nearest gram. Measurement of weight can be regarded as a useful general test of size, bulk or volume of an implement in certain broad senses; it fulfilled its purpose well enough in my study of British assemblages, but it is only fair to point out that a far greater variety of rock types was used at Olduvai than in Britain for handaxe and cleaver manufacture, and that some rocks are relatively heavy and some relatively light, so that weight and absolute size will not be as closely correlated as one might expect. Nevertheless the actual weight of an implement seems likely to have been of some importance to its user. There may of course have been some alteration in the weight of some tools between the time of their abandonment and the present, caused by chemical processes, and it has not been possible to make any allowance for this.

Not all the measurements listed above are used in isolation for the analyses; some are combined to form ratios, namely Th/B, T_1/L, B/L, B_1/B_2, L_1/L and CEL/B. In addition, use is made of various combinations of ratios to produce the shape diagrams for handaxes and for cleavers, as explained below. In the cleaver-shape diagrams, the measurements of the cleaver edge angle also play a part.

In the analysis of the Olduvai samples, for each measurement or ratio used, as applied to handaxes and cleavers together, handaxes alone and cleavers alone, a table is given (Tables 8.2–8.8, 8.17–8.23 and 8.32–8.40), showing sample details, means, standard deviations and frequency distributions. Each of these tables is supported by a table of statistical comparisons (Tables 8.9–8.15, 8.24–8.30 and 8.41–8.49), in which each one of the samples is compared with each other one. These Tables give both t-values (upper half) and significance estimates (lower half) for each of these comparisons. For those, like myself, to whom all this is somewhat unfamiliar ground, it should be explained that these tests of 'significant differences' take into account means, standard deviations and sample sizes. The difference between two means may appear dramatically large, but if for example one of the samples is very small, or if the standard deviations are large, the difference may not after all be 'significant' in the statistical sense – the two samples could perhaps have been drawn from a single population after all. It should be noted that Dr Paul Callow, in his program to obtain the t-values and significance estimates, first established by use of Fisher's F Ratio which pairs of sites showed significantly different variances, and the t-values were then computed using either pooled or separate variances as appropriate and the significance estimates were computed accordingly. The t-values (upper right triangle) and their associated probabilities (lower left triangle) are shown in bold type where the probability is 0.05 or less, indicating a significant difference between the sample means. Italic figures indicate that the samples have significantly different variances, and that a separate-variance estimate of t has been made.

The working of the shape diagrams for handaxes is explained in Figure 8.2, and 8.2(b) should be used as a key diagram for interpreting the handaxe-shape diagrams for the actual Olduvai sites, which are given in Figs 8.6–8.23. Essentially, the diagram is a scattergram in three parts, and the implements are each assigned to one of the three according to their values for the ratio L_1/L: those with a low position of maximum breadth are plotted on the right-hand section (L_1/L values up to 0·350), those with maximum breadth centrally placed on the centre section (L_1/L values in the range 0·351–0·550) and those with a high position of maximum breadth on the left-hand section (L_1/L values over 0·550). Then, on each section, values for B/L are plotted horizontally against values for B_1/B_2 vertically, so that each individual handaxe is represented by a dot whose position is further to the right according as the implement is broader (i.e. has a higher value for B/L), and lower down according as the implement is more pointed (i.e. has a lower value for B_1/B_2). Figure 8.2(b) shows how this operates in terms of actual plan-forms, which are here given as silhouettes drawn symmetrically. The large crosses, one on each section, are merely visual coordinates, always marked in in the same places, to assist comparison of shape diagrams.

The cleaver-shape diagrams are a little more complicated, but only because more metrical information has to be used; they are just an extension of the same ideas, and no sophisticated mathematics is involved. Figs 8.3 and 8.4 show how the diagrams work and should be used as a key to interpreting the actual Olduvai cleaver-shape diagrams, which are given in Figs 8.24–8.35.

The cleaver-shape diagrams are essentially scattergrams split into nine sections, rather than three as was the case with the handaxe diagrams. The top set of three sections include cleavers with edges that are set at a markedly oblique angle to the long axis: they are designated 'acutely angled' and defined by having a cleaver edge angle (see Fig. 8.1(b)) in the range up to 60°. The middle set of three sections, horizontally, has 'angled' cleavers, whose edges are only moderately oblique (cleaver edge angles in the range 61°–75°). The bottom set of three sections contains the 'transverse' cleavers, i.e. those whose cleaver edges are more or less perpendicular to the long axis, or only slightly oblique (cleaver edge angles in the range 76°–90°).

Within each set of three sections of the cleaver-shape diagram, procedure is much the same as with the handaxe-shape diagrams. First, the implements with low positions of maximum breadth go to the right-hand section, those with centrally placed maximum breadth to the centre section and those with a high position of maximum breadth to the left-hand section (the defining ranges of L_1/L values are as for the handaxes).

METRICAL ANALYSIS OF HANDAXES AND CLEAVERS

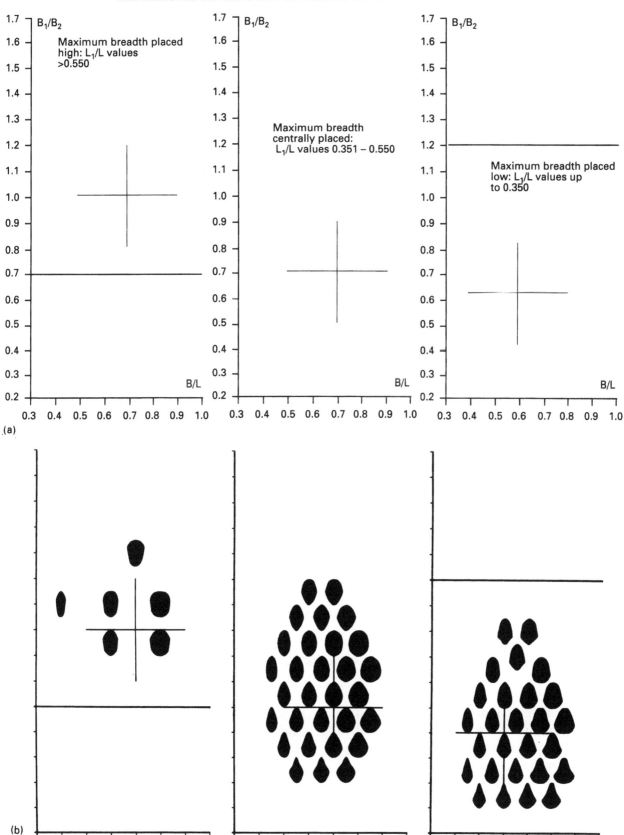

Fig. 8.2 (a) Framework for the handaxe shape diagrams. (b) Array of plan-forms on the handaxe-shape diagrams

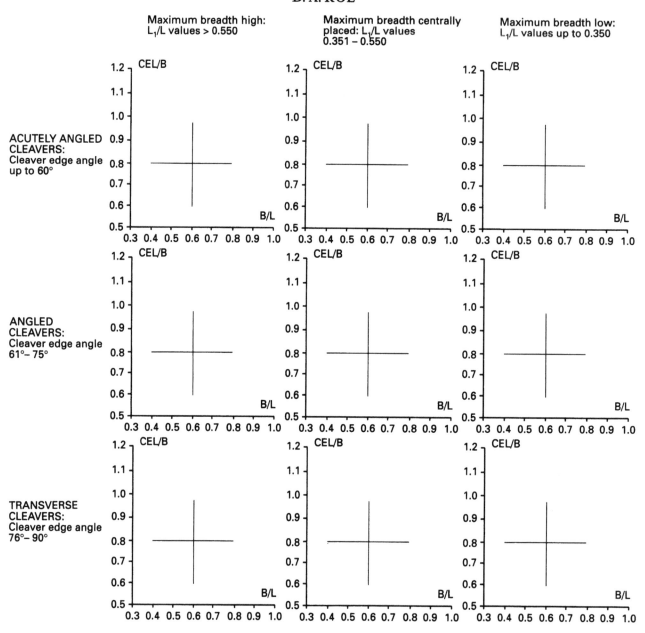

Fig. 8.3 Framework for the cleaver-shape diagrams

Then each cleaver is individually plotted, on its correct section out of the nine, the plot being in terms of B/L values on the horizontal axis and CEL/B values on the vertical axis. The effect of this is to throw the broader shapes (higher values for B/L) to the right, as was the case with the handaxes, and those with the longer cleaver edges (higher values for CEL/B) to the top of each section, as Fig. 8.4 shows. Visual coordinate crosses are again provided. There is, however, one further aspect of cleaver shape to be dealt with. The combination of metrical data used so

METRICAL ANALYSIS OF HANDAXES AND CLEAVERS

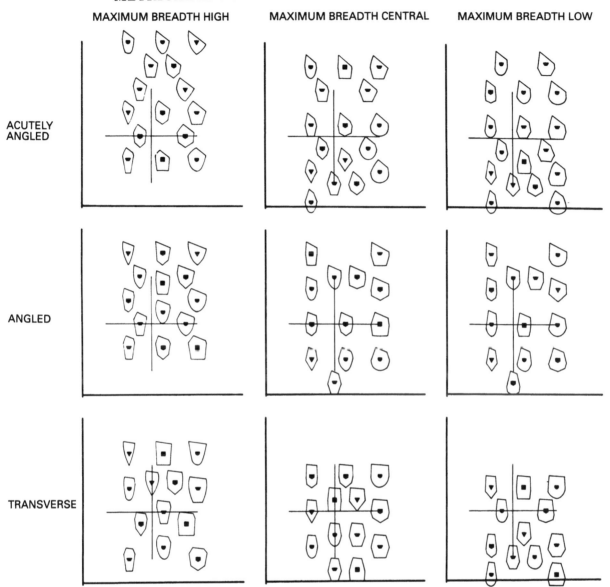

Fig. 8.4 Array of plan-forms on the cleaver-shape diagrams (random selection of butt shapes)

far does not express clearly enough the shape of the butt end of the cleaver, which may be a very important component of its overall plan-form. To cover this deficiency, when the plotting of the cleavers is done on each appropriate section, instead of a mere dot a symbol is used to represent any of five standard butt shapes, whichever is most appropriate for that particular cleaver. The butt shapes and symbols are shown in Fig. 8.5. On the key diagram, a random selection of butt

- rounded (U-shaped) base
- truncated convergent base
- squared base
- short pointed base
- long pointed base

Fig. 8.5 Cleaver butt-shape symbols

Table 8.2 *Metrical analysis of bifaces: handaxes and cleavers: length*

Site	N	% value of each biface	Mean	S.D.	Length (mm)												
					31–40	41–50	51–60	61–70	71–80	81–90	91–100	101–110	111–120	121–130	131–140	141–150	151–160
HK	360	0.3	127.8	21.1	–	–	–	–	0.3	1.9	4.7	9.2	22.8	20.6	18.9	9.4	6.4
TK FG	69	1.4	162.4	44.9	–	–	–	–	–	4.3	5.8	2.9	1.4	4.3	14.5	5.8	13.0
FLK Masek	22	4.5	181.2	56.8	–	–	–	–	–	–	4.5	–	9.1	4.5	9.1	18.2	4.5
PDK Trenches I–III	17	5.9	72.5	17.4	–	–	29.4	29.4	17.6	5.9	5.9	5.9	5.9	–	–	–	–
WK East A	42	2.4	91.9	25.8	–	–	9.5	14.3	16.7	7.1	16.7	11.9	14.3	2.4	2.4	2.4	–
WK East C	9	11.1	107.0	29.2	–	–	–	–	22.2	11.1	22.2	11.1	11.1	–	11.1	–	–
WK	146	0.7	133.4	26.3	–	–	4.1	2.1	1.4	–	–	4.1	11.6	12.3	18.5	20.5	15.8
HEB West 2a	35	2.9	153.6	25.2	–	–	–	–	–	–	2.9	5.7	–	8.6	11.4	17.1	11.4
HEB West 2b	43	2.3	156.6	28.9	–	–	–	–	–	–	4.7	4.7	4.7	4.7	9.3	11.6	7.0
HEB West 3	54	1.9	137.9	26.6	–	–	–	–	1.9	–	3.7	7.4	11.1	20.4	14.8	11.1	13.0
HEB East	44	2.3	94.8	31.3	–	–	13.6	13.6	9.1	15.9	9.1	9.1	6.8	6.8	6.8	4.5	–
PDK Trench IV	18	5.6	158.4	24.0	–	–	–	–	–	–	–	5.6	–	5.6	16.7	11.1	11.1
BK	62	1.6	79.0	31.4	4.8	16.1	11.3	14.5	12.9	12.9	11.3	3.2	–	3.2	1.6	4.8	1.6
TK Upper	15	6.7	110.7	59.5	–	–	–	40.0	6.7	6.7	–	6.7	6.7	6.7	–	–	6.7
TK Lower	9	11.1	151.0	57.2	–	–	–	–	–	–	11.1	22.2	11.1	–	–	11.1	11.1
SHK	47	2.1	105.7	42.1	4.3	4.3	10.6	4.3	6.4	8.5	10.6	8.5	6.4	8.5	4.3	4.3	8.5
EF–HR	35	2.9	146.6	27.2	–	–	–	–	–	–	–	5.7	5.7	22.9	5.7	20.0	14.3
MLK	29	3.4	167.9	36.2	–	–	–	–	–	–	3.4	–	3.4	–	17.2	–	17.2

shapes has been used. It is not quite true to say that any one of the five butt shapes could fall at any point on any of the nine sections of the diagram, but there is nevertheless a great deal of interchangeability. In actual practice, my impression is that clear preferences for particular cleaver butt shapes are to be found within at least some Lower Palaeolithic industries.

All this may sound very complicated, but I have found no simpler way to express the essentials of African cleaver plan-forms on a diagram of this nature, and those who are willing to follow step by step the description of how the diagrams are made, with reference to Figs. 8.3 and 8.4, should find nothing difficult about it. However, the Olduvai cleaver diagrams require careful study rather than a casual glance if their full potential for yielding information is to be realised. The handaxe-shape diagrams can be digested rather more rapidly.

Through the programming skills of Dr Callow, the handaxe and cleaver-shape diagrams as presented in this report have actually been drawn by computer, another major saving of effort for which I am very grateful.

This section may conveniently be concluded and summarised with a straightforward list of the full set of metrical analyses used in this report:

Measurement, ratio, etc.	Aim	Presentation of results
(a) *For whole sets of bifaces* (18 samples)		
Length (L)	Test of overall size of implement	Frequency table with means and standard deviations (Table 8.2); statistical comparison table for all pairs of sites, with t-values and estimates of significance (Table 8.9)
Weight (Wt)	ditto	ditto (Tables 8.3, 8.10)
Th/B	Test of relative thickness of implement cross section (short profile); related to technology and hence to 'refinement' in a broad sense	ditto (Tables 8.4, 8.11)
T_1/L	Test of flatness of tip in long profile view. Related to technology and hence to 'refinement' in a broad sense	ditto (Tables 8.5, 8.12)

METRICAL ANALYSIS OF HANDAXES AND CLEAVERS

Table 8.2 (cont.)

	Length (mm)														
Site	161-170	171-180	181-190	191-200	201-210	211-220	221-230	231-240	241-250	251-260	261-270	271-280	281-290	291-300	301-310
HK	1.4	2.2	0.8	0.8	–	–	0.6	–	–	–	–	–	–	–	–
TK FG	13.0	7.2	5.8	4.3	2.9	4.3	4.3	1.4	1.4	–	–	–	–	1.4	1.4
FLK Masek	4.5	–	4.5	9.1	4.5	4.5	–	–	4.5	–	9.1	9.1	–	–	–
PDK Trenches I–III	–	–	–	–	–	–	–	–	–	–	–	–	–	–	–
WK East A	2.4	–	–	–	–	–	–	–	–	–	–	–	–	–	–
WK East C	11.1	–	–	–	–	–	–	–	–	–	–	–	–	–	–
WK	6.8	2.1	0.7	–	–	–	–	–	–	–	–	–	–	–	–
HEB West 2a	14.3	11.4	11.4	5.7	–	–	–	–	–	–	–	–	–	–	–
HEB West 2b	23.3	7.0	11.6	4.7	7.0	–	–	–	–	–	–	–	–	–	–
HEB West 3	5.6	1.9	7.4	–	–	1.9	–	–	–	–	–	–	–	–	–
HEB East	4.5	–	–	–	–	–	–	–	–	–	–	–	–	–	–
PDK Trench IV	11.1	22.2	11.1	5.6	–	–	–	–	–	–	–	–	–	–	–
BK	–	1.6	–	–	–	–	–	–	–	–	–	–	–	–	–
TK Upper	6.7	–	–	6.7	–	–	–	–	–	6.7	–	–	–	–	–
TK Lower	11.1	–	–	11.1	–	–	–	–	–	–	–	11.1	–	–	–
SHK	2.1	2.1	4.3	2.1	–	–	–	–	–	–	–	–	–	–	–
EF-HR	8.6	5.7	5.7	2.9	–	–	–	2.9	–	–	–	–	–	–	–
MLK	17.2	17.2	3.4	6.9	3.4	3.4	–	–	–	3.4	3.4	–	–	–	–

Measurement, ratio, etc.	Aim	Presentation of results
B/L	General test of shape, reflecting overall broadness or narrowness of plan-form	ditto (Tables 8.6, 8.13)
B_1/B_2	Specific test of plan view shape, reflecting 'pointedness' or 'blunt-endedness'	ditto (Tables 8.7, 8.14)
L_1/L	Specific test of plan view shape, indicating position of maximum breadth	ditto (Tables 8.8, 8.15)
(b) *For handaxes alone (18 samples)*		
L, Wt, Th/B, T_1/L, B/L, B_1/B_2, L_1/L	as before	as before (Tables 8.17–8.30)
Handaxe-shape diagrams	Array of handaxe plan-forms, using the ratios L_1/L, B/L and B_1/B_2	One diagram per site (Figs 8.6–8.23; refer to Fig. 8.2 for key)
(c) *For cleavers alone (9 samples)*		
L, Wt, Th/B, T_1/L, B/L, B_1/B_2, L_1/L	as before	as before (Tables 8.32–8.38, 8.41–8.47)
Cleaver edge angle	Specific test of plan view shape: examines degree of inclination of cleaver edge to long axis	Frequency table with means and standard deviations (Table 8.39); statistical comparison table for all pairs of sites, with t-values and estimates of significance (Table 8.48)
CEL/B	Specific test of plan view shape: examines length of cleaver edge, relative to overall breadth of implement	ditto (Tables 8.40, 8.49)
Cleaver-shape diagrams	Array of cleaver plan-forms, using cleaver edge angle and the ratios L_1/L, B/L and CEL/B, with butt shape symbols	One diagram for each site producing one or more cleavers (Figs 8.24–8.35; refer to Figs 8.3 and 8.4 for key)

Table 8.3 *Metrical analysis of bifaces: handaxes and cleavers: weight*

Site	N	% value of each biface	Mean	S.D.	Weight (g)														
					0–100	101–200	201–300	301–400	401–500	501–600	601–700	701–800	801–900	901–1000	1001–1100	1101–1200	1201–1300	1301–1400	1401–1500
HK	345	0.3	405.6	212.1	0.3	8.1	23.2	29.3	19.4	9.0	3.8	1.4	1.2	2.0	0.9	0.6	–	–	–
TK FG	68	1.5	715.4	477.5	–	8.8	5.9	5.9	11.8	14.7	11.8	10.3	8.8	4.4	2.9	2.9	4.4	2.9	–
FLK Masek	22	4.5	1158.7	783.8	–	–	4.5	4.5	18.2	13.6	13.6	–	–	–	–	–	–	9.1	–
PDK Trenches I–III	8	12.5	170.3	128.6	25.0	50.0	–	12.5	12.5	–	–	–	–	–	–	–	–	–	–
WK East A	32	3.1	212.8	147.1	21.9	40.6	12.5	12.5	9.4	–	3.1	–	–	–	–	–	–	–	–
WK East C	9	11.1	314.2	237.9	11.1	22.2	33.3	–	11.1	11.1	–	–	11.1	–	–	–	–	–	–
WK	133	0.8	539.5	190.3	3.0	2.3	5.3	5.3	23.3	27.1	15.0	12.0	3.8	1.5	1.5	–	–	–	–
HEB West 2a	35	2.9	621.2	270.0	–	5.7	2.9	14.3	14.3	17.1	11.4	8.6	8.6	5.7	8.6	–	2.9	–	–
HEB West 2b	42	2.4	616.9	267.3	–	4.8	7.1	11.9	11.9	14.3	11.9	14.3	11.9	2.4	2.4	7.1	–	–	–
HEB West 3	52	1.9	387.1	162.1	1.9	11.5	21.2	17.3	26.9	13.5	1.9	1.9	3.8	–	–	–	–	–	–
HEB East	34	2.9	292.9	220.7	14.7	35.3	14.7	5.9	14.7	5.9	5.9	–	–	–	2.9	–	–	–	–
PDK Trench IV	18	5.6	485.3	148.9	–	–	5.6	27.8	16.7	22.2	27.8	–	–	–	–	–	–	–	–
BK	59	1.7	185.4	201.4	42.4	27.1	13.6	8.5	1.7	3.4	1.7	–	–	–	–	–	1.7	–	–
TK Upper	0	–	–	–	–	–	–	–	–	–	–	–	–	–	–	–	–	–	–
TK Lower	0	–	–	–	–	–	–	–	–	–	–	–	–	–	–	–	–	–	–
SHK	0	–	–	–	–	–	–	–	–	–	–	–	–	–	–	–	–	–	–
EF–HR	29	3.4	593.9	197.5	–	–	3.4	10.3	24.1	17.2	17.2	13.8	6.9	3.4	3.4	–	–	–	–
MLK	28	3.6	1072.8	526.5	–	–	3.6	3.6	–	3.6	14.3	21.4	–	3.6	10.7	14.3	–	–	3.6

Table 8.4 *Metrical analysis of bifaces: handaxes and cleavers: ratio Th/B*

Site	N	% value of each biface	Mean	S.D.	Ratio Th/B					
					0.201–0.250	0.251–0.300	0.301–0.350	0.351–0.400	0.401–0.450	0.451–0.500
HK	382	0.3	0.543	0.104	–	0.8	1.0	5.2	10.5	21.2
TK FG	71	1.4	0.580	0.115	–	1.4	–	2.8	2.8	15.5
FLK Masek	23	4.3	0.612	0.079	–	–	–	–	–	8.7
PDK Trenches I–III	9	11.1	0.603	0.107	–	–	–	–	11.1	–
WK East A	42	2.4	0.590	0.147	–	–	2.4	4.8	11.9	9.5
WK East C	9	11.1	0.601	0.177	–	–	11.1	–	–	11.1
WK	133	0.8	0.529	0.089	–	0.8	0.8	2.3	10.5	30.1
HEB West 2a	35	2.9	0.561	0.107	–	–	11.4	2.9	17.1
HEB West 2b	43	2.3	0.526	0.107	–	–	9.3	–	9.3	20.9
HEB West 3	58	1.7	0.513	0.089	–	–	3.4	5.2	13.8	27.6
HEB East	44	2.3	0.598	0.120	–	–	2.3	2.3	4.5	13.6
PDK Trench IV	18	5.6	0.481	0.087	–	–	–	22.2	22.2	16.7
BK	67	1.5	0.659	0.138	–	–	–	–	3.0	13.4
TK Upper	13	7.7	0.611	0.138	–	–	–	7.7	7.7	7.7
TK Lower	9	11.1	0.512	0.078	–	–	–	11.1	11.1	11.1
SHK	46	2.2	0.618	0.113	–	–	–	2.2	6.5	8.7
EF–HR	37	2.7	0.621	0.139	–	–	–	–	5.4	13.5
MLK	28	3.6	0.684	0.096	–	–	–	–	–	3.6

METRICAL ANALYSIS OF HANDAXES AND CLEAVERS

Table 8.3 (*cont.*)

	Weight (g)																	
te	1501–1600	1601–1700	1701–1800	1801–1900	1901–2000	2001–2100	2101–2200	2201–2300	2301–2400	2401–2500	2501–2600	2601–2700	2701–2800	2801–2900	2901–3000	3001–3100	3101–3200
K	0.6	–	0.3	–	–	–	–	–	–	–	–	–	–	–	–	–	–
K FG	–	–	–	1.5	1.5	–	–	–	–	–	–	–	–	–	–	–	1.5
LK Masek	4.5	4.5	–	–	4.5	4.5	–	4.5	9.1	–	–	4.5	–	–	–	–	–
OK Trenches I–III	–	–	–	–	–	–	–	–	–	–	–	–	–	–	–	–	–
K East A	–	–	–	–	–	–	–	–	–	–	–	–	–	–	–	–	–
K East C	–	–	–	–	–	–	–	–	–	–	–	–	–	–	–	–	–
K	–	–	–	–	–	–	–	–	–	–	–	–	–	–	–	–	–
EB West 2a	–	–	–	–	–	–	–	–	–	–	–	–	–	–	–	–	–
EB West 2b	–	–	–	–	–	–	–	–	–	–	–	–	–	–	–	–	–
EB West 3	–	–	–	–	–	–	–	–	–	–	–	–	–	–	–	–	–
EB East	–	–	–	–	–	–	–	–	–	–	–	–	–	–	–	–	–
OK Trench IV	–	–	–	–	–	–	–	–	–	–	–	–	–	–	–	–	–
K	–	–	–	–	–	–	–	–	–	–	–	–	–	–	–	–	–
K Upper	–	–	–	–	–	–	–	–	–	–	–	–	–	–	–	–	–
K Lower	–	–	–	–	–	–	–	–	–	–	–	–	–	–	–	–	–
HK	–	–	–	–	–	–	–	–	–	–	–	–	–	–	–	–	–
F–HR	–	–	–	–	–	–	–	–	–	–	–	–	–	–	–	–	–
LK	–	7.1	3.6	3.6	–	–	–	3.6	3.6	–	–	–	–	–	–	–	–

Table 8.4 (*cont.*)

	Ratio Th/B									
te	0.501–0.550	0.551–0.600	0.601–0.650	0.651–0.700	0.701–0.750	0.751–0.800	0.801–0.850	0.851–0.900	0.901–0.950	0.951–1.000
K	17.8	15.4	14.9	5.8	3.4	2.6	0.8	0.3	0.3	–
K FG	19.7	22.5	14.1	8.5	8.5	–	–	2.8	–	1.4
LK Masek	17.4	21.7	21.7	13.0	8.7	8.7	–	–	–	–
OK Trenches I–III	11.1	44.4	–	11.1	11.1	11.1	–	–	–	–
K East A	11.9	16.7	14.3	7.1	9.5	4.8	–	4.8	–	2.4
K East C	22.2	11.1	–	33.3	–	–	–	–	–	11.1
K	22.6	14.3	8.3	6.8	1.5	0.8	1.5	–	–	–
EB West 2a	11.4	17.1	20.0	8.6	8.6	2.9	–	–	–	–
EB West 2b	25.6	11.6	14.0	4.7	2.3	–	2.3	–	–	–
EB West 3	17.2	15.5	10.3	3.4	3.4	–	–	–	–	–
EB East	15.9	11.4	9.1	22.7	13.6	–	2.3	–	2.3	–
OK Trench IV	11.1	16.7	11.1	–	–	–	–	–	–	–
K	11.9	6.0	14.9	14.9	7.5	17.9	1.5	3.0	1.5	4.5
K Upper	–	30.8	–	30.8	–	–	15.4	–	–	–
K Lower	22.2	33.3	11.1	–	–	–	–	–	–	–
HK	8.7	10.9	21.7	23.9	4.3	8.7	2.2	2.2	–	–
F–HR	21.6	13.5	8.1	13.5	5.4	8.1	2.7	2.7	5.4	–
LK	7.1	7.1	10.7	32.1	17.9	10.7	3.6	7.1	–	–

Table 8.5 Metrical analysis of bifaces: handaxes and cleavers: ratio T_1/L

Site	N	% value of each biface	Mean	S.D.	0.076–0.100	0.101–0.125	0.126–0.150	0.151–0.175	0.176–0.200	0.201–0.225	0.226–0.250	0.251–0.275	0.276–0.300	0.301–0.325	0.326–0.350	0.351–0.375	0.376–0.400	0.401–0.425	0.426–0.450	0.451–0.475	0.476–0.500
HK	353	0.3	0.179	0.030	–	1.4	15.6	29.7	32.0	13.0	6.2	1.7	0.3	–	–	–	–	–	–	–	–
TK FG	66	1.5	0.172	0.056	4.5	15.2	15.2	30.3	16.7	9.1	–	3.0	–	3.0	1.5	–	1.5	–	–	–	–
FLK Masek	21	4.8	0.185	0.040	–	–	19.0	28.6	14.3	19.0	14.3	4.8	–	–	–	–	–	–	–	–	–
PDK Trenches I–III	8	12.5	0.344	0.069	–	–	–	–	–	–	12.5	12.5	–	12.5	12.5	12.5	25.0	–	–	12.5	–
WK East A	32	3.1	0.232	0.061	–	–	3.1	12.5	28.1	12.5	6.3	15.6	3.1	9.4	6.3	–	3.1	–	–	–	–
WK East C	9	11.1	0.230	0.075	–	11.1	–	11.1	11.1	11.1	22.2	22.2	–	–	–	–	11.1	–	–	–	–
WK	132	0.8	0.177	0.040	–	6.1	18.9	28.8	26.5	12.1	3.0	0.8	2.3	0.8	0.8	–	–	–	–	–	–
HEB West 2a	35	2.9	0.154	0.040	2.9	25.7	17.1	31.4	17.1	2.9	–	–	–	2.9	–	–	–	–	–	–	–
HEB West 2b	42	2.4	0.163	0.035	–	9.5	33.3	21.4	16.7	11.9	7.1	–	–	–	–	–	–	–	–	–	–
HEB West 3	52	1.9	0.152	0.041	5.8	25.0	23.1	25.0	7.7	3.8	5.8	3.8	–	–	–	–	–	–	–	–	–
HEB East	34	2.9	0.245	0.077	–	11.8	5.9	8.8	23.5	14.7	5.9	11.8	5.9	2.9	8.8	2.9	5.9	2.9	–	–	–
PDK Trench IV	17	5.9	0.144	0.042	17.6	–	35.3	11.8	17.6	–	5.9	–	–	–	–	–	–	–	–	–	–
BK	59	1.7	0.300	0.105	–	6.8	–	6.8	5.1	13.6	5.1	3.4	11.9	6.8	10.2	6.8	3.4	1.7	6.8	6.8	5.1
TK Upper	0	–	–	–	–	–	–	–	–	–	–	–	–	–	–	–	–	–	–	–	–
TK Lower	0	–	–	–	–	–	–	–	–	–	–	–	–	–	–	–	–	–	–	–	–
SHK	0	–	–	–	–	–	–	–	–	–	–	–	–	–	–	–	–	–	–	–	–
EF–HR	28	3.6	0.193	0.057	–	10.7	17.9	10.7	25.0	7.1	7.1	17.9	–	3.6	–	–	–	–	–	–	–
MLK	28	3.6	0.244	0.068	–	–	7.1	7.1	14.3	14.3	14.3	14.3	14.3	3.6	–	3.6	3.6	3.6	–	–	–

Table 8.6 *Metrical analysis of bifaces: handaxes and cleavers: ratio B/L*

Site	N	% value of each biface	Mean	S.D.	0.251–0.300	0.301–0.350	0.351–0.400	0.401–0.450	0.451–0.500	0.501–0.550	0.551–0.600	0.601–0.650	0.651–0.700	0.701–0.750	0.751–0.800	0.801–0.850	0.851–0.900	0.901–0.950	0.951–1.000
HK	360	0.3	0.602	0.066	–	–	–	0.3	3.9	18.9	29.4	23.6	16.7	4.4	2.5	0.3	–	–	–
TK FG	69	1.4	0.545	0.080	–	–	1.4	5.8	30.4	18.8	23.2	10.1	5.8	1.4	2.9	–	–	–	–
FLK Masek	22	4.5	0.577	0.090	–	–	–	4.5	27.3	13.6	9.1	22.7	13.6	9.1	–	–	–	–	–
PDK Trenches I–III	17	5.9	0.742	0.103	–	–	–	–	–	–	5.9	17.6	11.8	11.8	29.4	5.9	11.8	–	5.9
WK East A	42	2.4	0.648	0.102	–	–	–	2.4	7.1	4.8	21.4	16.7	16.7	19.0	4.8	5.9	–	2.4	–
WK East C	9	11.1	0.648	0.120	–	–	–	11.1	–	–	–	22.2	22.2	11.1	–	4.8	11.1	–	–
WK	146	0.7	0.653	0.083	–	–	–	–	2.7	4.1	22.2	31.5	16.4	15.1	7.5	0.7	2.7	–	0.7
HEB West 2a	35	2.9	0.579	0.065	–	–	–	–	5.7	31.4	18.5	25.7	2.9	–	–	2.9	–	–	–
HEB West 2b	43	2.3	0.580	0.077	–	–	–	2.3	16.3	18.6	32.6	14.0	9.3	2.3	4.7	–	–	–	–
HEB West 3	54	1.9	0.593	0.102	–	–	1.9	1.9	7.4	27.8	25.9	7.4	11.1	11.1	–	3.7	1.9	–	–
HEB East	44	2.3	0.706	0.110	–	–	–	–	2.3	9.1	9.1	13.6	6.8	15.9	25.0	9.1	9.1	–	–
PDK Trench IV	18	5.6	0.539	0.081	–	5.6	–	–	11.1	44.4	22.2	11.1	5.6	–	–	–	–	–	–
BK	62	1.6	0.718	0.125	–	–	–	–	3.2	8.1	3.2	16.1	11.3	25.8	11.3	8.1	3.2	3.2	6.5
TK Upper	15	6.7	0.624	0.125	–	–	–	13.3	–	20.0	13.3	6.7	26.7	–	6.7	13.3	–	–	–
TK Lower	9	11.1	0.670	0.169	–	–	–	–	11.1	22.2	–	–	–	11.1	11.1	–	11.1	11.1	–
SHK	47	2.1	0.703	0.120	–	–	–	–	2.1	10.6	10.6	8.5	12.8	14.9	19.1	10.6	8.5	2.1	–
EF-HR	35	2.9	0.583	0.073	–	–	–	2.9	11.4	20.0	28.6	22.9	5.7	8.6	–	–	–	–	–
MLK	29	3.4	0.578	0.092	–	–	–	10.3	6.9	27.6	20.7	6.9	17.2	6.9	–	3.4	–	–	–

163

Table 8.7 Metrical analysis of bifaces: handaxes and cleavers: ratio B_1/B_2

Site	N	% value of each biface	Mean	S.D.	0.101–0.200	0.201–0.300	0.301–0.400	0.401–0.500	0.501–0.600	0.601–0.700	0.701–0.800	0.801–0.900	0.901–1.000	1.001–1.100	1.101–1.200	1.201–1.300	1.301–1.400	1.401–1.500	1.501–1.600	1.601–1.700	1.701–1.800	1.801–1.900	1.901–2.000	2.001–2.100
HK	359	0.3	0.767	0.168	–	–	0.3	1.1	11.7	25.6	27.0	17.3	8.1	4.5	2.2	1.4	0.3	0.3	–	0.3	–	–	–	–
TK FG	69	1.4	0.696	0.121	–	–	–	4.3	15.9	33.3	23.2	18.8	4.3	–	–	–	–	–	–	–	–	–	–	–
FLK Masek	22	4.5	0.746	0.107	–	–	–	–	4.5	36.4	36.4	13.6	9.1	–	–	–	–	–	–	–	–	–	–	–
PDK Trenches I–III	17	5.9	0.754	0.185	–	–	5.9	5.9	5.9	17.6	11.8	29.4	17.6	5.9	–	–	–	–	–	–	–	–	–	–
WK East A	42	2.4	0.713	0.159	–	–	–	7.1	19.0	23.8	23.8	19.0	2.4	2.4	–	2.4	–	–	–	–	–	–	–	–
WK East C	9	11.1	0.635	0.141	–	–	11.1	11.1	11.1	33.3	22.2	11.1	–	–	–	–	–	–	–	–	–	–	–	–
WK	146	0.7	0.866	0.247	–	–	0.7	1.4	10.3	10.3	20.5	19.9	13.0	11.6	3.4	4.1	2.7	0.7	–	–	0.7	–	–	0.7
HEB West 2a	35	2.9	0.694	0.173	–	–	–	2.9	28.6	40.0	5.7	5.7	11.4	2.9	2.9	–	–	–	–	–	–	–	–	–
HEB West 2b	43	2.3	0.803	0.216	–	–	–	7.0	9.3	18.6	18.6	23.3	4.7	4.7	9.3	2.3	2.3	–	–	–	–	–	–	–
HEB West 3	54	1.9	0.714	0.233	–	–	–	18.5	18.5	25.9	7.4	9.3	7.4	5.6	–	5.6	1.9	–	–	–	–	–	–	–
HEB East	44	2.3	0.786	0.240	–	–	–	2.3	15.9	25.0	22.7	11.4	13.6	4.5	–	–	2.3	–	–	–	–	–	–	–
PDK Trench IV	17	5.9	0.841	0.303	–	–	–	5.9	29.4	5.9	11.8	11.8	5.9	11.8	5.9	5.9	–	–	–	–	–	–	–	–
BK	62	1.6	0.760	0.206	–	–	1.6	8.1	14.5	17.7	19.4	11.3	14.5	9.7	1.6	–	1.6	–	–	–	–	–	–	–
TK Upper	15	6.7	0.763	0.151	–	–	–	–	20.0	13.3	33.3	13.3	13.3	6.7	–	–	–	–	–	–	–	–	–	–
TK Lower	9	11.1	0.677	0.193	–	–	–	22.2	11.1	33.3	11.1	11.1	–	11.1	–	–	–	–	–	–	–	–	–	–
SHK	47	2.1	0.781	0.177	–	–	–	12.8	8.5	8.5	17.0	23.4	19.1	10.6	–	–	–	–	–	–	–	–	–	–
EF–HR	35	2.9	0.735	0.196	–	–	–	11.4	17.1	14.3	28.6	8.6	5.7	11.4	–	2.9	–	–	–	–	–	–	–	–
MLK	29	3.4	0.732	0.183	–	–	–	3.4	20.7	27.6	20.7	10.3	3.4	6.9	6.9	–	–	–	–	–	–	–	–	–

Table 8.8 Metrical analysis of bifaces: handaxes and cleavers: ratio L_1/L

Site	N	% value of each biface	Mean	S.D.	0.001–0.050	0.051–0.100	0.101–0.150	0.151–0.200	0.201–0.250	0.251–0.300	0.301–0.350	0.351–0.400	0.401–0.450	0.451–0.500	0.501–0.550	0.551–0.600	0.601–0.650	0.651–0.700	0.701–0.750	0.751–0.800	0.801–0.850	0.851–0.900	0.901–0.950	0.951–1.000
HK	357	0.3	0.407	0.085	–	–	–	–	0.8	5.6	16.8	27.5	26.3	13.7	4.2	1.7	0.8	1.7	–	0.6	0.3	–	–	–
TK FG	69	1.4	0.391	0.072	–	–	–	–	1.4	11.6	13.0	29.0	24.6	11.6	7.2	1.4	–	–	–	–	–	–	–	–
FLK Masek	22	4.5	0.426	0.101	–	–	–	–	–	13.6	4.5	22.7	27.3	13.6	4.5	9.1	–	4.5	–	–	–	–	–	–
PDK Trenches I–III	17	5.9	0.401	0.087	–	–	–	–	–	23.5	5.9	23.5	11.8	23.5	5.9	5.9	–	–	–	–	–	–	–	–
WK East A	42	2.4	0.418	0.094	–	–	–	2.4	2.4	9.5	11.9	2.4	26.2	31.0	9.5	2.4	2.4	–	–	–	–	–	–	–
WK East C	9	11.1	0.411	0.053	–	–	–	–	–	11.1	–	33.3	33.3	22.2	–	–	–	–	–	–	–	–	–	–
WK	146	0.7	0.436	0.131	–	–	–	0.7	1.4	4.8	16.4	24.7	20.5	12.3	4.1	3.4	4.1	1.4	2.1	0.7	1.4	0.7	1.4	–
HEB West 2a	35	2.9	0.386	0.073	–	–	–	–	–	14.3	20.0	25.7	14.3	20.0	5.7	–	–	–	–	–	–	–	–	–
HEB West 2b	43	2.3	0.435	0.075	–	–	–	–	2.3	2.3	4.7	20.9	27.9	25.6	7.0	7.0	2.3	–	–	–	–	–	–	–
HEB West 3	54	1.9	0.388	0.103	–	–	–	–	3.7	7.4	31.5	24.1	14.8	5.6	5.6	3.7	1.9	–	–	–	1.9	–	–	–
HEB East	44	2.3	0.438	0.094	–	–	–	–	2.3	2.3	4.5	18.2	29.5	27.3	11.4	–	–	2.3	–	–	2.3	–	–	–
PDK Trench IV	17	5.9	0.420	0.118	–	–	–	–	–	5.9	23.5	17.6	29.4	5.9	5.9	5.9	–	–	5.9	–	–	–	–	–
BK	62	1.6	0.408	0.094	–	–	–	–	1.6	11.3	14.5	19.4	24.2	11.3	6.5	11.3	–	–	–	–	–	–	–	–
TK Upper	15	6.7	0.412	0.049	–	–	–	–	–	–	13.3	20.0	53.3	6.7	6.7	–	–	–	–	–	–	–	–	–
TK Lower	9	11.1	0.357	0.075	–	–	–	–	11.1	11.1	11.1	33.3	33.3	–	–	–	2.1	–	–	–	–	–	–	–
SHK	47	2.1	0.402	0.086	–	–	–	–	4.3	8.5	17.0	19.1	17.0	25.5	6.4	–	2.1	–	–	–	–	–	–	–
EF–HR	35	2.9	0.392	0.087	–	–	–	–	2.9	8.6	22.9	22.9	22.9	5.7	11.4	2.9	–	–	–	–	–	–	–	–
MLK	29	3.4	0.403	0.102	–	–	–	3.4	–	6.9	20.7	20.7	13.8	17.2	13.8	–	–	3.4	–	–	–	–	–	–

Table 8.9 *Metrical analysis of bifaces: handaxes and cleavers: t-values and estimates of significance: length (for an explanation of Tables 8.9–8.15 see p. 154)*

Site	HK (360)	TK Fish Gully (69)	FLK Masek (22)	PDK Trenches I–III (17)	WK East A (42)	WK East C (9)	WK (146)	HEB West 2a (35)	HEB West 2b (43)	HEB West 3 (54)	HEB East (44)	PDK Trench IV (18)	BK (62)	TK Upper (15)	TK Lower (9)	SHK (47)	EF–HR (35)	MLK (29)	Site
HK (360)		10.0	9.9	10.7	10.2	2.9	2.5	6.8	8.1	3.2	9.2	6.0	15.5	2.7	3.1	5.9	4.9	9.2	HK
TK Fish Gully (69)	0.00		1.6	8.1	9.3	3.6	6.0	1.1	0.8	3.6	8.7	0.4	12.2	3.8	0.7	6.9	1.9	0.6	TK Fish Gully
FLK Masek (22)	0.00	0.11		7.6	8.7	3.7	6.6	2.5	2.3	4.5	8.0	1.6	10.4	3.6	1.3	6.2	3.1	1.0	FLK Masek
PDK Trenches I–III (17)	0.00	0.00	0.00		2.9	3.8	9.3	11.9	11.2	9.5	2.8	12.1	0.8	2.5	5.3	3.1	10.2	10.2	PDK Trenches I–III
WK East A (42)	0.00	0.00	0.00	0.01		1.6	9.0	10.6	10.9	8.5	0.5	9.3	2.2	1.7	4.9	1.8	9.0	10.3	WK East A
WK East C (9)	0.00	0.00	0.00	0.00	0.13		2.9	4.8	4.7	3.2	1.1	4.9	2.5	0.2	2.1	0.1	3.8	4.6	WK East C
WK (146)	0.02	0.00	0.00	0.00	0.00	0.00		4.1	5.0	1.1	8.2	3.8	12.9	2.7	1.8	5.4	2.7	6.0	WK
HEB West 2a (35)	0.00	0.20	0.04	0.00	0.00	0.00	0.00		0.5	2.8	9.0	0.7	12.0	3.6	0.2	6.0	1.1	1.9	HEB West 2a
HEB West 2b (43)	0.00	0.40	0.07	0.00	0.00	0.00	0.00	0.64		3.3	9.6	0.2	12.9	3.9	0.4	6.6	1.5	1.5	HEB West 2b
HEB West 3 (54)	0.01	0.00	0.00	0.00	0.00	0.00	0.28	0.01	0.00		7.4	2.9	10.8	2.6	1.1	4.7	1.5	4.3	HEB West 3
HEB East (44)	0.00	0.00	0.00	0.00	0.00	0.64	0.00	0.00	0.00	0.00		7.7	2.6	1.3	4.2	1.4	7.7	9.2	HEB East
PDK Trench IV (18)	0.00	0.61	0.10	0.00	0.00	0.29	0.00	0.51	0.81	0.00	0.00		9.9	3.1	0.5	5.0	1.5	1.0	PDK Trench IV
BK (62)	0.00	0.00	0.00	0.00	0.00	0.00	0.00	0.00	0.00	0.00	0.01	0.00		2.9	5.7	3.8	10.7	12.0	BK
TK Upper (15)	0.29	0.00	0.00	0.27	0.25	0.84	0.17	0.02	0.01	0.11	0.34	0.01	0.06		1.6	0.4	2.9	4.0	TK Upper
TK Lower (9)	0.26	0.49	0.19	0.03	0.02	0.06	0.39	0.90	0.78	0.52	0.02	0.72	0.01	0.12		2.8	0.3	1.1	TK Lower
SHK (47)	0.00	0.00	0.00	0.00	0.00	0.93	0.00	0.00	0.00	0.00	0.00	0.00	0.00	0.72	0.01		5.0	6.6	SHK
EF–HR (35)	0.00	0.03	0.01	0.00	0.00	0.00	0.01	0.27	0.13	0.14	0.00	0.13	0.00	0.04	0.83	0.00		2.7	EF–HR
MLK (29)	0.00	0.56	0.35	0.00	0.00	0.00	0.00	0.08	0.14	0.00	0.00	0.33	0.00	0.00	0.30	0.00	0.01		MLK
Site	HK	TK Fish Gully	FLK Masek	PDK Trenches I–III	WK East A	WK East C	WK	HEB West 2a	HEB West 2b	HEB West 3	HEB East	PDK Trench IV	BK	TK Upper	TK Lower	SHK	EF–HR	MLK	

Table 8.10 *Metrical analysis of bifaces: handaxes and cleavers: t-values and estimates of significance: weight*

Site	HK (345)	TK Fish Gully (68)	FLK Masek (22)	PDK Trenches I–III (8)	WK East A (32)	WK East C (9)	WK (133)	HEB West 2a (35)	HEB West 2b (42)	HEB West 3 (52)	HEB East (34)	PDK Trench IV (18)	BK (59)	EF-HR (29)	MLK (28)	Site
HK (345)		8.5	12.3	3.1	5.0	1.3	6.4	5.6	5.9	0.6	2.9	1.6	7.4	4.6	13.6	HK
TK Fish Gully (68)	0.00		3.2	3.2	5.8	2.5	3.7	1.1	1.2	4.7	4.9	2.0	7.9	1.3	3.2	TK Fish Gully
FLK Masek (22)	0.00	0.02		3.5	6.7	3.1	7.9	3.7	4.1	6.3	6.1	3.6	8.9	3.7	0.5	FLK Masek
PDK Trenches I–III (8)	0.00	0.00	0.00		0.7	1.5	5.4	4.6	4.6	3.5	1.5	5.2	0.2	5.7	4.8	PDK Trenches I–III
WK East A (32)	0.00	0.00	0.00	0.46		1.6	9.1	7.6	7.7	5.3	1.7	6.3	0.7	8.6	8.9	WK East A
WK East C (9)	0.20	0.00	0.00	0.15	0.12		3.4	3.1	3.1	1.2	0.3	2.3	1.7	3.5	4.2	WK East C
WK (133)	0.00	0.00	0.00	0.00	0.00	0.00		2.1	2.1	5.1	6.5	1.2	11.7	1.4	9.2	WK
HEB West 2a (35)	0.00	0.20	0.00	0.00	0.00	0.00	0.10		0.1	5.2	5.5	2.0	8.9	0.5	4.4	HEB West 2a
HEB West 2b (42)	0.00	0.17	0.00	0.00	0.00	0.00	0.09	0.94		5.2	5.7	2.0	9.3	0.4	4.8	HEB West 2b
HEB West 3 (52)	0.46	0.00	0.00	0.00	0.00	0.25	0.00	0.00	0.00			2.3	5.8	5.1	8.7	HEB West 3
HEB East (34)	0.00	0.00	0.00	0.14	0.09	0.80	0.00	0.00	0.00	0.04		3.3	2.4	5.7	7.9	HEB East
PDK Trench IV (18)	0.12	0.00	0.00	0.00	0.00	0.03	0.25	0.02	0.02	0.03	0.00		5.8	2.0	4.6	PDK Trench IV
BK (59)	0.00	0.00	0.00	0.84	0.50	0.09	0.00	0.02	0.00	0.00	0.02	0.00		9.0	11.4	BK
EF-HR (29)	0.00	0.08	0.00	0.00	0.00	0.00	0.17	0.65	0.69	0.00	0.00	0.05	0.00		4.6	EF-HR
MLK (28)	0.00	0.00	0.65	0.00	0.00	0.00	0.00	0.00	0.00	0.00	0.00	0.00	0.00	0.00		MLK

Table 8.11 Metrical analysis of bifaces: handaxes and cleavers: t-values and estimates of significance: ratio Th/B

Site	HK (382)	TK Fish Gully (71)	FLK Masek (23)	PDK Trenches I-III (9)	WK East A (42)	WK East C (9)	WK (133)	HEB West 2a (35)	HEB West 2b (43)	HEB West 3 (58)	HEB East (44)	PDK Trench IV (18)	BK (67)	TK Upper (13)	TK Lower (9)	SHK (46)	EF-HR (37)	MLK (28)	Site
HK (382)		2.8	3.1	1.7	2.7	1.6	1.3	1.0	1.0	2.1	3.3	2.5	8.0	2.3	0.9	4.6	4.2	7.0	HK
TK Fish Gully (71)	0.01		1.2	0.6	0.4	0.5	3.5	0.8	2.5	3.7	0.8	3.4	3.6	0.9	1.7	1.8	1.6	4.2	TK Fish Gully
FLK Masek (23)	0.00	0.22		0.2	0.7	0.2	4.2	1.9	3.4	4.7	0.5	5.0	1.5	0.0	3.2	0.2	0.3	2.9	FLK Masek
PDK Trenches I-III (9)	0.08	0.57	0.81		0.3	0.0	2.4	1.1	2.0	2.8	0.1	3.2	1.2	0.1	2.1	0.4	0.4	2.1	PDK Trenches I-III
WK East A (42)	0.05	0.69	0.44	0.80		0.2	3.3	1.0	2.3	3.3	0.3	2.9	2.5	0.5	1.5	1.0	1.0	3.0	WK East A
WK East C (9)	0.35	0.63	0.87	0.98	0.84		2.2	0.9	1.7	2.4	0.1	2.4	1.1	0.1	1.4	0.4	0.4	1.8	WK East C
WK (133)	0.15	0.00	0.00	0.02	0.01	0.26		1.8	0.2	1.2	4.1	2.2	8.0	3.0	0.6	5.5	4.9	8.2	WK
HEB West 2a (35)	0.31	0.41	0.06	0.30	0.33	0.53	0.07		1.4	2.4	1.4	2.8	3.6	1.3	1.3	2.3	2.0	4.7	HEB West 2a
HEB West 2b (43)	0.33	0.01	0.00	0.05	0.02	0.25	0.86	0.15		0.7	2.9	1.6	5.3	2.4	0.4	3.9	3.5	6.3	HEB West 2b
HEB West 3 (58)	0.04	0.00	0.00	0.01	0.00	0.18	0.24	0.02	0.49		4.1	1.3	6.9	3.2	0.0	5.3	4.7	8.1	HEB West 3
HEB East (44)	0.00	0.43	0.57	0.90	0.79	0.94	0.00	0.16	0.00	0.00		3.7	2.4	0.3	2.1	0.8	0.8	3.2	HEB East
PDK Trench IV (18)	0.01	0.00	0.00	0.00	0.00	0.08	0.03	0.01	0.12	0.19	0.00		5.2	3.2	0.9	4.6	3.9	7.2	PDK Trench IV
BK (67)	0.00	0.00	0.05	0.25	0.02	0.26	0.00	0.00	0.00	0.00	0.02	0.00		1.1	3.1	1.6	1.3	0.9	BK
TK Upper (13)	0.02	0.39	0.99	0.89	0.65	0.88	0.06	0.19	0.02	0.03	0.74	0.00	0.26		2.0	0.2	0.2	1.9	TK Upper
TK Lower (9)	0.38	0.09	0.00	0.05	0.13	0.19	0.57	0.20	0.70	0.97	0.04	0.38	0.00	0.06		2.7	2.3	4.8	TK Lower
SHK (46)	0.00	0.08	0.80	0.71	0.31	0.71	0.00	0.02	0.00	0.00	0.41	0.00	0.11	0.85	0.01		0.1	2.5	SHK
EF-HR (37)	0.00	0.11	0.74	0.72	0.34	0.72	0.00	0.04	0.00	0.00	0.42	0.00	0.19	0.82	0.03	0.92		2.0	EF-HR
MLK (28)	0.00	0.00	0.01	0.04	0.00	0.22	0.00	0.00	0.00	0.00	0.00	0.00	0.32	0.06	0.00	0.01	0.05		MLK

Table 8.12 *Metrical analysis of bifaces: handaxes and cleavers: t-values and estimates of significance: ratio T_1/L*

Site	HK (353)	TK Fish Gully (66)	FLK Masek (21)	PDK Trenches I–III (8)	WK East A (32)	WK East C (9)	WK (132)	HEB West 2a (35)	HEB West 2b (42)	HEB West 3 (52)	HEB East (34)	PDK Trench IV (17)	BK (59)	EF–HR (28)	MLK (28)	Site
HK (353)		1.6	0.8	14.6	8.5	4.7	0.7	4.6	3.2	5.9	10.0	4.6	17.8	2.0	9.6	HK
TK Fish Gully (66)	0.28		1.0	8.0	4.9	2.8	0.8	1.7	0.9	2.2	5.5	1.9	8.7	1.7	5.4	TK Fish Gully
FLK Masek (21)	0.44	0.32		7.8	3.2	2.2	0.8	2.8	2.2	3.2	3.3	3.1	4.9	0.5	3.6	FLK Masek
PDK Trenches I–III (8)	0.00	0.00	0.00		4.5	3.2	11.0	10.5	11.1	11.1	3.3	9.0	1.1	6.3	3.6	PDK Trenches I–III
WK East A (32)	0.00	0.00	0.00	0.00		0.1	6.3	6.3	6.2	7.3	0.8	5.4	3.4	2.6	0.7	WK East A
WK East C (9)	0.08	0.01	0.12	0.01	0.94		3.6	4.2	4.1	4.6	0.5	3.8	1.9	1.6	0.5	WK East C
WK (132)	0.53	0.50	0.40	0.00	0.00	0.07		3.1	2.0	3.9	7.2	3.2	11.8	1.7	7.1	WK
HEB West 2a (35)	0.00	0.07	0.01	0.00	0.00	0.02	0.00		1.1	0.3	6.2	0.8	7.9	3.2	6.6	HEB West 2a
HEB West 2b (42)	0.00	0.32	0.03	0.00	0.00	0.03	0.04	0.29		1.4	6.1	1.8	8.1	2.7	6.5	HEB West 2b
HEB West 3 (52)	0.00	0.02	0.00	0.00	0.00	0.02	0.00	0.79	0.15		7.3	0.7	9.6	3.7	7.6	HEB West 3
HEB East (34)	0.00	0.00	0.00	0.00	0.00	0.61	0.00	0.00	0.00	0.00		5.0	2.7	3.0	0.1	HEB East
PDK Trench IV (17)	0.00	0.06	0.00	0.00	0.00	0.01	0.00	0.40	0.08	0.50	0.00		6.0	3.1	5.5	PDK Trench IV
BK (59)	0.00	0.00	0.00	0.00	0.00	0.06	0.00	0.00	0.00	0.00	0.01	0.00		5.1	2.6	BK
EF–HR (28)	0.23	0.10	0.59	0.00	0.01	0.12	0.17	0.00	0.02	0.00	0.00	0.01	0.00		3.1	EF–HR
MLK (28)	0.00	0.00	0.00	0.00	0.48	0.61	0.00	0.00	0.00	0.00	0.94	0.00	0.00	0.00		MLK
	HK	TK Fish Gully	FLK Masek	PDK Trenches I–III	WK East A	WK East C	WK	HEB West 2a	HEB West 2b	HEB West 3	HEB East	PDK Trench IV	BK	EF–HR	MLK	

Table 8.13 *Metrical analysis of bifaces: handaxes and cleavers: t-values and estimates of significance: ratio B/L*

Site	HK (360)	TK Fish Gully (69)	FLK Masek (22)	PDK Trenches I-III (17)	WK East A (42)	WK East C (9)	WK (146)	HEB West 2a (35)	HEB West 2b (43)	HEB West 3 (54)	HEB East (44)	PDK Trench IV (18)	BK (62)	TK Upper (15)	TK Lower (9)	SHK (47)	EF-HR (35)	MLK (29)	Site
HK (360)		6.3	1.7	8.3		4.0	2.0	7.3	2.0	2.1	0.8	9.0	3.9	10.8	1.2	2.9	8.7	1.7	1.9	HK
TK Fish Gully (69)	0.00		1.6	8.6	5.3	5.9	3.4	9.0	2.2	2.3	2.9	9.0	0.3	9.5	3.1	3.8	8.5	2.3	1.7	TK Fish Gully
FLK Masek (22)	0.20	0.12		5.3		2.8	1.8	4.0	0.1	0.2	0.7	4.8	1.4	4.8	1.3	2.0	4.4	0.3	0.0	FLK Masek
PDK Trenches I-III (17)	0.00	0.00	0.00		3.2	2.1	4.1	6.9	6.7	5.2	1.1	6.5	0.7	2.9	1.4	1.2	6.4	5.6	PDK Trenches I-III
WK East A (42)	0.01	0.00	0.01	0.00		0.0	0.3	3.5	3.5	2.6	2.5	4.0	3.0	0.8	0.5	2.3	3.2	3.0	WK East A
WK East C (9)	0.29	0.00	0.08	0.05	0.98		0.2	2.3	2.2	1.4	1.4	2.8	1.6	0.5	0.3	1.3	2.1	1.8	WK East C
WK (146)	0.00	0.00	0.00	0.00	0.75	0.85		4.9	5.2	4.2	3.4	5.5	4.4	1.2	0.5	3.2	4.6	4.4	WK
HEB West 2a (35)	0.05	0.03	0.91	0.00	0.00	0.13	0.00		0.1	0.7	6.1	1.9	6.1	1.7	2.6	5.5	0.2	0.1	HEB West 2a
HEB West 2b (43)	0.04	0.03	0.88	0.00	0.00	0.04	0.00	0.95		0.7	6.2	1.9	6.4	1.6	2.5	5.7	0.2	0.1	HEB West 2b
HEB West 3 (54)	0.54	0.00	0.51	0.00	0.00	0.16	0.00	0.42	0.48		5.3	2.0	5.8	1.0	1.9	4.9	0.5	0.7	HEB West 3
HEB East (44)	0.00	0.00	0.00	0.26	0.00	0.16	0.00	0.00	0.00	0.00		5.8	0.5	2.4	0.8	0.2	5.7	5.2	HEB East
PDK Trench IV (18)	0.00	0.78	0.18	0.00	0.00	0.01	0.00	0.06	0.07	0.05	0.00		5.7	2.3	2.7	5.3	2.0	1.4	PDK Trench IV
BK (62)	0.00	0.00	0.00	0.47	0.00	0.12	0.00	0.00	0.00	0.00	0.64	0.00		2.6	1.0	0.6	5.8	5.4	BK
TK Upper (15)	0.52	0.03	0.19	0.01	0.45	0.65	0.39	0.21	0.22	0.34	0.02	0.03	0.01		0.8	2.2	1.5	1.4	TK Upper
TK Lower (9)	0.27	0.06	0.15	0.19	0.72	0.75	0.78	0.15	0.16	0.23	0.41	0.06	0.31	0.45		0.7	2.4	2.1	TK Lower
SHK (47)	0.00	0.00	0.00	0.24	0.02	0.21	0.01	0.00	0.00	0.00	0.88	0.00	0.53	0.03	0.48		5.2	4.8	SHK
EF-HR (35)	0.10	0.02	0.78	0.00	0.00	0.16	0.00	0.82	0.87	0.57	0.00	0.05	0.00	0.25	0.17	0.00		0.2	EF-HR
MLK (29)	0.17	0.08	0.97	0.00	0.00	0.07	0.00	0.95	0.91	0.49	0.00	0.15	0.00	0.17	0.15	0.00	0.81		MLK

Table 8.14 Metrical analysis of bifaces: handaxes and cleavers: t-values and estimates of significance: ratio B_1/B_2

Site	HK (359)	TK Fish Gully (69)	FLK Masek (22)	PDK Trenches I-III (17)	WK East A (42)	WK East C (9)	WK (146)	HEB West 2a (35)	HEB West 2b (43)	HEB West 3 (54)	HEB East (44)	PDK Trench IV (17)	BK (62)	TK Upper (15)	TK Lower (9)	SHK (47)	EF-HR (35)	MLK (29)	Site
HK (359)		3.4	0.6	0.3	2.3	2.0	5.2	2.5	1.3	2.1	0.5	1.7	0.3	0.1	1.6	0.5	1.1	1.1	HK
TK Fish Gully (69)	0.00		1.7	1.6	1.4	0.6	5.4	0.1	3.4	0.5	2.0	3.1	2.2	1.9	0.4	3.1	1.2	1.1	TK Fish Gully
FLK Masek (22)	0.40	0.09		0.2	2.4	0.9	2.3	1.3	1.2	0.6	0.7	1.4	0.3	0.4	1.3	0.9	0.2	0.3	FLK Masek
PDK Trenches I-III (17)	0.76	0.23	0.87		1.7	0.9	1.8	1.2	0.8	0.7	0.5	1.0	0.1	0.1	1.0	0.5	0.3	0.4	PDK Trenches I-III
WK East A (42)	0.05	0.55	0.39	0.39			3.8	0.5	2.2	0.0	1.6	2.1	1.3	1.1	0.6	1.9	0.5	0.5	WK East A
WK East C (9)	0.02	0.17	0.02	0.11		1.4	2.8	0.9	2.2	1.0	1.3	1.9	1.8	2.1	0.5	2.3	1.4	1.4	WK East C
WK (146)	0.00	0.00	0.00	0.07	0.00	0.01		3.9	1.5	4.0	1.9	0.4	3.0	1.6	2.3	2.2	2.9	2.8	WK
HEB West 2a (35)	0.01	0.95	0.17	0.26	0.62	0.35	0.00		2.4	0.4	1.3	2.2	1.6	1.3	0.3	2.2	0.9	0.8	HEB West 2a
HEB West 2b (43)	0.30	0.00	0.16	0.42	0.03	0.03	0.13	0.02		1.9	0.4	0.5	1.0	0.7	1.6	0.5	1.4	1.5	HEB West 2b
HEB West 3 (54)	0.11	0.62	0.41	0.51	0.99	0.33	0.00	0.67	0.06		1.5	1.8	1.1	0.8	0.4	1.6	0.4	0.4	HEB West 3
HEB East (44)	0.62	0.03	0.36	0.63	0.10	0.08	0.06	0.06	0.73	0.14		0.45	0.6	0.3	1.3	0.1	1.0	1.0	HEB East
PDK Trench IV (17)	0.34	0.07	0.23	0.32	0.12	0.03	0.69	0.08	0.59	0.07	0.45		0.7		1.3	0.9	1.5	1.0	1.5	PDK Trench IV
BK (62)	0.80	0.03	0.69	0.92	0.21	0.08	0.00	0.11	0.31	0.25	0.55	0.31		0.1	1.1	0.6	0.6	0.6	BK
TK Upper (15)	0.93	0.07	0.69	0.88	0.29	0.05	0.03	0.19	0.51	0.44	0.74	0.36	0.96		1.2	0.4	0.5	0.6	TK Upper
TK Lower (9)	0.11	0.78	0.33	0.33	0.55	0.61	0.02	0.79	0.11	0.65	0.21	0.16	0.26	0.23		1.6	0.8	0.8	TK Lower
SHK (47)	0.59	0.01	0.31	0.60	0.06	0.02	0.01	0.03	0.60	0.11	0.52	0.45	0.58	0.73	0.12		1.1	1.2	SHK
EF-HR (35)	0.28	0.29	0.78	0.73	0.60	0.16	0.00	0.36	0.15	0.66	0.31	0.20	0.55	0.62	0.43	0.27		0.1	EF-HR
MLK (29)	0.28	0.34	0.73	0.69	0.65	0.16	0.01	0.40	0.15	0.71	0.31	0.19	0.53	0.57	0.44	0.25	0.95		MLK
Site	HK	TK Fish Gully	FLK Masek	PDK Trenches I-III	WK East A	WK East C	WK	HEB West 2a	HEB West 2b	HEB West 3	HEB East	PDK Trench IV	BK	TK Upper	TK Lower	SHK	EF-HR	MLK	

Table 8.15 Metrical analysis of bifaces: handaxes and cleavers: t-values and estimates of significance: ratio L_1/L

Site	HK (357)	TK Fish Gully (69)	FLK Masek (22)	PDK Trenches I-III (17)	WK East A (42)	WK East C (9)	WK (146)	HEB West 2a (35)	HEB West 2b (43)	HEB West 3 (54)	HEB East (44)	PDK Trench IV (17)	BK (62)	TK Upper (15)	TK Lower (9)	SHK (47)	EF-HR (35)	MLK (29)	Site
HK (357)		1.4	1.0	0.3	0.8	0.2	3.0	1.4	2.1	1.5	2.3	0.6	0.1	0.2	1.7	0.4	1.0	0.2	HK
TK Fish Gully (69)	0.15		1.8	0.5	1.7	0.8	2.7	0.3	3.1	0.2	3.0	1.3	1.2	1.1	1.3	0.7	0.1	0.7	TK Fish Gully
FLK Masek (22)	0.30	0.14		0.8	0.3	0.4	0.3	1.7	0.4	1.5	0.5	0.2	0.7	0.5	1.8	1.0	1.3	0.8	FLK Masek
PDK Trenches I-III (17)	0.77	0.64	0.41		0.7	0.3	1.1	0.6	1.5	0.5	1.4	0.5	0.3	0.4	1.3	0.1	0.3	0.1	PDK Trenches I-III
WK East A (42)	0.41	0.11	0.76	0.51		0.2	0.8	1.7	0.9	1.5	1.0	0.1	0.5	0.3	1.8	0.9	1.2	0.6	WK East A
WK East C (9)	0.87	0.41	0.69	0.74	0.84		0.6	1.0	0.9	0.7	0.8	0.2	0.1	0.0	1.8	0.3	0.6	0.2	WK East C
WK (146)	*0.01*	*0.00*	0.73	0.28	0.33	0.26		2.2	0.0	2.5	0.1	0.5	1.5	0.7	1.8	*1.7*	*1.9*	*1.3*	WK
HEB West 2a (35)	0.16	0.73	0.09	0.52	0.10	0.33	*0.00*		2.9	0.1	2.7	1.3	1.2	1.3	1.0	0.9	0.3	0.8	HEB West 2a
HEB West 2b (43)	*0.03*	*0.00*	0.68	0.13	0.36	0.37	0.96	*0.00*		2.5	0.1	0.6	1.6	1.1	*2.8*	2.0	*2.3*	1.5	HEB West 2b
HEB West 3 (54)	0.20	0.84	0.14	0.64	0.14	0.50	*0.01*	0.92	*0.01*		2.5	1.1	1.1	0.9	0.8	0.7	0.2	0.7	HEB West 3
HEB East (44)	*0.02*	*0.00*	0.64	0.16	0.34	0.42	0.92	*0.01*	0.90	*0.01*		0.6	1.6	1.0	2.4	1.9	2.2	1.5	HEB East
PDK Trench IV (17)	0.65	0.34	0.87	0.59	0.95	0.80	0.63	0.28	0.62	0.28	0.54		0.4	0.3	1.4	0.7	1.0	0.5	PDK Trench IV
BK (62)	0.89	0.24	0.46	0.76	0.60	0.92	0.09	0.22	0.12	0.26	0.11	0.67		*0.1*	1.6	0.4	0.8	0.2	BK
TK Upper (15)	0.71	0.29	0.57	0.65	0.74	0.99	0.15	0.22	0.26	0.21	0.18	0.79	0.85		2.2	0.4	0.8	0.3	TK Upper
TK Lower (9)	0.08	0.19	0.08	0.22	0.08	0.10	0.08	0.31	*0.01*	0.40	*0.02*	0.16	0.12	*0.04*		1.4	1.1	1.2	TK Lower
SHK (47)	0.72	0.46	0.31	0.96	0.39	0.75	*0.04*	0.38	0.05	0.46	0.06	0.50	0.71	0.58	0.15		0.5	0.1	SHK
EF-HR (35)	0.34	0.94	0.19	0.75	0.22	0.53	*0.02*	0.74	*0.02*	0.83	*0.03*	0.34	0.40	0.32	0.28	0.62		0.5	EF-HR
MLK (29)	0.84	0.56	0.43	0.93	0.53	0.82	0.20	0.43	0.13	0.51	0.14	0.61	0.81	0.71	0.22	0.95	0.64		MLK

Table 8.16 *Metrical analysis of bifaces: summary of all the statistical comparisons of the handaxe and cleaver samples. Counts of positive scores out of a possible 5 (SHK, TK Upper, TK Lower) or 7 (the rest) (for an explanation of this table, see page 202)*

Site	HK	TK Fish Gully	FLK Masek	PDK Trenches I–III	WK East A	WK East C	WK	HEB West 2a	HEB West 2b	HEB West 3	HEB East	PDK Trench IV	BK	TK Upper	TK Lower	SHK	EF-HR	MLK
HK		4	3	4	6	2	5	5	5	3	6	4	5	1	0	3	3	4
TK Fish Gully	5		0	4	4	4	6	1	4	5	6	2	6	2	0	2	0	3
FLK Masek	3	0		4	3	2	5	3	3	4	4	3	4	1	1	2	1	1
PDK Trenches I–III	4	4	4		2	2	4	4	4	4	2	5	0	0	1	1	4	4
WK East A	6	4	4			0	4	4	4	4	0	5	1	0	1	0	3	3
WK East C	2	4	3	3	0		1	2	2	2	0	2	0	0	0	0	2	2
WK	5	6	5	5	5	3		5	2	4	4	3	5	0	0	2	2	5
HEB West 2a	5	1	3	4	4	3	5		0	1	4	0	5	5	0	2	1	3
HEB West 2b	5	4	3	4	6	5	3	0		2	4	0	5	0	0	3	1	3
HEB West 3	3	5	4	5	5	2	5	3	2		4	0	5	0	0	3	3	4
HEB East	6	6	4	2	1	0	5	5	5	6		5	0	0	0	0	4	4
PDK Trench IV	4	2	3	5	5	5	4	2	1	3	5		5	1	0	3	2	3
BK	5	6	4	0	4	1	6	5	5	5	4	5		0	2	1	4	4
TK Upper	1	2	1	2	0	0	0	1	2	1	1	3	1		0	0	0	1
TK Lower	0	0	0	1	1	0	1	0	1	0	3	0	2	1		0	0	1
SHK	3	3	2	1	1	1	5	4	3	3	0	3	1	1	2		2	2
EF-HR	3	2	2	4	4	2	5	2	3	3	5	2	4	1	1	2		2
MLK	4	3	2	5	4	2	6	3	3	4	4	3	4	1	1	3	4	

Table 8.17 *Metrical analysis of bifaces: handaxes: length*

Site	N	% value of each biface	Mean	S.D.	31-40	41-50	51-60	61-70	71-80	81-90	91-100	101-110	111-120	121-130	131-140	141-150	151-160
HK	322	0.3	127.3	20.8	–	–	–	–	0.3	2.2	4.7	9.3	23.6	21.1	18.3	8.7	6.5
TK FG	69	1.4	162.4	44.9	–	–	–	–	–	4.3	5.8	2.9	1.4	4.3	14.5	5.8	13.0
FLK Masek	20	5.0	186.5	56.8	–	–	–	–	–	–	5.0	–	5.0	5.0	5.0	20.0	5.0
PDK Trenches I–III	17	5.9	72.5	17.4	–	–	29.4	29.4	17.6	5.9	5.9	5.9	5.9	–	–	–	–
WK East A	42	2.4	91.9	25.8	–	–	9.5	14.3	16.7	7.1	16.7	11.9	14.3	2.4	2.4	2.4	–
WK East C	9	11.1	107.0	29.2	–	–	–	–	22.2	11.1	22.2	11.1	11.1	–	11.1	–	–
WK	84	1.2	129.7	32.0	–	–	7.1	3.6	2.4	–	–	7.1	8.3	11.9	15.5	16.7	13.1
HEB West 2a	28	3.6	153.0	25.3	–	–	–	–	–	–	–	3.6	7.1	–	7.1	7.1	17.9	14.3
HEB West 2b	28	3.6	151.9	30.0	–	–	–	–	–	–	–	7.1	7.1	3.6	3.6	10.7	17.9	–
HEB West 3	44	2.3	138.2	28.5	–	–	–	–	2.3	–	4.5	6.8	11.4	25.0	9.1	6.8	15.9
HEB East	42	2.4	94.8	31.9	–	–	14.3	14.3	9.5	14.3	9.5	7.1	7.1	7.1	7.1	4.8	–
PDK Trench IV	11	9.1	153.8	26.2	–	–	–	–	–	–	–	9.1	–	9.1	18.2	9.1	18.2
BK	61	1.6	77.9	30.5	4.9	16.4	11.5	14.8	13.1	13.1	11.5	3.3	–	3.3	1.6	3.3	1.6
TK Upper	15	6.7	110.7	59.5	–	–	–	40.0	6.7	6.7	–	6.7	6.7	6.7	–	–	6.7
TK Lower	9	11.1	151.0	57.2	–	–	–	–	–	–	–	11.1	22.2	11.1	–	–	11.1	11.1
SHK	46	2.2	105.4	42.5	4.3	4.3	10.9	4.3	6.5	8.7	10.9	8.7	4.3	8.7	4.3	4.3	8.7
EF–HR	29	3.4	146.8	29.5	–	–	–	–	–	–	–	6.9	6.9	24.1	6.9	13.8	13.8
MLK	28	3.6	164.9	33.0	–	–	–	–	–	–	3.6	–	3.6	–	17.9	–	17.9

Table 8.18 *Metrical analysis of bifaces: handaxes: weight*

Site	N	% value of each biface	Mean	S.D.	0-100	101-200	201-300	301-400	401-500	501-600	601-700	701-800	801-900	901-1000	1001-1100	1101-1200	1201-1300	1301-1400	1401-1500
HK	308	0.3	392.7	201.8	0.3	9.1	23.1	31.8	17.9	8.4	3.2	1.6	1.0	1.6	0.6	0.6	–	–	–
TK FG	68	1.5	715.4	477.5	–	8.8	5.9	5.9	11.8	14.7	11.8	10.3	8.8	4.4	2.9	2.9	4.4	2.9	–
FLK Masek	20	5.0	1224.0	793.3	–	–	5.0	5.0	15.0	10.0	15.0	–	–	–	–	–	–	10.0	–
PDK Trenches I–III	8	12.5	170.3	128.6	25.0	50.0	–	12.5	12.5	–	–	–	–	–	–	–	–	–	–
WK East A	32	3.1	212.8	147.1	21.9	40.6	12.5	12.5	9.4	–	3.1	–	–	–	–	–	–	–	–
WK East C	9	11.1	314.2	237.9	11.1	22.2	33.3	–	11.1	11.1	–	–	11.1	–	–	–	–	–	–
WK	72	1.4	490.1	212.7	5.6	4.2	9.7	8.3	19.4	29.2	12.5	4.2	2.8	2.8	1.4	–	–	–	–
HEB West 2a	28	3.6	601.4	257.2	–	7.1	3.6	14.3	10.7	21.4	10.7	7.1	10.7	3.6	10.7	–	–	–	–
HEB West 2b	27	3.7	576.7	293.7	–	7.4	7.4	14.8	14.8	18.5	14.8	3.7	3.7	–	3.7	11.1	–	–	–
HEB West 3	42	2.4	374.6	163.7	2.4	14.3	23.8	14.3	23.8	14.3	2.4	2.4	2.4	–	–	–	–	–	–
HEB East	32	3.1	303.5	223.4	15.6	31.3	15.6	6.3	15.6	6.3	6.3	–	–	–	3.1	–	–	–	–
PDK Trench IV	11	9.1	449.8	147.0	–	–	9.1	27.3	27.3	18.2	18.2	–	–	–	–	–	–	–	–
BK	58	1.7	178.8	196.7	43.1	27.6	13.8	8.6	1.7	1.7	1.7	–	–	–	–	–	1.7	–	–
TK Upper	0	–	–	–	–	–	–	–	–	–	–	–	–	–	–	–	–	–	–
TK Lower	0	–	–	–	–	–	–	–	–	–	–	–	–	–	–	–	–	–	–
SHK	0	–	–	–	–	–	–	–	–	–	–	–	–	–	–	–	–	–	–
EF–HR	24	4.2	601.8	210.2	–	–	4.2	8.3	25.0	20.8	8.3	16.7	8.3	4.2	4.2	–	–	–	–
MLK	27	3.7	1026.4	474.7	–	–	3.7	3.7	–	3.7	14.8	22.2	–	3.7	11.1	14.8	–	–	3.7

METRICAL ANALYSIS OF HANDAXES AND CLEAVERS

Table 8.17 (*cont.*)

Site	Length (mm)														
	161–170	171–180	181–190	191–200	201–210	211–220	221–230	231–240	241–250	251–260	261–270	271–280	281–290	291–300	301–310
HK	1.2	2.5	0.3	0.6	–	–	0.6	–	–	–	–	–	–	–	–
TK FG	13.0	7.2	5.8	4.3	2.9	4.3	4.3	1.4	1.4	–	–	–	–	1.4	1.4
FLK Masek	5.0	–	5.0	10.0	5.0	5.0	–	–	5.0	–	10.0	10.0	–	–	–
PDK Trenches I–III	–	–	–	–	–	–	–	–	–	–	–	–	–	–	–
WK East A	2.4	–	–	–	–	–	–	–	–	–	–	–	–	–	–
WK East C	11.1	–	–	–	–	–	–	–	–	–	–	–	–	–	–
WK	10.7	2.4	1.2	–	–	–	–	–	–	–	–	–	–	–	–
HEB West 2a	14.3	14.3	10.7	3.6	–	–	–	–	–	–	–	–	–	–	–
HEB West 2b	28.6	7.1	–	7.1	7.1	–	–	–	–	–	–	–	–	–	–
HEB West 3	4.5	2.3	9.1	–	–	2.3	–	–	–	–	–	–	–	–	–
HEB East	4.8	–	–	–	–	–	–	–	–	–	–	–	–	–	–
PDK Trench IV	9.1	9.1	9.1	9.1	–	–	–	–	–	–	–	–	–	–	–
BK	–	1.6	–	–	–	–	–	–	–	–	–	–	–	–	–
TK Upper	6.7	–	–	6.7	–	–	–	–	–	6.7	–	–	–	–	–
TK Lower	11.1	–	–	11.1	–	–	–	–	–	–	–	11.1	–	–	–
SHK	2.2	2.2	4.3	2.2	–	–	–	–	–	–	–	–	–	–	–
EF–HR	6.9	6.9	6.9	3.4	–	–	–	3.4	–	–	–	–	–	–	–
MLK	17.9	17.9	3.6	7.1	3.6	3.6	–	–	–	–	3.6	–	–	–	–

Table 8.18 (*cont.*)

Site	Weight (g)																
	1501–1600	1601–1700	1701–1800	1801–1900	1901–2000	2001–2100	2101–2200	2201–2300	2301–2400	2401–2500	2501–2600	2601–2700	2701–2800	2801–2900	2901–3000	3001–3100	3101–3200
HK	0.3	–	0.3	–	–	–	–	–	–	–	–	–	–	–	–	–	–
TK FG	–	–	–	1.5	1.5	–	–	–	–	–	–	–	–	–	–	–	1.5
FLK Masek	5.0	5.0	–	–	5.0	5.0	–	5.0	10.0	–	–	5.0	–	–	–	–	–
PDK Trenches I–III	–	–	–	–	–	–	–	–	–	–	–	–	–	–	–	–	–
WK East A	–	–	–	–	–	–	–	–	–	–	–	–	–	–	–	–	–
WK East C	–	–	–	–	–	–	–	–	–	–	–	–	–	–	–	–	–
WK	–	–	–	–	–	–	–	–	–	–	–	–	–	–	–	–	–
HEB West 2a	–	–	–	–	–	–	–	–	–	–	–	–	–	–	–	–	–
HEB West 2b	–	–	–	–	–	–	–	–	–	–	–	–	–	–	–	–	–
HEB West 3	–	–	–	–	–	–	–	–	–	–	–	–	–	–	–	–	–
HEB East	–	–	–	–	–	–	–	–	–	–	–	–	–	–	–	–	–
PDK Trench IV	–	–	–	–	–	–	–	–	–	–	–	–	–	–	–	–	–
BK	–	–	–	–	–	–	–	–	–	–	–	–	–	–	–	–	–
TK Upper	–	–	–	–	–	–	–	–	–	–	–	–	–	–	–	–	–
TK Lower	–	–	–	–	–	–	–	–	–	–	–	–	–	–	–	–	–
SHK	–	–	–	–	–	–	–	–	–	–	–	–	–	–	–	–	–
EF–HR	–	–	–	–	–	–	–	–	–	–	–	–	–	–	–	–	–
MLK	–	7.4	3.7	3.7	–	–	–	3.7	–	–	–	–	–	–	–	–	–

Table 8.19 Metrical analysis of bifaces: handaxes: ratio Th/B

Site	N	% value of each biface	Mean	S.D.	0.201-0.250	0.251-0.300	0.301-0.350	0.351-0.400	0.401-0.450	0.451-0.500	0.501-0.550	0.551-0.600	0.601-0.650	0.651-0.700	0.701-0.750	0.751-0.800	0.801-0.850	0.851-0.900	0.901-0.950	0.951-1.000
HK	344	0.3	0.542	0.105	–	0.9	1.2	5.2	10.8	21.5	18.0	14.5	14.2	6.1	3.5	2.9	0.6	0.3	0.3	–
TK FG	71	1.4	0.580	0.115	–	1.4	–	2.8	2.8	15.5	19.7	22.5	14.1	8.5	8.5	–	–	2.8	–	1.4
FLK Masek	21	4.8	0.606	0.080	–	–	–	–	–	9.5	19.0	23.8	19.0	14.3	4.8	9.5	–	–	–	–
PDK Trenches I–III	9	11.1	0.603	0.107	–	–	–	–	11.1	–	11.1	44.4	–	11.1	11.1	11.1	–	–	–	–
WK East A	42	2.4	0.590	0.147	–	–	2.4	4.8	11.9	9.5	11.9	16.7	14.3	7.1	9.5	4.8	–	4.8	–	2.4
WK East C	9	11.1	0.601	0.177	–	–	11.1	–	–	11.1	22.2	11.1	–	33.3	–	–	–	–	–	11.1
WK	73	1.4	0.539	0.098	–	1.4	–	2.7	12.3	20.5	27.4	15.1	6.8	8.2	2.7	–	2.7	–	–	–
HEB West 2a	28	3.6	0.562	0.112	–	–	–	14.3	3.6	14.3	7.1	17.9	21.4	10.7	7.1	3.6	–	–	–	–
HEB West 2b	28	3.6	0.537	0.097	–	–	7.1	–	7.1	21.4	21.4	14.3	17.9	7.1	3.6	–	–	–	–	–
HEB West 3	48	2.1	0.520	0.089	–	–	2.1	4.2	12.5	31.3	14.6	14.6	12.5	4.2	4.2	–	–	–	–	–
HEB East	42	2.4	0.606	0.117	–	–	2.4	2.4	2.4	11.9	16.7	11.9	9.5	23.8	14.3	–	2.4	–	2.4	–
PDK Trench IV	11	9.1	0.499	0.096	–	–	–	27.3	9.1	9.1	18.2	18.2	18.2	–	–	–	–	–	–	–
BK	66	1.5	0.662	0.136	–	–	–	–	1.5	13.6	12.1	6.1	15.2	15.2	7.6	18.2	1.5	3.0	1.5	4.5
TK Upper	13	7.7	0.611	0.138	–	–	–	7.7	7.7	7.7	–	30.8	30.8	30.8	–	–	15.4	–	–	–
TK Lower	9	11.1	0.512	0.078	–	–	–	11.1	11.1	11.1	22.2	33.3	11.1	–	–	–	–	–	–	–
SHK	45	2.2	0.617	0.114	–	–	–	2.2	6.7	8.9	8.9	11.1	22.2	22.2	4.4	8.9	2.2	2.2	–	–
EF–HR	31	3.2	0.637	0.140	–	–	–	–	3.2	9.7	22.6	12.9	9.7	16.1	3.2	9.7	3.2	3.2	6.5	–
MLK	27	3.7	0.685	0.098	–	–	–	–	–	3.7	7.4	7.4	7.4	33.3	18.5	11.1	3.7	7.4	–	–

Table 8.20 *Metrical analysis of bifaces: handaxes: ratio T_1/L*

Site	N	% value of each biface	Mean	S.D.	0.076–0.100	0.101–0.125	0.126–0.150	0.151–0.175	0.176–0.200	0.201–0.225	0.226–0.250	0.251–0.275	0.276–0.300	0.301–0.325	0.326–0.350	0.351–0.375	0.376–0.400	0.401–0.425	0.426–0.450	0.451–0.475	0.476–0.500
HK	315	0.3	0.179	0.030	–	1.3	14.9	31.1	32.7	11.4	6.7	1.9	–	–	–	–	–	–	–	–	–
TK FG	66	1.5	0.172	0.056	4.5	15.2	15.2	30.3	16.7	9.1	–	3.0	–	3.0	1.5	–	1.5	–	–	–	–
FLK Masek	19	5.3	0.181	0.039	–	–	21.1	31.6	15.8	15.8	10.5	5.3	–	–	–	–	–	–	–	–	–
PDK Trenches I–III	8	12.5	0.344	0.069	–	–	–	–	–	–	12.5	12.5	–	12.5	12.5	12.5	25.0	–	–	12.5	–
WK East A	32	3.1	0.232	0.061	–	–	3.1	12.5	28.1	12.5	6.3	15.6	3.1	9.4	6.3	3.1	3.1	–	–	–	–
WK East C	9	11.1	0.230	0.075	–	11.1	–	11.1	11.1	11.1	22.2	22.2	4.2	–	–	–	11.1	–	–	–	–
WK	71	1.4	0.178	0.049	–	9.9	23.9	22.5	19.7	12.7	2.8	1.4	–	1.4	1.4	–	–	–	–	–	–
HEB West 2a	28	3.6	0.157	0.042	3.6	21.4	21.4	28.6	17.9	3.6	1.1	–	–	3.6	–	–	–	–	–	–	–
HEB West 2b	27	3.7	0.170	0.039	–	11.1	25.9	18.5	14.8	18.5	7.1	–	–	–	–	–	–	–	–	–	–
HEB West 3	42	2.4	0.151	0.044	7.1	–	21.4	26.2	2.4	4.8	6.3	4.8	6.3	3.1	9.4	3.1	6.3	3.1	–	–	–
HEB East	32	3.1	0.250	0.077	–	26.2	3.1	9.4	21.9	15.6	9.1	12.5	–	–	–	–	–	–	–	–	–
PDK Trench IV	11	9.1	0.145	0.048	18.2	9.1	45.5	–	18.2	–	–	–	–	–	–	–	–	–	–	–	–
BK	58	1.7	0.302	0.104	–	6.9	–	5.2	5.2	13.8	5.2	3.4	12.1	6.9	10.3	6.9	3.4	1.7	6.9	6.9	5.2
TK Upper	0	–	–	–	–	–	–	–	–	–	–	–	–	–	–	–	–	–	–	–	–
TK Lower	0	–	–	–	–	–	–	–	–	–	–	–	–	–	–	–	–	–	–	–	–
SHK	0	–	–	–	–	–	–	–	–	–	–	–	–	–	–	–	–	–	–	–	–
EF–HR	23	4.3	0.193	0.062	–	13.0	21.7	8.7	17.4	8.7	4.3	21.7	–	4.3	–	–	–	–	–	–	–
MLK	27	3.7	0.247	0.068	–	–	7.4	3.7	14.8	14.8	14.8	14.8	14.8	3.7	–	3.7	3.7	3.7	–	–	–

Table 8.21 Metrical analysis of bifaces: handaxes: ratio B/L

Site	N	% value of each biface	Mean	S.D.	0.251-0.300	0.301-0.350	0.351-0.400	0.401-0.450	0.451-0.500	0.501-0.550	0.551-0.600	0.601-0.650	0.651-0.700	0.701-0.750	0.751-0.800	0.801-0.850	0.851-0.900	0.901-0.950	0.951-1.000
HK	322	0.3	0.600	0.066	–	–	–	0.3	4.0	18.9	31.1	23.9	15.5	3.4	2.5	0.3	–	–	–
TK FG	69	1.4	0.545	0.080	–	–	1.4	5.8	30.4	18.8	23.2	10.1	5.8	1.4	2.9	–	–	–	–
FLK Masek	20	5.0	0.570	0.092	–	–	–	5.0	30.0	15.0	10.0	15.0	15.0	10.0	–	–	–	–	–
PDK Trenches I–III	17	5.9	0.742	0.103	–	–	–	–	–	–	5.9	17.6	11.8	11.8	29.4	5.9	11.8	–	5.9
WK East A	42	2.4	0.648	0.102	–	–	–	2.4	7.1	4.8	21.4	16.7	16.7	19.0	4.8	4.8	–	2.4	–
WK East C	9	11.1	0.648	0.120	–	–	–	11.1	–	–	22.2	22.2	22.2	11.1	–	–	11.1	–	–
WK	84	1.2	0.644	0.092	–	–	–	–	4.8	7.1	16.7	34.5	15.5	10.7	4.8	1.2	3.6	–	1.2
HEB West 2a	28	3.6	0.579	0.069	–	–	–	–	3.6	35.7	28.6	25.0	3.6	–	–	3.6	–	–	–
HEB West 2b	28	3.6	0.585	0.081	–	–	–	–	21.4	14.3	32.1	17.9	3.6	3.6	7.1	–	–	–	–
HEB West 3	44	2.3	0.582	0.103	–	–	2.3	2.3	9.1	34.1	22.7	6.8	4.5	13.6	–	2.3	2.3	–	–
HEB East	42	2.4	0.709	0.109	–	–	–	–	2.4	7.1	9.5	14.3	7.1	16.7	23.8	9.5	9.5	–	–
PDK Trench IV	11	9.1	0.536	0.097	–	9.1	–	–	–	54.5	18.2	9.1	9.1	–	–	–	–	–	–
BK	61	1.6	0.718	0.126	–	–	–	–	3.3	8.2	3.3	16.4	9.8	26.2	11.5	8.2	3.3	3.3	6.6
TK Upper	15	6.7	0.624	0.125	–	–	–	13.3	–	20.0	13.3	6.7	26.7	–	6.7	13.3	–	–	–
TK Lower	9	11.1	0.670	0.169	–	–	–	–	11.1	22.2	–	22.2	–	11.1	11.1	–	11.1	11.1	–
SHK	46	2.2	0.703	0.121	–	–	–	–	2.2	10.9	10.9	8.7	13.0	13.0	19.6	10.9	8.7	2.2	–
EF–HR	29	3.4	0.582	0.078	–	–	–	3.4	13.8	17.2	31.0	17.2	6.9	10.3	–	–	–	–	–
MLK	28	3.6	0.580	0.093	–	–	–	10.7	7.1	25.0	21.4	7.1	17.9	7.1	–	3.6	–	–	–

Table 8.22 *Metrical analysis of bifaces: handaxes: ratio B_1/B_2*

Site	N	% value of each biface	Mean	S.D.	Ratio B_1/B_2																	
					0.101–0.200	0.201–0.300	0.301–0.400	0.401–0.500	0.501–0.600	0.601–0.700	0.701–0.800	0.801–0.900	0.901–1.000	1.001–1.100	1.101–1.200	1.201–1.300	1.301–1.400	1.401–1.500	1.501–1.600	1.601–1.700	1.701–1.800	1.801–1.900	1.901–2.000	2.001–2.100
HK	321	0.3	0.735	0.129	–	–	0.3	1.2	13.1	28.3	29.0	17.8	6.9	2.8	0.3	0.3	–	–	–	–	–	–	–	–
TK FG	69	1.4	0.696	0.121	–	–	–	4.3	15.9	33.3	23.2	18.8	4.3	–	–	–	–	–	–	–	–	–	–	–
FLK Masek	20	5.0	0.736	0.106	–	–	–	–	5.0	40.0	40.0	5.0	10.0	–	–	–	–	–	–	–	–	–	–	–
PDK Trenches I–III	17	5.9	0.754	0.185	–	–	5.9	5.9	5.9	17.6	11.8	29.4	17.6	5.9	–	–	–	–	–	–	–	–	–	–
WK East A	42	2.4	0.713	0.159	–	–	–	7.1	19.0	23.8	23.8	19.0	2.4	2.4	–	2.4	–	–	–	–	–	–	–	–
WK East C	9	11.1	0.635	0.141	–	–	11.1	11.1	11.1	33.3	22.2	11.1	–	–	–	–	–	–	–	–	–	–	–	–
WK	84	1.2	0.723	0.130	–	–	1.2	2.4	17.9	16.7	33.3	19.0	8.3	1.2	–	–	–	–	–	–	–	–	–	–
HEB West 2a	28	3.6	0.636	0.117	–	–	–	3.6	35.7	46.4	3.6	3.6	7.1	–	–	–	–	–	–	–	–	–	–	–
HEB West 2b	28	3.6	0.722	0.194	–	–	–	10.7	14.3	25.0	21.4	21.4	–	–	3.6	–	3.6	–	–	–	–	–	–	–
HEB West 3	44	2.3	0.629	0.144	–	–	–	22.7	22.7	31.8	6.8	11.4	4.5	4.8	–	–	–	–	–	–	–	–	–	–
HEB East	42	2.4	0.746	0.149	–	–	–	2.4	16.7	26.2	23.8	11.9	14.3	4.8	–	–	–	–	–	–	–	–	–	–
PDK Trench IV	11	9.1	0.670	0.142	–	–	–	9.1	45.5	9.1	18.2	9.1	9.1	–	–	–	–	–	–	–	–	–	–	–
BK	61	1.6	0.755	0.203	–	–	1.6	8.2	14.8	18.0	19.7	11.5	14.8	8.2	1.6	–	–	–	1.6	–	–	–	–	–
TK Upper	15	6.7	0.763	0.151	–	–	–	–	20.0	13.3	33.3	13.3	13.3	6.7	–	–	–	–	–	–	–	–	–	–
TK Lower	9	11.1	0.677	0.193	–	–	–	22.2	11.1	33.3	11.1	11.1	–	11.1	–	–	–	–	–	–	–	–	–	–
SHK	46	2.2	0.778	0.178	–	–	–	13.0	8.7	8.7	17.4	23.9	17.4	10.9	–	–	–	–	–	–	–	–	–	–
EF–HR	29	3.4	0.676	0.150	–	–	–	13.8	20.7	17.2	31.0	10.3	3.4	3.4	–	–	–	–	–	–	–	–	–	–
MLK	28	3.6	0.720	0.174	–	–	–	3.6	21.4	28.6	21.4	10.7	3.6	3.6	7.1	–	–	–	–	–	–	–	–	–

Table 8.23 *Metrical analysis of bifaces: handaxes: ratio L_1/L*

Site	N	% value of each biface	Mean	S.D.	0.001–0.050	0.051–0.100	0.101–0.150	0.151–0.200	0.201–0.250	0.251–0.300	0.301–0.350	0.351–0.400
HK	319	0.3	0.394	0.066	–	–	–	–	0.9	6.0	17.9	28.8
TK FG	69	1.4	0.391	0.072	–	–	–	–	1.4	11.6	13.0	29.0
FLK Masek	20	5.0	0.428	0.106	–	–	–	–	–	15.0	5.0	20.0
PDK Trenches I–III	17	5.9	0.401	0.087	–	–	–	–	–	23.5	5.9	23.5
WK East A	42	2.4	0.418	0.094	–	–	–	2.4	2.4	9.5	11.9	2.4
WK East C	9	11.1	0.411	0.053	–	–	–	–	–	11.1	–	33.3
WK	84	1.2	0.382	0.068	–	–	–	1.2	2.4	8.3	20.2	28.6
HEB West 2a	28	3.6	0.379	0.069	–	–	–	–	–	14.3	21.4	28.6
HEB West 2b	28	3.6	0.410	0.063	–	–	–	–	3.6	3.6	7.1	25.0
HEB West 3	44	2.3	0.362	0.074	–	–	–	–	4.5	9.1	36.4	25.0
HEB East	42	2.4	0.423	0.066	–	–	–	–	2.4	2.4	4.8	19.0
PDK Trench IV	11	9.1	0.381	0.077	–	–	–	–	–	9.1	27.3	27.3
BK	61	1.6	0.409	0.094	–	–	–	–	1.6	11.5	14.8	19.7
TK Upper	15	6.7	0.412	0.049	–	–	–	–	–	–	13.3	20.0
TK Lower	9	11.1	0.357	0.075	–	–	–	–	11.1	11.1	11.1	33.3
SHK	46	2.2	0.401	0.087	–	–	–	–	4.3	8.7	17.4	19.6
EF–HR	29	3.4	0.380	0.084	–	–	–	–	3.4	10.3	24.1	27.6
MLK	28	3.6	0.393	0.088	–	–	–	3.6	–	7.1	21.4	21.4

Table 8.24 *Metrical analysis of bifaces: handaxes: t-values and estimates of significance: length (for explanation of tables 8.24–8.30 see p. 154)*

Site	HK (322)	TK Fish Gully (69)	FLK Masek (20)	PDK Trenches I–III (17)	WK East A (42)	WK East C (9)	WK (84)	HEB West 2a (28)	HEB West 2b (28)	HEB West 3 (44)	HEB East (42)	PDK Trench IV (11)	BK (61)	TK Upper (15)	TK Lower (9)	SHK (46)	EF–HR (29)	MLK (28)	Site
HK (322)		10.0	10.6	10.7	10.1	2.9	0.8	6.2	5.8	3.1	8.9	4.1	15.7	2.6	3.1	5.7	4.7	8.7	HK
TK Fish Gully (69)	0.00		2.0	8.1	9.3	3.6	5.3	1.0	1.1	3.2	8.5	0.6	12.4	3.8	0.7	6.8	1.7	0.3	TK Fish Gully
FLK Masek (20)	0.00	0.05		7.9	9.1	3.9	6.0	2.8	2.7	4.5	8.1	1.8	10.9	3.8	1.6	6.4	3.2	1.7	FLK Masek
PDK Trenches I–III (17)	0.00	0.00	0.00		2.9	3.8	7.1	11.6	9.9	8.9	2.7	9.9	0.7	2.5	5.3	3.1	9.4	10.7	PDK Trenches I–III
WK East A (42)	0.00	0.00	0.00	0.01		1.6	6.6	9.8	8.9	7.9	0.5	7.1	2.4	1.7	4.9	1.8	8.3	10.4	WK East A
WK East C (9)	0.00	0.00	0.00	0.00	0.13		2.0	4.6	3.9	3.0	1.1	3.8	2.7	0.2	2.1	0.1	3.5	4.7	WK East C
WK (84)	0.51	0.00	0.00	0.00	0.00	0.04		3.5	3.2	1.5	5.8	2.4	9.8	1.8	1.7	3.7	2.5	5.0	WK
HEB West 2a (28)	0.00	0.19	0.02	0.00	0.00	0.00	0.00		0.1	2.2	8.1	0.1	11.3	3.3	0.1	5.4	0.8	1.5	HEB West 2a
HEB West 2b (28)	0.00	0.18	0.02	0.00	0.00	0.00	0.00	0.89		1.9	7.5	0.2	10.7	3.0	0.1	5.1	0.6	1.5	HEB West 2b
HEB West 3 (44)	0.02	0.00	0.00	0.00	0.00	0.00	0.14	0.03	0.06		6.7	1.6	10.3	2.4	1.0	4.3	1.2	3.6	HEB West 3
HEB East (42)	0.00	0.00	0.00	0.00	0.65	0.30	0.00	0.00	0.00	0.00		5.6	2.7	1.3	4.1	1.3	7.0	8.9	HEB East
PDK Trench IV (11)	0.00	0.54	0.04	0.00	0.00	0.00	0.02	0.93	0.85	0.11	0.00		7.7	2.2	0.1	3.6	0.7	1.0	PDK Trench IV
BK (61)	0.00	0.00	0.00	0.35	0.02	0.01	0.00	0.00	0.00	0.00	0.01	0.00		3.0	5.9	3.9	10.1	12.2	BK
TK Upper (15)	0.30	0.00	0.00	0.03	0.25	0.84	0.25	0.02	0.02	0.10	0.34	0.02	0.06		1.6	0.4	2.7	3.9	TK Upper
TK Lower (9)	0.25	0.49	0.13	0.00	0.02	0.06	0.30	0.92	0.97	0.53	0.02	0.89	0.01	0.12		2.8	0.3	0.9	TK Lower
SHK (46)	0.00	0.00	0.00	0.00	0.07	0.92	0.00	0.00	0.00	0.00	0.19	0.00	0.00	0.71	0.01		4.6	6.3	SHK
EF–HR (29)	0.00	0.05	0.01	0.00	0.00	0.00	0.01	0.40	0.52	0.22	0.00	0.49	0.00	0.04	0.84	0.00		2.2	EF–HR
MLK (28)	0.00	0.79	0.14	0.00	0.00	0.00	0.00	0.13	0.13	0.00	0.00	0.32	0.00	0.00	0.50	0.00	0.03		MLK

METRICAL ANALYSIS OF HANDAXES AND CLEAVERS

Table 8.23 (cont.)

Site	Ratio L_1/L												
	0.401–0.450	0.451–0.500	0.501–0.550	0.551–0.600	0.601–0.650	0.651–0.700	0.701–0.750	0.751–0.800	0.801–0.850	0.851–0.900	0.901–0.950	0.951–1.000	
HK	27.9	13.5	3.8	0.9	–	0.3	–	–	–	–	–	–	
TK FG	24.6	11.6	7.2	1.4	–	–	–	–	–	–	–	–	
FLK Masek	25.0	15.0	5.0	10.0	–	5.0	–	–	–	–	–	–	
PDK Trenches I–III	11.8	23.5	5.9	5.9	–	–	–	–	–	–	–	–	
WK East A	26.2	31.0	9.5	2.4	2.4	–	–	–	–	–	–	–	
WK East C	33.3	22.2	–	–	–	–	–	–	–	–	–	–	
WK	23.8	11.9	2.4	1.2	–	–	–	–	–	–	–	–	
HEB West 2a	17.9	14.3	3.6	–	–	–	–	–	–	–	–	–	
HEB West 2b	32.1	25.0	3.6	–	–	–	–	–	–	–	–	–	
HEB West 3	15.9	4.5	2.3	–	2.3	–	–	–	–	–	–	–	
HEB East	31.0	28.6	11.9	–	–	–	–	–	–	–	–	–	
PDK Trench IV	18.2	9.1	9.1	–	–	–	–	–	–	–	–	–	
BK	23.0	11.5	6.6	11.5	–	–	–	–	–	–	–	–	
TK Upper	53.3	6.7	6.7	–	–	–	–	–	–	–	–	–	
TK Lower	33.3	–	–	–	–	–	–	–	–	–	–	–	
SHK	15.2	26.1	6.5	–	2.2	–	–	–	–	–	–	–	
EF–HR	17.2	6.9	10.3	–	–	–	–	–	–	–	–	–	
MLK	14.3	17.9	14.3	–	–	–	–	–	–	–	–	–	

Table 8.25 *Metrical analysis of bifaces: handaxes: t-values and estimates of significance: weight*

Site	HK (308)	TK Fish Gully (68)	FLK Masek (20)	PDK Trenches I–III (8)	WK East A (32)	WK East C (9)	WK (72)	HEB West 2a (28)	HEB West 2b (27)	HEB West 3 (42)	HEB East (32)	PDK Trench IV (11)	BK (58)	EF–HR (24)	MLK (27)	Site
HK (308)		8.8	13.2	3.1	4.9	1.1	3.6	5.1	4.4	0.6	2.4	0.9	7.4	4.9	13.4	HK
TK Fish Gully (68)	0.00		3.6	3.2	5.8	2.5	3.6	12	14	4.5	4.6	1.8	8.0	1.1	2.9	TK Fish Gully
FLK Masek (20)	0.00	0.01		3.7	7.1	3.3	7.1	3.9	3.9	6.7	6.2	3.2	9.3	3.7	1.1	FLK Masek
PDK Trenches I–III (8)	0.00	0.00	0.00		0.7	1.5	4.2	4.5	3.8	3.3	1.6	4.3	0.1	5.4	5.0	PDK Trenches I–III
WK East A (32)	0.00	0.00	0.00	0.46		1.6	6.7	7.3	6.2	4.4	1.9	4.6	0.9	8.2	9.2	WK East A
WK East C (9)	0.25	0.00	0.00	0.15	0.12		2.3	3.0	2.4	0.9	0.1	1.6	1.9	3.4	4.3	WK East C
WK (72)	0.00	0.00	0.00	0.00	0.00	0.02		2.2	1.6	3.0	4.1	0.6	8.6	2.2	7.8	WK
HEB West 2a (28)	0.00	0.14	0.00	0.00	0.00	0.01	0.03		0.3	4.5	4.8	1.8	8.4	0.0	4.1	HEB West 2a
HEB West 2b (27)	0.00	0.09	0.00	0.00	0.00	0.02	0.17	0.74		3.7	4.1	1.4	7.4	0.3	4.2	HEB West 2b
HEB West 3 (42)	0.58	0.00	0.00	0.00	0.00	0.36	0.00	0.00	0.00		1.6	1.4	5.3	4.9	8.2	HEB West 3
HEB East (32)	0.02	0.00	0.00	0.12	0.06	0.90	0.00	0.00	0.00	0.12		2.0	2.7	5.1	7.7	HEB East
PDK Trench IV (11)	0.35	0.00	0.00	0.00	0.00	0.14	0.55	0.08	0.09	0.17	0.05		4.3	2.2	3.9	PDK Trench IV
BK (58)	0.00	0.00	0.00	0.91	0.40	0.07	0.00	0.00	0.00	0.00	0.01	0.00		8.7	11.7	BK
EF–HR (24)	0.00	0.12	0.00	0.00	0.00	0.00	0.03	0.99	0.73	0.00	0.00	0.04	0.00		4.0	EF–HR
MLK (27)	0.00	0.01	0.33	0.00	0.00	0.00	0.00	0.00	0.00	0.00	0.00	0.00	0.00	0.00		MLK

Table 8.26 Metrical analysis of bifaces: handaxes: t-values and estimates of significance: ratio Th/B

Site	HK (344)	TK Fish Gully (71)	FLK Masek (21)	PDK Trenches I-III (9)	WK East A (42)	WK East C (9)	WK (73)	HEB West 2a (28)	HEB West 2b (28)	HEB West 3 (48)	HEB East (42)	PDK Trench IV (11)	BK (66)	TK Upper (13)	TK Lower (9)	SHK (45)	EF-HR (31)	MLK (27)	Site
HK (344)		*2.8*	*2.7*	*1.7*	*1.6*	*2.7*	0.2	1.0	0.2	1.3	*3.7*	1.3	*8.1*	*2.3*	0.8	*4.4*	*4.7*	*6.8*	HK
TK Fish Gully (71)	0.01		1.0	0.6	0.5	0.4	*2.3*	0.7	1.8	*3.1*	1.1	*2.2*	*3.8*	0.9	1.7	1.7	*2.1*	*4.2*	TK Fish Gully
FLK Masek (21)	0.01	0.34		0.1	0.1	0.5	*2.9*	*1.5*	*2.7*	*3.8*	0.0	*3.4*	*1.8*	0.1	*3.0*	0.4	0.9	*3.0*	FLK Masek
PDK Trenches I-III (9)	0.08	0.57	0.94		0.3	0.0	*1.8*	1.0	*1.8*	*2.5*	0.1	*2.3*	1.2	0.1	*2.1*	0.3	0.7	*2.1*	PDK Trenches I-III
WK East A (42)	0.04	0.69	0.58	0.80		0.2	*2.2*	0.9	*1.7*	*2.8*	0.5	*2.0*	*2.6*	0.5	1.5	0.9	1.4	*3.0*	WK East A
WK East C (9)	0.34	0.63	0.94	0.98	0.84		*1.6*	0.8	*1.4*	*2.1*	0.1	1.7	1.2	0.1	1.4	0.3	0.6	*1.8*	WK East C
WK (73)	0.83	0.02	0.01	0.07	0.05	0.33		1.0	0.1	1.0	*3.3*	1.3	*6.1*	*2.3*	0.8	*3.9*	*4.1*	*6.6*	WK
HEB West 2a (28)	0.34	0.46	0.13	0.33	0.39	0.43	0.32		0.9	*1.8*	1.6	1.6	*3.4*	1.2	1.2	*2.0*	*2.3*	*4.3*	HEB West 2a
HEB West 2b (28)	0.82	0.08	0.01	0.09	0.07	0.32	0.93	0.38		0.8	*2.6*	1.1	*4.4*	*2.0*	0.7	*3.1*	*3.2*	*5.7*	HEB West 2b
HEB West 3 (48)	0.18	0.00	0.00	0.02	0.01	0.22	0.30	0.08	0.45		*3.9*	0.7	*6.3*	*2.9*	0.3	*4.6*	*4.5*	*7.4*	HEB West 3
HEB East (42)	0.00	0.27	0.99	0.96	0.60	0.93	0.00	0.12	0.01	0.00		*2.8*	*2.2*	0.1	*2.3*	0.4	1.0	*2.9*	HEB East
PDK Trench IV (11)	0.18	0.03	0.00	0.03	0.03	0.12	0.21	0.11	0.27	0.01	0.01		*3.8*	*2.3*	0.3	*3.2*	*3.0*	*5.3*	PDK Trench IV
BK (66)	0.00	0.00	0.00	0.22	0.01	0.23	0.00	0.02	0.00	0.00	0.03	0.00		1.2	*3.2*	1.8	0.8	0.8	BK
TK Upper (13)	0.02	0.39	0.90	0.89	0.65	0.88	0.02	0.23	0.05	0.02	0.04	0.03	0.23		*2.0*	0.1	0.6	*2.0*	TK Upper
TK Lower (9)	0.40	0.09	0.01	0.05	0.13	0.19	0.43	0.23	0.48	0.23	0.78	0.03	0.75	0.00		*2.6*	*2.6*	*4.8*	TK Lower
SHK (45)	0.00	0.10	0.70	0.75	0.35	0.74	0.00	0.05	0.00	0.00	0.66	0.03	0.00	0.00	0.01		0.7	*2.6*	SHK
EF-HR (31)	0.00	0.04	0.32	0.51	0.18	0.53	0.00	0.03	0.00	0.00	0.31	0.00	0.40	0.07	0.01	0.50		1.5	EF-HR
MLK (27)	0.00	0.00	0.00	0.04	0.00	0.21	0.00	0.00	0.00	0.00	0.00	0.00	0.42	0.06	0.00	0.01	0.14		MLK

Table 8.27 Metrical analysis of bifaces: handaxes: t-values and estimates of significance: ratio T_1/L

Site	HK (315)	TK Fish Gully (66)	FLK Masek (19)	PDK Trenches I-III (8)	WK East A (32)	WK East C (9)	WK (71)	HEB West 2a (28)	HEB West 2b (27)	HEB West 3 (42)	HEB East (32)	PDK Trench IV (11)	BK (58)	EF-HR (23)	MLK (27)	Site
HK (315)		*1.5*	*0.2*	*14.7*	*8.5*	*4.8*	*0.3*	*3.7*	*1.6*	*5.4*	*10.4*	*3.6*	*17.5*	*1.9*	*9.8*	HK
TK Fish Gully (66)	0.30		*0.7*	*8.0*	*4.9*	*2.8*	*0.7*	*1.3*	*0.2*	*2.0*	*5.8*	*1.5*	*8.8*	*1.5*	*5.5*	TK Fish Gully
FLK Masek (19)	0.81	0.51		*7.8*	*3.3*	*2.3*	*0.3*	*2.0*	*1.0*	*2.5*	*3.6*	*2.2*	*4.9*	*0.7*	*3.8*	FLK Masek
PDK Trenches I-III (8)	0.00	0.00	0.00		*4.5*	*3.2*	*8.7*	*9.6*	*9.2*	*10.3*	*3.1*	*7.4*	*1.1*	*5.8*	*3.5*	PDK Trenches I-III
WK East A (32)	0.00	0.00	0.00	0.00		*0.1*	*4.8*	*5.5*	*4.6*	*6.7*	*1.0*	*4.3*	*3.5*	*2.4*	*0.9*	WK East A
WK East C (9)	0.08	0.01	0.09	0.01	0.94		*2.8*	*3.7*	*3.2*	*4.3*	*0.7*	*3.1*	*2.0*	*1.5*	*0.6*	WK East C
WK (71)	0.82	0.50	0.80	0.00	0.00	0.01		*2.0*	*0.8*	*2.9*	*5.8*	*2.0*	*8.9*	*1.2*	*5.6*	WK
HEB West 2a (28)	0.01	0.21	0.05	0.00	0.00	0.02	0.05		*1.2*	*0.5*	*5.7*	*0.7*	*7.1*	*2.5*	*5.9*	HEB West 2a
HEB West 2b (27)	0.22	0.83	0.34	0.00	0.00	0.04	0.44	0.25		*1.8*	*4.9*	*1.6*	*6.4*	*1.6*	*5.1*	HEB West 2b
HEB West 3 (42)	0.00	0.04	0.01	0.00	0.00	0.01	0.00	0.59	0.08		*7.0*	*0.4*	*8.8*	*3.2*	*7.1*	HEB West 3
HEB East (32)	0.00	0.00	0.00	0.00	0.31	0.50	0.00	0.00	0.00	0.00		*4.2*	*2.5*	*3.0*	*0.2*	HEB East
PDK Trench IV (11)	0.04	0.14	0.04	0.00	0.00	0.01	0.04	0.46	0.11	0.00	0.00		*4.9*	*2.2*	*4.5*	PDK Trench IV
BK (58)	0.00	0.00	0.00	0.00	0.00	0.05	0.00	0.00	0.13	0.00	0.02	0.00		*4.7*	*2.5*	BK
EF-HR (23)	0.31	0.13	0.47	0.00	0.02	0.15	0.24	0.02	0.00	0.00	0.00	0.03	0.00		*2.9*	EF-HR
MLK (27)	0.00	0.00	0.00	0.00	0.39	0.55	0.00	0.00	0.00	0.00	0.85	0.00	0.00	0.01		MLK

Table 8.28 Metrical analysis of bifaces: handaxes: t-values and estimates of significance: ratio B/L

Site	HK (322)	TK Fish Gully (69)	FLK Masek (20)	PDK Trenches I-III (17)	WK East A (42)	WK East C (9)	WK (84)	HEB West 2a (28)	HEB West 2b (28)	HEB West 3 (44)	HEB East (42)	PDK Trench IV (11)	BK (61)	TK Upper (15)	TK Lower (9)	SHK (46)	EF-HR (29)	MLK (28)
HK (322)		6.0	1.9	8.4	4.2	2.1	5.0	1.6	1.1	1.6	9.3	3.1	10.8	1.3	3.0	8.7	1.4	1.5
TK Fish Gully (69)	0.00		1.2	8.6	5.9	3.4	7.0	1.9	2.2	2.1	9.1	0.3	9.4	3.1	3.8	8.4	2.1	1.8
FLK Masek (20)	0.18	0.24		5.3	2.9	1.9	3.2	0.4	0.6	0.4	4.9	1.0	4.8	1.5	2.1	4.4	0.5	0.4
PDK Trenches I-III (17)	0.00	0.00	0.00		3.2	2.1	3.9	6.4	5.7	5.4	1.1	5.3	0.7	2.9	1.4	1.2	5.9	5.4
WK East A (42)	0.00	0.00	0.01	0.00		0.0	0.3	3.2	2.8	3.0	2.6	3.3	3.0	0.8	0.5	2.3	2.9	2.8
WK East C (9)	0.27	0.00	0.07	0.05	0.98		0.1	2.1	1.8	1.7	1.5	2.3	1.6	0.5	0.3	1.2	1.9	1.8
WK (84)	0.00	0.00	0.00	0.00	0.80	0.91		3.4	3.0	3.5	3.5	3.6	4.1	0.7	0.7	3.1	3.2	3.2
HEB West 2a (28)	0.11	0.06	0.72	0.00	0.00	0.14	0.00		0.3	0.1	5.6	1.5	5.5	1.5	2.3	4.9	0.2	0.1
HEB West 2b (28)	0.27	0.03	0.55	0.00	0.01	0.08	0.00	0.74		0.2	5.1	1.6	5.1	1.2	2.0	4.5	0.1	0.2
HEB West 3 (44)	0.26	0.04	0.68	0.00	0.00	0.09	0.00	0.89	0.87		5.6	1.3	5.9	1.3	2.1	5.1	0.0	0.1
HEB East (42)	0.00	0.00	0.00	0.29	0.01	0.14	0.00	0.00	0.00	0.00		4.8	0.4	2.5	0.9	0.3	5.4	5.1
PDK Trench IV (11)	0.06	0.74	0.34	0.00	0.00	0.03	0.00	0.13	0.12	0.19	0.00		4.5	1.9	2.2	4.2	1.6	1.3
BK (61)	0.00	0.00	0.00	0.48	0.00	0.12	0.00	0.00	0.00	0.00	0.71	0.00		2.6	1.0	0.6	5.3	5.2
TK Upper (15)	0.47	0.03	0.16	0.01	0.45	0.65	0.46	0.21	0.30	0.20	0.02	0.07	0.01		0.8	2.2	1.4	1.3
TK Lower (9)	0.25	0.06	0.13	0.19	0.72	0.75	0.66	0.16	0.18	0.17	0.38	0.04	0.31	0.45		0.7	2.2	2.0
SHK (46)	0.00	0.00	0.00	0.24	0.03	0.22	0.01	0.00	0.00	0.00	0.80	0.00	0.53	0.03	0.49		4.8	4.6
EF-HR (29)	0.17	0.04	0.63	0.00	0.00	0.06	0.00	0.86	0.88	0.98	0.00	0.13	0.00	0.26	0.17	0.00		0.1
MLK (28)	0.28	0.07	0.73	0.00	0.01	0.09	0.00	0.96	0.82	0.94	0.00	0.20	0.00	0.20	0.16	0.00	0.92	

Table 8.29 Metrical analysis of bifaces: handaxes: t-values and estimates of significance: ratio B_1/B_2

Site	HK (321)	TK Fish Gully (69)	FLK Masek (20)	PDK Trenches I–III (17)	WK East A (42)	WK East C (9)	WK (84)	HEB West 2a (28)	HEB West 2b (28)	HEB West 3 (44)	HEB East (42)	PDK Trench IV (11)	BK (61)	TK Upper (15)	TK Lower (9)	SHK (46)	EF–HR (29)	MLK (28)	Site
HK (321)		2.3	0.0	0.6	1.0	2.3	0.7	3.9	0.5	5.0	0.5	1.6	1.0	0.8	1.3	2.0	2.3	0.6	HK
TK Fish Gully (69)	0.02		1.3	1.6	0.6	1.4	1.3	2.2	0.8	2.7	1.9	0.6	2.0	1.9	0.4	2.9	0.7	0.8	TK Fish Gully
FLK Masek (20)	0.96	0.19		0.4	0.6	2.1	0.4	3.0	0.3	3.0	0.3	1.5	0.4	0.6	1.1	1.0	1.5	0.4	FLK Masek
PDK Trenches I–III (17)	0.67	0.23	0.72		0.9	1.7	0.8	2.6	0.6	2.8	0.2	1.3	0.0	0.1	1.0	0.5	1.6	0.6	PDK Trenches I–III
WK East A (42)	0.32	0.55	0.56	0.39		1.4	0.4	2.2	0.2	2.6	1.0	0.8	1.1	1.1	0.6	1.8	1.0	0.2	WK East A
WK East C (9)	0.02	0.17	0.04	0.11	0.18		1.9	0.0	1.2	0.1	2.0	0.5	1.7	2.1	0.5	2.3	0.7	1.3	WK East C
WK (84)	0.46	0.19	0.68	0.51	0.71	0.06		3.2	0.0	3.7	0.9	1.3	1.2	1.1	1.0	2.0	1.6	0.1	WK
HEB West 2a (28)	0.00	0.03	0.00	0.03	0.03	1.00	0.00		2.0	0.2	3.3	0.8	2.9	3.1	0.8	3.8	1.1	2.1	HEB West 2a
HEB West 2b (28)	0.74	0.51	0.75	0.58	0.83	0.22	0.98	0.05		2.3	0.6	0.8	0.7	0.7	0.6	1.3	1.0	0.0	HEB West 2b
HEB West 3 (44)	0.00	0.01	0.00	0.01	0.01	0.90	0.00	0.84	0.02		3.7	0.8	3.5	3.1	0.9	4.4	1.3	2.4	HEB West 3
HEB East (42)	0.59	0.06	0.79	0.86	0.33	0.05	0.37	0.00	0.56	0.00		1.5	0.2	0.4	1.2	0.9	1.9	0.7	HEB East
PDK Trench IV (11)	0.10	0.52	0.15	0.21	0.42	0.59	0.21	0.44	0.42	0.40	0.13		1.3	1.6	0.1	1.9	0.1	0.8	PDK Trench IV
BK (61)	0.45	0.05	0.59	0.99	0.26	0.09	0.28	0.00	0.47	0.00	0.80	0.19		0.1	1.1	0.6	1.9	0.8	BK
TK Upper (15)	0.40	0.07	0.53	0.88	0.29	0.05	0.28	0.00	0.48	0.00	0.71	0.12	0.88		1.2	0.3	1.8	0.8	TK Upper
TK Lower (9)	0.40	0.78	0.41	0.33	0.55	0.61	0.34	0.56	0.54	0.40	0.23	0.93	0.28	0.23		1.5	0.0	0.6	TK Lower
SHK (46)	0.12	0.01	0.24	0.65	0.08	0.03	0.07	0.00	0.21	0.00	0.37	0.07	0.55	0.78	0.13		2.6	1.4	SHK
EF–HR (29)	0.02	0.49	0.13	0.12	0.33	0.48	0.11	0.26	0.32	0.18	0.06	0.91	0.07	0.08	0.99	0.01		1.0	EF–HR
MLK (28)	0.66	0.51	0.69	0.53	0.87	0.20	0.93	0.04	0.96	0.02	0.50	0.40	0.43	0.42	0.53	0.17	0.31		MLK
	HK	TK Fish Gully	FLK Masek	PDK Trenches I–III	WK East A	WK East C	WK	HEB West 2a	HEB West 2b	HEB West 3	HEB East	PDK Trench IV	BK	TK Upper	TK Lower	SHK	EF–HR	MLK	

185

Table 8.30 Metrical analysis of bifaces: handaxes: t-values and estimates of significance: ratio L_1/L

Site	HK (319)	TK Fish Gully (69)	FLK Masek (20)	PDK Trenches I-III (17)	WK East A (42)	WK East C (9)	WK (84)	HEB West 2a (28)	HEB West 2b (28)	HEB West 3 (44)	HEB East (42)	PDK Trench IV (11)	BK (61)	TK Upper (15)	TK Lower (9)	SHK (46)	EF-HR (29)	MLK (28)	Site
HK (319)		0.3	2.2	0.4	2.1	0.8	1.4	1.2	1.3	2.9	2.7	0.7	1.5	1.0	1.7	0.7	1.1	0.1	HK
TK Fish Gully (69)	0.74		1.8	0.5	1.7	0.8	0.8	0.8	1.2	2.0	2.4	0.4	1.2	1.1	1.3	0.7	0.7	0.1	TK Fish Gully
FLK Masek (20)	0.17	0.15		0.9	0.4	0.4	2.4	2.0	0.7	2.9	0.2	1.3	0.8	0.6	1.8	1.1	1.8	1.3	FLK Masek
PDK Trenches I-III (17)	0.69	0.64	0.40		0.7	0.3	1.0	0.9	0.4	1.7	1.1	0.6	0.3	0.4	1.3	0.0	0.8	0.3	PDK Trenches I-III
WK East A (42)	0.11	0.11	0.71	0.51		0.2	2.4	1.9	0.4	3.1	0.3	1.2	0.5	0.3	1.8	0.9	1.8	1.1	WK East A
WK East C (9)	0.43	0.41	0.66	0.74	0.84		1.2	1.3	0.1	1.9	0.5	1.0	0.1	0.0	1.8	0.3	1.1	0.6	WK East C
WK (84)	0.15	0.45	0.08	0.34	0.03	0.22		0.2	1.9	1.5	3.2	0.1	1.9	1.6	1.0	1.4	0.2	0.7	WK
HEB West 2a (28)	0.24	0.45	0.08	0.36	0.06	0.20	0.81		1.8	0.9	2.7	0.1	1.5	1.6	0.8	1.1	0.0	0.7	HEB West 2a
HEB West 2b (28)	0.21	0.22	0.50	0.68	0.66	0.95	0.06	0.08		2.8	0.8	1.2	0.1	0.1	2.1	0.5	1.5	0.8	HEB West 2b
HEB West 3 (44)	0.00	0.04	0.01	0.09	0.00	0.07	0.13	0.35	0.01		4.0	0.7	2.7	2.4	0.2	2.3	0.9	1.6	HEB West 3
HEB East (42)	0.01	0.02	0.86	0.28	0.77	0.61	0.00	0.01	0.40	0.00		1.9	0.9	0.6	2.7	1.4	2.5	1.7	HEB East
PDK Trench IV (11)	0.51	0.66	0.20	0.54	0.23	0.32	0.94	0.95	0.23	0.47	0.07		0.9	1.3	0.7	0.7	0.0	0.4	PDK Trench IV
BK (61)	0.25	0.24	0.43	0.76	0.61	0.93	0.07	0.14	0.93	0.01	0.34	0.36		0.1	1.6	0.4	1.4	0.7	BK
TK Upper (15)	0.30	0.29	0.55	0.65	0.74	0.99	0.11	0.11	0.93	0.02	0.53	0.22	0.85		2.2	0.5	1.4	0.8	TK Upper
TK Lower (9)	0.10	0.19	0.08	0.22	0.08	0.10	0.30	0.43	0.04	0.85	0.01	0.50	0.12	0.04		1.4	0.7	1.1	TK Lower
SHK (46)	0.60	0.50	0.28	0.98	0.38	0.73	0.18	0.25	0.64	0.03	0.18	0.48	0.68	0.56	0.16		1.0	0.4	SHK
EF-HR (29)	0.28	0.50	0.08	0.43	0.08	0.29	0.87	0.96	0.13	0.35	0.02	0.98	0.17	0.12	0.48	0.30		0.6	EF-HR
MLK (28)	0.96	0.90	0.22	0.78	0.27	0.56	0.50	0.50	0.41	0.11	0.10	0.68	0.47	0.38	0.28	0.71	0.56		MLK

Table 8.31 *Metrical analysis of bifaces: summary of all the statistical comparisons of the handaxe samples. Counts of positive scores out of a possible 5 (SHK, TK Upper, TK Lower) or 7 (the rest) (for an explanation of this table, see p. 202)*

Site	HK	TK Fish Gully	FLK Masek	PDK Trenches I–III	WK East A	WK East C	WK	HEB West 2a	HEB West 2b	HEB West 3	HEB East	PDK Trench IV	BK	TK Upper	TK Lower	SHK	EF-HR	MLK	Site
HK		3	2	4	4	0	2	3	2	2	4	1	5	0	0	3	3	4	HK
TK Fish Gully	5		0	4	4	3	3	0	0	3	4	1	5	1	0	2	0	2	TK Fish Gully
FLK Masek	3	2		4	3	2	3	2	1	3	4	2	4	1	0	2	1	2	FLK Masek
PDK Trenches I–III	4	4	4		2	2	4	4	4	4	2	4	0	0	1	1	4	4	PDK Trenches I–III
WK East A	5	4	4	3		0	3	4	3	5	0	4	1	0	1	0	2	3	WK East A
WK East C	2	4	3	3	0		0	1	1	1	0	2	0	0	0	0	2	2	WK East C
WK	2	4	4	4	5	3		3	1	3	6	1	5	0	0	2	2	5	WK
HEB West 2a	4	1	3	5	5	3	5		0	1	5	0	6	1	0	3	0	3	HEB West 2a
HEB West 2b	2	1	3	4	4	3	2	1		1	4	0	5	0	0	3	1	3	HEB West 2b
HEB West 3	4	7	6	6	7	2	4	2	3		6	0	6	1	0	4	3	4	HEB West 3
HEB East	6	5	4	2	1	1	6	6	5	6		3	0	0	0	0	3	3	HEB East
PDK Trench IV	2	2	4	5	4	3	3	0	0	0	4		5	0	0	3	1	3	PDK Trench IV
BK	5	5	5	0	4	1	5	6	5	7	4	5		0	2	1	4	3	BK
TK Upper	1	2	1	2	0	0	1	2	1	3	1	2	1		0	0	0	1	TK Upper
TK Lower	0	0	1	1	1	0	0	0	1	0	3	1	2	1		0	0	1	TK Lower
SHK	3	3	2	1	1	1	3	4	3	5	5	3	3	1	2		2	2	SHK
EF-HR	4	3	2	4	4	2	4	2	1	3	5	3	4	1	1	3		1	EF-HR
MLK	4	3	2	5	4	2	5	4	3	5	4	3	4	1	1	3	3		MLK
Site	HK	TK Fish Gully	FLK Masek	PDK Trenches I–III	WK East A	WK East C	WK	HEB West 2a	HEB West 2b	HEB West 3	HEB East	PDK Trench IV	BK	TK Upper	TK Lower	SHK	EF-HR	MLK	Site

Table 8.32 *Metrical analysis of bifaces: cleavers: length*

Site	N	% value of each biface	Mean	S.D.	Length (mm)												
					31–40	41–50	51–60	61–70	71–80	81–90	91–100	101–110	111–120	121–130	131–140	141–150	151–160
HK	38	2.6	132.7	23.4	–	–	–	–	–	–	5.3	7.9	15.8	15.8	23.7	15.8	5.3
TK FG	0	–	–	–	–	–	–	–	–	–	–	–	–	–	–	–	–
FLK Masek	2	50.0	128.0	14.1	–	–	–	–	–	–	–	–	50.0	–	50.0	–	–
PDK Trenches I–III	0	–	–	–	–	–	–	–	–	–	–	–	–	–	–	–	–
WK East A	0	–	–	–	–	–	–	–	–	–	–	–	–	–	–	–	–
WK East C	0	–	–	–	–	–	–	–	–	–	–	–	–	–	–	–	–
WK	62	1.6	138.4	14.5	–	–	–	–	–	–	–	–	16.1	12.9	22.6	25.8	19.4
HEB West 2a	7	14.3	156.1	26.6	–	–	–	–	–	–	–	–	–	14.3	28.6	14.3	–
HEB West 2b	15	6.7	165.3	25.2	–	–	–	–	–	–	–	–	6.7	6.7	6.7	–	20.0
HEB West 3	10	10.0	136.6	16.8	–	–	–	–	–	–	–	10.0	10.0	–	40.0	30.0	–
HEB East	2	50.0	94.0	15.6	–	–	–	–	–	50.0	–	50.0	–	–	–	–	–
PDK Trench IV	7	14.3	165.6	19.6	–	–	–	–	–	–	–	–	–	–	14.3	14.3	–
BK	1	100.0	–	–	–	–	–	–	–	–	–	–	–	–	100.0	–	–
TK Upper	0	–	–	–	–	–	–	–	–	–	–	–	–	–	–	–	–
TK Lower	0	–	–	–	–	–	–	–	–	–	–	–	–	–	–	–	–
SHK	1	100.0	–	–	–	–	–	–	–	–	–	–	100.0	–	–	–	–
EF–HR	6	16.7	145.8	13.6	–	–	–	–	–	–	–	–	–	–	16.7	–	50.0	16.7
MLK	1	100.0	–	–	–	–	–	–	–	–	–	–	–	–	–	–	–

Table 8.33 *Metrical analysis of bifaces: cleavers: weight*

Site	N	% value of each biface	Mean	S.D.	Weight (g)														
					0–100	101–200	201–300	301–400	401–500	501–600	601–700	701–800	801–900	901–1000	1001–1100	1101–1200	1201–1300	1301–1400	1401–1500
HK	37	2.7	513.4	263.0	–	–	24.3	8.1	32.4	13.5	8.1	–	2.7	5.4	2.7	–	–	–	–
TK FG	0	–	–	–	–	–	–	–	–	–	–	–	–	–	–	–	–	–	–
FLK Masek	2	50.0	505.0	77.8	–	–	–	–	50.0	50.0	–	–	–	–	–	–	–	–	–
PDK Trenches I–III	0	–	–	–	–	–	–	–	–	–	–	–	–	–	–	–	–	–	–
WK East A	0	–	–	–	–	–	–	–	–	–	–	–	–	–	–	–	–	–	–
WK East C	0	–	–	–	–	–	–	–	–	–	–	–	–	–	–	–	–	–	–
WK	61	1.6	597.8	140.6	–	–	–	1.6	27.9	24.6	18.0	21.3	4.9	–	1.6	–	–	–	–
HEB West 2a	7	14.3	700.6	326.2	–	–	–	14.3	28.6	–	14.3	14.3	–	14.3	–	–	14.3	–	–
HEB West 2b	15	6.7	689.3	200.8	–	–	6.7	6.7	6.7	6.7	6.7	33.3	26.7	6.7	–	–	–	–	–
HEB West 3	10	10.0	439.2	151.9	–	–	10.0	30.0	40.0	10.0	–	–	10.0	–	–	–	–	–	–
HEB East	2	50.0	123.5	16.3	–	100.0	–	–	–	–	–	–	–	–	–	–	–	–	–
PDK Trench IV	7	14.3	541.1	144.3	–	–	–	28.6	–	28.6	42.9	–	–	–	–	–	–	–	–
BK	1	100.0	–	–	–	–	–	–	–	100.0	–	–	–	–	–	–	–	–	–
TK Upper	0	–	–	–	–	–	–	–	–	–	–	–	–	–	–	–	–	–	–
TK Lower	0	–	–	–	–	–	–	–	–	–	–	–	–	–	–	–	–	–	–
SHK	0	–	–	–	–	–	–	–	–	–	–	–	–	–	–	–	–	–	–
EF–HR	5	20.0	555.8	129.5	–	–	–	20.0	20.0	–	60.0	–	–	–	–	–	–	–	–
MLK	1	100.0	–	–	–	–	–	–	–	–	–	–	–	–	–	–	–	–	–

METRICAL ANALYSIS OF HANDAXES AND CLEAVERS

Table 8.32 (*cont.*)

Site	\multicolumn{15}{c}{Length (mm)}
	161–170	171–180	181–190	191–200	201–210	211–220	221–230	231–240	241–250	251–260	261–270	271–280	281–290	291–300	301–310
HK	2.6	–	5.3	2.6	–	–	–	–	–	–	–	–	–	–	–
TK FG	–	–	–	–	–	–	–	–	–	–	–	–	–	–	–
FLK Masek	–	–	–	–	–	–	–	–	–	–	–	–	–	–	–
PDK Trenches I–III	–	–	–	–	–	–	–	–	–	–	–	–	–	–	–
WK East A	–	–	–	–	–	–	–	–	–	–	–	–	–	–	–
WK East C	–	–	–	–	–	–	–	–	–	–	–	–	–	–	–
WK	1.6	1.6	–	–	–	–	–	–	–	–	–	–	–	–	–
HEB West 2a	14.3	–	14.3	14.3	–	–	–	–	–	–	–	–	–	–	–
HEB West 2b	13.3	6.7	33.3	–	6.7	–	–	–	–	–	–	–	–	–	–
HEB West 3	10.0	–	–	–	–	–	–	–	–	–	–	–	–	–	–
HEB East	–	–	–	–	–	–	–	–	–	–	–	–	–	–	–
PDK Trench IV	14.3	42.9	14.3	–	–	–	–	–	–	–	–	–	–	–	–
BK	–	–	–	–	–	–	–	–	–	–	–	–	–	–	–
TK Upper	–	–	–	–	–	–	–	–	–	–	–	–	–	–	–
TK Lower	–	–	–	–	–	–	–	–	–	–	–	–	–	–	–
SHK	–	–	–	–	–	–	–	–	–	–	–	–	–	–	–
EF–HR	16.7	–	–	–	–	–	–	–	–	–	–	–	–	–	–
MLK	–	–	–	–	–	–	–	–	100.0	–	–	–	–	–	–

Table 8.33 (*cont.*)

Site	\multicolumn{17}{c}{Weight (g)}
	1501–1600	1601–1700	1701–1800	1801–1900	1901–2000	2001–2100	2101–2200	2201–2300	2301–2400	2401–2500	2501–2600	2601–2700	2701–2800	2801–2900	2901–3000	3001–3100	3101–3200
HK	2.7	–	–	–	–	–	–	–	–	–	–	–	–	–	–	–	–
TK FG	–	–	–	–	–	–	–	–	–	–	–	–	–	–	–	–	–
FLK Masek	–	–	–	–	–	–	–	–	–	–	–	–	–	–	–	–	–
PDK Trenches I–III	–	–	–	–	–	–	–	–	–	–	–	–	–	–	–	–	–
WK East A	–	–	–	–	–	–	–	–	–	–	–	–	–	–	–	–	–
WK East C	–	–	–	–	–	–	–	–	–	–	–	–	–	–	–	–	–
WK	–	–	–	–	–	–	–	–	–	–	–	–	–	–	–	–	–
HEB West 2a	–	–	–	–	–	–	–	–	–	–	–	–	–	–	–	–	–
HEB West 2b	–	–	–	–	–	–	–	–	–	–	–	–	–	–	–	–	–
HEB West 3	–	–	–	–	–	–	–	–	–	–	–	–	–	–	–	–	–
HEB East	–	–	–	–	–	–	–	–	–	–	–	–	–	–	–	–	–
PDK Trench IV	–	–	–	–	–	–	–	–	–	–	–	–	–	–	–	–	–
BK	–	–	–	–	–	–	–	–	–	–	–	–	–	–	–	–	–
TK Upper	–	–	–	–	–	–	–	–	–	–	–	–	–	–	–	–	–
TK Lower	–	–	–	–	–	–	–	–	–	–	–	–	–	–	–	–	–
SHK	–	–	–	–	–	–	–	–	–	–	–	–	–	–	–	–	–
EF–HR	–	–	–	–	–	–	–	–	–	–	–	–	–	–	–	–	–
MLK	–	–	–	–	–	–	–	–	100.0	–	–	–	–	–	–	–	–

Table 8.34 Metrical analysis of bifaces: cleavers: ratio Th/B

Site	N	% value of each biface	Mean	S.D.	0.201-0.250	0.251-0.300	0.301-0.350	0.351-0.400	0.401-0.450	0.451-0.500	0.501-0.550	0.551-0.600	0.601-0.650	0.651-0.700	0.701-0.750	0.751-0.800	0.801-0.850	0.851-0.900	0.901-0.950	0.951-1.000
HK	38	2.6	0.552	0.092	–	–	–	5.3	7.9	18.4	15.8	23.7	21.1	2.6	2.6	–	2.6	–	–	–
TK FG	0	–	–	–	–	–	–	–	–	–	–	–	–	–	–	–	–	–	–	–
FLK Masek	2	50.0	0.673	0.052	–	–	–	–	–	–	–	–	50.0	–	50.0	–	–	–	–	–
PDK Trenches I–III	0	–	–	–	–	–	–	–	–	–	–	–	–	–	–	–	–	–	–	–
WK East A	0	–	–	–	–	–	–	–	–	–	–	–	–	–	–	–	–	–	–	–
WK East C	0	–	–	–	–	–	–	–	–	–	–	–	–	–	–	–	–	–	–	–
WK	60	1.7	0.517	0.075	–	–	1.7	1.7	8.3	41.7	16.7	13.3	10.0	5.0	–	1.7	–	–	–	–
HEB West 2a	7	14.3	0.560	0.087	–	–	–	–	–	28.6	28.6	14.3	14.3	–	14.3	–	–	–	–	–
HEB West 2b	15	6.7	0.506	0.124	–	–	13.3	–	13.3	20.0	33.3	6.7	6.7	–	–	–	6.7	–	–	–
HEB West 3	10	10.0	0.476	0.083	–	–	10.0	10.0	20.0	10.0	30.0	20.0	–	–	–	–	–	–	–	–
HEB East	2	50.0	0.440	0.051	–	–	–	–	50.0	50.0	–	–	–	–	–	–	–	–	–	–
PDK Trench IV	7	14.3	0.453	0.069	–	–	–	14.3	42.9	28.6	–	14.3	–	–	–	–	–	–	–	–
BK	1	100.0	–	–	–	–	–	–	100.0	–	–	–	–	–	–	–	–	–	–	–
TK Upper	0	–	–	–	–	–	–	–	–	–	–	–	–	–	–	–	–	–	–	–
TK Lower	0	–	–	–	–	–	–	–	–	–	–	–	–	–	–	–	–	–	–	–
SHK	1	100.0	–	–	–	–	–	–	–	–	–	–	–	100.0	–	–	–	–	–	–
EF-HR	6	16.7	0.542	0.113	–	–	–	–	16.7	33.3	16.7	16.7	–	–	16.7	–	–	–	–	–
MLK	1	100.0	–	–	–	–	–	–	–	–	–	–	100.0	–	–	–	–	–	–	–

Table 8.35 Metrical analysis of bifaces. cleavers: ratio T_1/L

Site	N	% value of each biface	Mean	S.D.	Ratio T_1/L														
					0.076–0.100	0.101–0.125	0.126–0.150	0.151–0.175	0.176–0.200	0.201–0.225	0.226–0.250	0.251–0.275	0.276–0.300	0.301–0.325	0.326–0.350	0.351–0.375	0.376–0.400	0.401–0.425	0.426–0.450	0.451–0.475	0.476–0.500
HK	38	2.6	0.181	0.034	–	2.6	21.1	18.4	26.3	26.3	2.6	–	2.6	–	–	–	–	–	–	–	–
TK FG	0	–	–	–	–	–	–	–	–	–	–	–	–	–	–	–	–	–	–	–	–
FLK Masek	2	50.0	0.221	0.025	–	–	–	–	–	50.0	50.0	–	–	–	–	–	–	–	–	–	–
PDK Trenches I–III	0	–	–	–	–	–	–	–	–	–	–	–	–	–	–	–	–	–	–	–	–
WK East A	0	–	–	–	–	–	–	–	–	–	–	–	–	–	–	–	–	–	–	–	–
WK East C	0	–	–	–	–	–	–	–	–	–	–	–	–	–	–	–	–	–	–	–	–
WK	61	1.6	0.176	0.025	–	1.6	13.1	36.1	34.4	11.5	3.3	–	–	–	–	–	–	–	–	–	–
HEB West 2a	7	14.3	0.142	0.031	–	42.9	–	42.9	14.3	–	–	–	–	–	–	–	–	–	–	–	–
HEB West 2b	15	6.7	0.151	0.025	–	6.7	46.7	26.7	20.0	–	–	–	–	–	–	–	–	–	–	–	–
HEB West 3	10	10.0	0.153	0.029	–	20.0	30.0	20.0	30.0	–	–	–	–	–	–	–	–	–	–	–	–
HEB East	2	50.0	0.168	0.035	–	–	50.0	–	50.0	–	–	–	–	–	–	–	–	–	–	–	–
PDK Trench IV	6	16.7	0.140	0.034	16.7	16.7	16.7	33.3	16.7	–	–	–	–	–	–	–	–	–	–	–	–
BK	1	100.0	–	–	–	–	–	100.0	–	–	–	–	–	–	–	–	–	–	–	–	–
TK Upper	0	–	–	–	–	–	–	–	–	–	–	–	–	–	–	–	–	–	–	–	–
TK Lower	0	–	–	–	–	–	–	–	–	–	–	–	–	–	–	–	–	–	–	–	–
SHK	0	–	–	–	–	–	–	–	–	–	–	–	–	–	–	–	–	–	–	–	–
EF–HR	5	20.0	0.192	0.028	–	–	–	20.0	60.0	–	20.0	–	–	–	–	–	–	–	–	–	–
MLK	1	100.0	–	–	–	–	–	100.0	–	–	–	–	–	–	–	–	–	–	–	–	–

Table 8.36 *Metrical analysis of bifaces: cleavers: ratio B/L*

Site	N	% value of each biface	Mean	S.D.	0.251-0.300	0.301-0.350	0.351-0.400	0.401-0.450	0.451-0.500	0.501-0.550	0.551-0.600	0.601-0.650	0.651-0.700	0.701-0.750	0.751-0.800	0.801-0.850	0.851-0.900	0.901-0.950	0.951-1.000
HK	38	2.6	0.623	0.067	–	–	–	–	2.6	18.4	15.8	21.1	26.3	13.2	2.6	–	–	–	–
TK FG	0	–	–	–	–	–	–	–	–	–	–	–	–	–	–	–	–	–	–
FLK Masek	2	50.0	0.641	0.005	–	–	–	–	–	–	–	100.0	–	–	–	–	–	–	–
PDK Trenches I–III	0	–	–	–	–	–	–	–	–	–	–	–	–	–	–	–	–	–	–
WK East A	0	–	–	–	–	–	–	–	–	–	–	–	–	–	–	–	–	–	–
WK East C	0	–	–	–	–	–	–	–	–	–	–	–	–	–	–	–	–	–	–
WK	62	1.6	0.666	0.068	–	–	–	–	–	–	21.0	27.4	17.7	21.0	11.3	–	1.6	–	–
HEB West 2a	7	14.3	0.580	0.050	–	–	–	–	14.3	14.3	42.9	28.6	–	–	–	–	–	–	–
HEB West 2b	15	6.7	0.570	0.071	–	–	–	6.7	6.7	26.7	33.3	6.7	20.0	–	–	–	–	–	–
HEB West 3	10	10.0	0.645	0.082	–	–	–	–	–	–	40.0	10.0	40.0	–	–	10.0	–	–	–
HEB East	2	50.0	0.651	0.153	–	–	–	–	–	50.0	–	–	–	–	50.0	–	–	–	–
PDK Trench IV	7	14.3	0.544	0.055	–	–	–	–	28.6	28.6	28.6	14.3	–	–	–	–	–	–	–
BK	1	100.0	–	–	–	–	–	–	–	–	–	–	100.0	–	–	–	–	–	–
TK Upper	0	–	–	–	–	–	–	–	–	–	–	–	–	–	–	–	–	–	–
TK Lower	0	–	–	–	–	–	–	–	–	–	–	–	–	–	–	–	–	–	–
SHK	1	100.0	–	–	–	–	–	–	–	–	–	–	–	100.0	–	–	–	–	–
EF–HR	6	16.7	0.586	0.042	–	–	–	–	–	33.3	16.7	50.0	–	–	–	–	–	–	–
MLK	1	100.0	–	–	–	–	–	–	–	100.0	–	–	–	–	–	–	–	–	–

Table 8.37 Metrical analysis of bifaces: cleavers: ratio B_1/B_2

Site	N	% value of each biface	Mean	S.D.	Ratio B_1/B_2 0.101-0.200	0.201-0.300	0.301-0.400	0.401-0.500	0.501-0.600	0.601-0.700	0.701-0.800	0.801-0.900	0.901-1.000	1.001-1.100	1.101-1.200	1.201-1.300	1.301-1.400	1.401-1.500	1.501-1.600	1.601-1.700	1.701-1.800	1.801-1.900	1.901-2.000	2.001-2.100
HK	38	2.6	1.041	0.205	–	–	–	–	–	2.6	10.5	13.2	18.4	18.4	18.4	10.5	2.6	2.6	–	2.6	–	–	–	–
TK FG	0	–	–	–	–	–	–	–	–	–	–	–	–	–	–	–	–	–	–	–	–	–	–
FLK Masek	2	50.0	0.847	0.054	–	–	–	–	–	–	–	100.0	–	–	–	–	–	–	–	–	–	–	–
PDK Trenches I–III	0	–	–	–	–	–	–	–	–	–	–	–	–	–	–	–	–	–	–	–	–	–	–
WK East A	0	–	–	–	–	–	–	–	–	–	–	–	–	–	–	–	–	–	–	–	–	–	–
WK East C	0	–	–	–	–	–	–	–	–	–	–	–	–	–	–	–	–	–	–	–	–	–	–
WK	62	1.6	1.061	0.234	–	–	–	–	–	1.6	3.2	21.0	19.4	25.8	8.1	9.7	6.5	1.6	–	1.6	1.6	–	–	1.6
HEB West 2a	7	14.3	0.929	0.169	–	–	–	–	–	14.3	14.3	14.3	28.6	14.3	14.3	–	–	–	–	–	–	–	–	–
HEB West 2b	15	6.7	0.954	0.173	–	–	–	–	–	6.7	13.3	26.7	13.3	13.3	20.0	6.7	–	–	–	–	–	–	–	–
HEB West 3	10	10.0	1.086	0.178	–	–	–	–	–	–	10.0	–	20.0	30.0	–	30.0	10.0	–	–	–	–	–	–	–
HEB East	2	50.0	1.616	0.361	–	–	–	–	–	–	–	–	–	–	–	–	50.0	–	–	–	–	50.0	–	–
PDK Trench IV	6	16.7	1.154	0.268	–	–	–	–	–	–	–	16.7	–	33.3	16.7	16.7	–	–	–	16.7	–	–	–	–
BK	1	100.0	–	–	–	–	–	–	–	–	–	–	–	100.0	–	–	–	–	–	–	–	–	–
TK Upper	0	–	–	–	–	–	–	–	–	–	–	–	–	–	–	–	–	–	–	–	–	–	–
TK Lower	0	–	–	–	–	–	–	–	–	–	–	–	–	–	–	–	–	–	–	–	–	–	–
SHK	1	100.0	–	–	–	–	–	–	–	–	–	–	100.0	–	–	–	–	–	–	–	–	–	–
EF-HR	6	16.7	1.018	0.141	–	–	–	–	–	–	16.7	–	16.7	50.0	–	16.7	–	–	–	–	–	–	–	–
MLK	1	100.0	–	–	–	–	–	–	–	–	–	–	–	100.0	–	–	–	–	–	–	–	–	–

Table 8.38 *Metrical analysis of bifaces: cleavers: ratio L_1/L*

		% value of each biface			Ratio L_1/L							
Site	N		Mean	S.D.	0.001–0.050	0.051–0.100	0.101–0.150	0.151–0.200	0.201–0.250	0.251–0.300	0.301–0.350	0.351–0.400
HK	38	2.6	0.514	0.139	–	–	–	–	–	2.6	7.9	15.8
TK FG	0	–	–	–	–	–	–	–	–	–	–	–
FLK Masek	2	50.0	0.405	0.050	–	–	–	–	–	–	–	50.0
PDK Trenches I–III	0	–	–	–	–	–	–	–	–	–	–	–
WK East A	0	–	–	–	–	–	–	–	–	–	–	–
WK East C	0	–	–	–	–	–	–	–	–	–	–	–
WK	62	1.6	0.509	0.158	–	–	–	–	–	–	11.3	19.4
HEB West 2a	7	14.3	0.413	0.090	–	–	–	–	–	14.3	14.3	14.3
HEB West 2b	15	6.7	0.483	0.074	–	–	–	–	–	–	–	13.3
HEB West 3	10	10.0	0.498	0.141	–	–	–	–	–	–	10.0	20.0
HEB East	2	50.0	0.738	0.115	–	–	–	–	–	–	–	–
PDK Trench IV	6	16.7	0.492	0.151	–	–	–	–	–	–	16.7	–
BK	1	100.0	–	–	–	–	–	–	–	–	–	–
TK Upper	0	–	–	–	–	–	–	–	–	–	–	–
TK Lower	0	–	–	–	–	–	–	–	–	–	–	–
SHK	1	100.0	–	–	–	–	–	–	–	–	–	–
EF–HR	6	16.7	0.452	0.088	–	–	–	–	–	–	16.7	–
MLK	1	100.0	–	–	–	–	–	–	–	–	–	–

Table 8.39 *Metrical analysis of bifaces: cleavers: cleaver edge angle*

		% value of each biface			C.E. angle (degrees)											
Site	N		Mean	S.D.	31–35	36–40	41–45	46–50	51–55	56–60	61–65	66–70	71–75	76–80	81–85	86–90
HK	38	2.6	74.6	11.0	–	–	–	–	–	18.4	7.9	7.9	10.5	18.4	15.8	21.1
TK FG	0	–	–	–	–	–	–	–	–	–	–	–	–	–	–	–
FLK Masek	2	50.0	66.0	2.8	–	–	–	–	–	50.0	50.0	–	–	–	–	–
PDK Trenches I–III	0	–	–	–	–	–	–	–	–	–	–	–	–	–	–	–
WK East A	0	–	–	–	–	–	–	–	–	–	–	–	–	–	–	–
WK East C	0	–	–	–	–	–	–	–	–	–	–	–	–	–	–	–
WK	62	1.6	76.7	9.3	–	–	–	1.6	1.6	1.6	4.8	16.1	19.4	21.0	14.5	19.4
HEB West 2a	7	14.3	69.4	6.7	–	–	–	–	–	–	14.3	14.3	28.6	14.3	28.6	–
HEB West 2b	15	6.7	70.6	13.1	–	–	6.7	–	6.7	–	13.3	26.7	6.7	13.3	13.3	13.3
HEB West 3	10	10.0	72.6	8.6	–	–	–	–	–	10.0	20.0	10.0	10.0	30.0	20.0	–
HEB East	2	50.0	73.5	6.4	–	–	–	–	–	–	–	50.0	–	50.0	–	–
PDK Trench IV	6	16.7	80.8	8.0	–	–	–	–	–	–	–	16.7	16.7	16.7	16.7	33.3
BK	1	100.0	–	–	–	–	–	–	–	–	–	–	–	–	–	100.0
TK Upper	0	–	–	–	–	–	–	–	–	–	–	–	–	–	–	–
TK Lower	0	–	–	–	–	–	–	–	–	–	–	–	–	–	–	–
SHK	1	100.0	–	–	–	–	–	–	–	–	–	–	–	100.0	–	–
EF–HR	6	16.7	77.0	14.2	–	–	–	–	–	16.7	16.7	–	–	–	16.7	50.0
MLK	1	100.0	–	–	–	–	–	–	100.0	–	–	–	–	–	–	–

METRICAL ANALYSIS OF HANDAXES AND CLEAVERS

Table 8.38 (cont.)

	\multicolumn{11}{c}{Ratio L_1/L}
	0.401–0.450	0.451–0.500	0.501–0.550	0.551–0.600	0.601–0.650	0.651–0.700	0.701–0.750	0.751–0.800	0.801–0.850	0.851–0.900	0.901–0.950	0.951–1.000
FG	13.2	15.8	7.9	7.9	7.9	13.2	–	5.3	2.6	–	–	–
Masek	50.0	–	–	–	–	–	–	–	–	–	–	–
Trenches I–III	–	–	–	–	–	–	–	–	–	–	–	–
East A	–	–	–	–	–	–	–	–	–	–	–	–
East C	–	–	–	–	–	–	–	–	–	–	–	–
	16.1	12.9	6.5	6.5	9.7	3.2	4.8	1.6	3.2	1.6	3.2	–
West 2a	–	42.9	14.3	–	–	–	–	–	–	–	–	–
West 2b	20.0	26.7	13.3	20.0	6.7	–	–	–	–	–	–	–
West 3	10.0	10.0	20.0	20.0	–	–	–	–	10.0	–	–	–
East	–	–	–	–	–	50.0	–	–	50.0	–	–	–
Trench IV	50.0	–	–	16.7	–	–	16.7	–	–	–	–	–
	100.0	–	–	–	–	–	–	–	–	–	–	–
Upper	–	–	–	–	–	–	–	–	–	–	–	–
Lower	–	–	–	–	–	–	–	–	–	–	–	–
	100.0	–	–	–	–	–	–	–	–	–	–	–
HR	50.0	–	16.7	16.7	–	–	–	–	–	–	–	–
K	–	–	–	–	–	100.0	–	–	–	–	–	–

Table 8.40 *Metrical analysis of bifaces: cleavers: ratio CEL/B*

	N	% value of each biface	Mean	S.D.	0.501–0.550	0.551–0.600	0.601–0.650	0.651–0.700	0.701–0.750	0.751–0.800	0.801–0.850	0.851–0.900	0.901–0.950	0.951–1.000	1.001–1.050	1.051–1.100	1.101–1.150	1.151–1.200
	38	2.6	0.755	0.140	5.3	10.5	10.5	15.8	5.3	15.8	7.9	13.2	7.9	2.6	5.3	–	–	–
FG	0	–	–	–	–	–	–	–	–	–	–	–	–	–	–	–	–	–
Masek	2	50.0	0.798	0.254	–	–	50.0	–	–	–	–	–	–	50.0	–	–	–	–
Trenches I–III	0	–	–	–	–	–	–	–	–	–	–	–	–	–	–	–	–	–
East A	0	–	–	–	–	–	–	–	–	–	–	–	–	–	–	–	–	–
East C	0	–	–	–	–	–	–	–	–	–	–	–	–	–	–	–	–	–
	62	1.6	0.805	0.147	6.5	8.1	3.2	4.8	9.7	14.5	12.9	11.3	9.7	11.3	6.5	1.6	–	–
West 2a	7	14.3	0.782	0.191	14.3	14.3	–	14.3	–	–	14.3	14.3	–	14.3	14.3	–	–	–
West 2b	15	6.7	0.739	0.206	20.0	13.3	6.7	13.3	6.7	6.7	6.7	6.7	–	6.7	6.7	–	–	6.7
West 3	10	10.0	0.856	0.144	–	–	20.0	–	–	20.0	–	–	20.0	40.0	–	–	–	–
East	2	50.0	0.939	0.087	–	–	–	–	–	–	–	50.0	–	50.0	–	–	–	–
Trench IV	6	16.7	0.803	0.142	–	16.7	–	–	16.7	16.7	16.7	–	16.7	16.7	–	–	–	–
	1	100.0	–	–	–	–	–	–	–	–	100.0	–	–	–	–	–	–	–
Upper	0	–	–	–	–	–	–	–	–	–	–	–	–	–	–	–	–	–
Lower	0	–	–	–	–	–	–	–	–	–	–	–	–	–	–	–	–	–
	1	100.0	–	–	–	–	–	100.0	–	–	–	–	–	–	–	–	–	–
HR	6	16.7	0.737	0.109	–	16.7	16.7	–	16.7	33.3	–	16.7	–	–	–	–	–	–
K	1	100.0	–	–	–	–	–	–	–	–	–	–	–	–	–	–	–	100.0

Table 8.41 *Metrical analysis of bifaces: cleavers: t-values and estimates of significance: length (for an explanation of Tables 8.41–8.49 see p. 154)*

Site	HK (38)	FLK Masek (2)	WK (62)	HEB West 2a (7)	HEB West 2b (15)	HEB West 3 (10)	HEB East (2)	PDK Trench IV (7)	EF–HR (6)	Site
HK (38)		0.3	1.5	2.4	4.5	0.5	2.3	3.5	1.3	HK
FLK Masek (2)	0.78		1.0	1.4	2.0	0.7	2.3	2.5	1.6	FLK Masek
WK (62)	0.19	0.32		2.8	5.5	0.4	4.3	4.5	1.2	WK
HEB West 2a (7)	0.02	0.21	0.13		0.8	1.9	3.1	0.8	0.9	HEB West 2a
HEB West 2b (15)	0.00	0.06	0.00	0.45		3.1	3.8	0.0	1.8	HEB West 2b
HEB West 3 (10)	0.63	0.52	0.72	0.08	0.00		3.3	3.3	1.1	HEB West 3
HEB East (2)	0.03	0.15	0.00	0.02	0.00	0.01		4.7	4.6	HEB East
PDK Trench IV (7)	0.00	0.04	0.00	0.46	0.98	0.01	0.00		2.1	PDK Trench IV
EF–HR (6)	0.19	0.16	0.23	0.41	0.09	0.27	0.00	0.06		EF–HR

Table 8.42 *Metrical analysis of bifaces: cleavers: t-values and estimates of significance: weight*

Site	HK (37)	FLK Masek (2)	WK (61)	HEB West 2a (7)	HEB West 2b (15)	HEB West 3 (10)	HEB East (2)	PDK Trench IV (7)	EF–HR (5)	Site
HK (37)		0.0	2.1	1.7	2.3	0.8	2.1	0.3	0.4	HK
FLK Masek (2)	0.96		0.9	0.8	1.3	0.6	6.8	0.3	0.5	FLK Masek
WK (61)	0.08	0.36		1.5	2.1	3.3	4.7	1.0	0.6	WK
HEB West 2a (7)	0.10	0.45	0.44		0.1	2.2	2.4	1.2	0.9	HEB West 2a
HEB West 2b (15)	0.02	0.23	0.04	0.92		3.3	3.9	1.7	1.4	HEB West 2b
HEB West 3 (10)	0.40	0.57	0.00	0.09	0.00		2.8	1.4	1.5	HEB West 3
HEB East (2)	0.05	0.02	0.00	0.05	0.00	0.02		3.9	4.5	HEB East
PDK Trench IV (7)	0.79	0.75	0.32	0.26	0.10	0.19	0.01		0.2	PDK Trench IV
EF–HR (5)	0.73	0.64	0.52	0.37	0.18	0.17	0.01	0.86		EF–HR

METRICAL ANALYSIS OF HANDAXES AND CLEAVERS

Table 8.43 *Metrical analysis of bifaces: cleavers: t-values and estimates of significance: ratio Th/B*

Site	HK (38)	FLK Masek (2)	WK (60)	HEB West 2a (7)	HEB West 2b (15)	HEB West 3 (10)	HEB East (2)	PDK Trench IV (7)	EF–HR (6)	Site
HK (38)		1.8	2.0	0.2	1.5	2.4	1.7	2.7	0.2	HK
FLK Masek (2)	0.07		2.9	1.7	1.8	3.2	4.5	4.1	1.5	FLK Masek
WK (60)	0.04	0.01		1.4	0.4	1.6	1.4	2.2	0.7	WK
HEB West 2a (7)	0.83	0.13	0.17		1.0	2.0	1.8	2.6	0.3	HEB West 2a
HEB West 2b (15)	0.15	0.09	0.75	0.32		0.7	0.7	1.1	0.6	HEB West 2b
HEB West 3 (10)	0.02	0.01	0.12	0.06	0.50		0.6	0.6	1.3	HEB West 3
HEB East (2)	0.10	0.05	0.16	0.11	0.48	0.58		0.2	1.2	HEB East
PDK Trench IV (7)	0.01	0.00	0.03	0.02	0.30	0.56	0.81		1.7	PDK Trench IV
EF–HR (6)	0.81	0.18	0.47	0.75	0.55	0.20	0.28	0.11		EF–HR
	HK	FLK Masek	WK	HEB West 2a	HEB West 2b	HEB West 3	HEB East	PDK Trench IV	EF–HR	

Table 8.44 *Metrical analysis of bifaces: cleavers: t-values and estimates of significance: ratio T_1/L*

Site	HK (38)	FLK Masek (2)	WK (61)	HEB West 2a (7)	HEB West 2b (15)	HEB West 3 (10)	HEB East (2)	PDK Trench IV (6)	EF–HR (5)	Site
HK (38)		1.6	0.8	2.8	3.0	2.4	0.5	2.7	0.7	HK
FLK Masek (2)	0.11		2.5	3.2	3.7	3.1	1.7	3.0	1.3	FLK Masek
WK (61)	0.46	0.01		3.3	3.4	2.7	0.5	3.2	1.4	WK
HEB West 2a (7)	0.01	0.01	0.00		0.7	0.7	1.0	0.1	2.8	HEB West 2a
HEB West 2b (15)	0.00	0.00	0.00	0.47		0.1	0.8	0.8	3.1	HEB West 2b
HEB West 3 (10)	0.02	0.01	0.01	0.48	0.88		0.7	0.8	2.5	HEB West 3
HEB East (2)	0.60	0.22	0.65	0.35	0.41	0.53		1.0	1.0	HEB East
PDK Trench IV (6)	0.01	0.02	0.00	0.93	0.44	0.45	0.37		2.7	PDK Trench IV
EF–HR (5)	0.49	0.26	0.18	0.02	0.01	0.03	0.37	0.02		EF–HR
	HK	FLK Masek	WK	HEB West 2a	HEB West 2b	HEB West 3	HEB East	PDK Trench IV	EF–HR	

Table 8.45 *Metrical analysis of bifaces: cleavers: t-values and estimates of significance: ratio B/L*

Site	HK (38)	FLK Masek (2)	WK (62)	HEB West 2a (7)	HEB West 2b (15)	HEB West 3 (10)	HEB East (2)	PDK Trench IV (7)	EF–HR (6)	Site
HK (38)		0.4	3.1	1.6	2.6	0.9	0.5	3.0	1.3	HK
FLK Masek (2)	0.72		0.5	1.6	1.4	0.1	*0.1*	2.4	1.8	FLK Masek
WK (62)	**0.00**	0.61		3.2	4.9	0.9	0.3	4.6	2.8	WK
HEB West 2a (7)	0.11	0.15	**0.00**		0.3	1.8	*1.2*	1.3	0.2	HEB West 2a
HEB West 2b (15)	**0.01**	0.19	**0.00**	0.73		2.4	1.4	0.9	0.5	HEB West 2b
HEB West 3 (10)	0.38	0.94	0.39	0.08	**0.02**		0.1	2.8	1.6	HEB West 3
HEB East (2)	0.59	*0.94*	0.77	*0.64*	0.20	0.94		1.7	*1.1*	HEB East
PDK Trench IV (7)	**0.00**	0.05	**0.00**	0.22	0.40	**0.01**	0.13		1.5	PDK Trench IV
EF–HR (6)	0.19	0.13	**0.01**	0.84	0.62	0.13	*0.66*	0.15		EF–HR

Table 8.46 *Metrical analysis of bifaces: cleavers: t-values and estimates of significance: ratio B_1/B_2*

Site	HK (38)	FLK Masek (2)	WK (62)	HEB West 2a (7)	HEB West 2b (15)	HEB West 3 (10)	HEB East (2)	PDK Trench IV (6)	EF–HR (6)	Site
HK (38)		1.3	0.4	1.4	1.4	0.6	**3.8**	1.2	0.3	HK
FLK Masek (2)	0.20		1.3	0.6	0.8	1.8	3.0	1.5	1.6	FLK Masek
WK (62)	0.67	0.21		1.4	1.7	0.3	**3.3**	0.9	0.4	WK
HEB West 2a (7)	0.18	0.54	0.15		0.3	1.8	**4.1**	1.8	1.0	HEB West 2a
HEB West 2b (15)	0.15	0.41	0.10	0.75		1.8	**4.6**	2.0	0.8	HEB West 2b
HEB West 3 (10)	0.53	0.10	0.75	0.09	0.08		**3.4**	0.6	0.8	HEB West 3
HEB East (2)	**0.00**	0.10	**0.00**	**0.00**	**0.00**	**0.01**		2.0	3.7	HEB East
PDK Trench IV (6)	0.24	0.18	0.36	0.09	0.06	0.55	0.09		1.1	PDK Trench IV
EF–HR (6)	0.79	0.16	0.66	0.33	0.43	0.44	**0.01**	0.30		EF–HR

METRICAL ANALYSIS OF HANDAXES AND CLEAVERS

Table 8.47 *Metrical analysis of bifaces: cleavers: t-values and estimates of significance: ratio L_1/L*

Site	HK (38)	FLK Masek (2)	WK (62)	HEB West 2a (7)	HEB West 2b (15)	HEB West 3 (10)	HEB East (2)	PDK Trench IV (6)	EF–HR (6)	Site
HK (38)		1.1	0.2	1.8	*0.8*	0.3	**2.2**	0.3	1.1	HK
FLK Masek (2)	0.28		0.9	0.1	1.4	0.9	**3.8**	0.8	0.7	FLK Masek
WK (62)	0.88	0.36		1.6	*0.6*	0.2	**2.0**	0.2	0.9	WK
HEB West 2a (7)	0.07	0.91	0.12		1.9	1.4	**4.3**	1.2	0.8	HEB West 2a
HEB West 2b (15)	*0.30*	0.18	*0.35*	0.07		*0.4*	**4.4**	*0.2*	0.8	HEB West 2b
HEB West 3 (10)	0.75	0.39	0.84	0.18	0.75		**2.2**	0.1	0.7	HEB West 3
HEB East (2)	**0.03**	0.06	**0.05**	**0.00**	**0.00**	**0.05**		2.1	**3.8**	HEB East
PDK Trench IV (6)	0.73	0.47	0.81	0.27	0.89	0.94	0.08		0.6	PDK Trench IV
EF–HR (6)	0.30	0.52	0.39	0.45	0.43	0.48	**0.01**	0.58		EF–HR
	HK	FLK Masek	WK	HEB West 2a	HEB West 2b	HEB West 3	HEB East	PDK Trench IV	EF–HR	

Table 8.48 *Metrical analysis of bifaces: cleavers: t-values and estimates of significance: cleaver edge angle*

Site	HK (38)	FLK Masek (2)	WK (62)	HEB West 2a (7)	HEB West 2b (15)	HEB West 3 (10)	HEB East (2)	PDK Trench IV (6)	EF–HR (6)	Site
HK (38)		1.1	1.0	1.2	1.1	0.5	0.1	1.3	0.5	HK
FLK Masek (2)	0.29		1.6	0.7	0.5	1.0	1.5	**2.5**	1.0	FLK Masek
WK (62)	0.32	0.11		2.0	2.1	1.3	0.5	1.1	0.1	WK
HEB West 2a (7)	0.24	0.52	0.05		0.2	0.8	0.8	**2.8**	1.3	HEB West 2a
HEB West 2b (15)	0.27	0.64	**0.04**	0.83		0.4	0.3	1.8	1.0	HEB West 2b
HEB West 3 (10)	0.60	0.33	0.20	0.43	0.68		0.1	1.9	0.8	HEB West 3
HEB East (2)	0.89	0.27	0.64	0.47	0.77	0.89		1.2	0.3	HEB East
PDK Trench IV (6)	0.19	**0.05**	0.30	**0.02**	0.09	0.08	0.29		0.6	PDK Trench IV
EF–HR (6)	0.63	0.34	0.94	0.23	0.33	0.45	0.76	0.58		EF–HR
	HK	FLK Masek	WK	HEB West 2a	HEB West 2b	HEB West 3	HEB East	PDK Trench IV	EF–HR	

Table 8.49 *Metrical analysis of bifaces: cleavers: t-values and estimates of significance: ratio CEL/B*

Site	HK (38)	FLK Masek (2)	WK (62)	HEB West 2a (7)	HEB West 2b (15)	HEB West 3 (10)	HEB East (2)	PDK Trench IV (6)	EF–HR (6)	Site
HK (38)		0.4	1.7	0.4	0.3	2.0	1.8	0.8	0.3	HK
FLK Masek (2)	0.69		0.1	0.1	0.4	0.5	0.7	0.0	0.5	FLK Masek
WK (62)	0.10	0.95		0.4	1.4	1.0	1.3	0.0	1.1	WK
HEB West 2a (7)	0.66	0.93	0.71		0.5	0.9	1.1	0.2	0.5	HEB West 2a
HEB West 2b (15)	0.75	0.72	0.16	0.65		1.5	1.3	0.7	0.0	HEB West 2b
HEB West 3 (10)	0.05	0.65	0.31	0.38	0.14		0.8	0.7	1.7	HEB West 3
HEB East (2)	0.08	0.54	0.21	0.32	0.21	0.46		1.2	2.3	HEB East
PDK Trench IV (6)	0.45	0.97	0.98	0.83	0.50	0.49	0.26		0.9	PDK Trench IV
EF–HR (6)	0.76	0.62	0.28	0.62	0.98	0.11	0.06	0.39		EF–HR
	HK	FLK Masek	WK	HEB West 2a	HEB West 2b	HEB West 3	HEB East	PDK Trench IV	EF–HR	

Table 8.50 *Metrical analysis of bifaces: summary table for all the statistical comparisons of the cleaver samples: counts of positive scores out of a possible 8 (for explanation of Tables 8.50 and 8.51, see p. 202)*

Site	HK	FLK Masek	WK	HEB West 2a	HEB West 2b	HEB West 3	HEB East	PDK Trench IV	EF–HR	Site
HK		0	1	0	2	0	1	1	0	HK
FLK Masek	0		0	1	1	2	3	2	0	FLK Masek
WK	2	2		2	3	1	3	3	0	WK
HEB West 2a	2	1	2		0	0	3	0	0	HEB West 2a
HEB West 2b	4	1	5	0		2	4	0	1	HEB West 2b
HEB West 3	2	2	2	0	3		2	1	0	HEB West 3
HEB East	4	2	4	4	4	4		2	4	HEB East
PDK Trench IV	4	5	4	2	0	2	2		0	PDK Trench IV
EF–HR	0	0	1	1	1	1	4	1		EF–HR
	HK	FLK Masek	WK	HEB West 2a	HEB West 2b	HEB West 3	HEB East	PDK Trench IV	EF–HR	

Table 8.51 *Metrical analysis of bifaces: summary table for all the statistical comparisons made: positive scores expressed as a percentage, i.e. out of a possible 100*

Site	HK	TK Fish Gully	FLK Masek	PDK Trenches I-III	WK East A	WK East C	WK	HEB West 2a	HEB West 2b	HEB West 3	HEB East	PDK Trench IV	BK	TK Upper	TK Lower	SHK	EF-HR	MLK	Site
HK		50	22	57	50	0	26	26	30	13	39	22	71	0	0	60	26	57	HK
TK Fish Gully	71		0	57	57	43	57	0	7	43	57	21	71	20	0	40	0	36	TK Fish Gully
FLK Masek	26	21		57	43	29	35	17	17	39	48	30	57	20	10	40	9	21	FLK Masek
PDK Trenches I-III	57	57	57		29	29	57	57	57	57	29	64	0	0	20	20	57	57	PDK Trenches I-III
WK East A	79	57	57	29		0	50	57	50	64	0	64	14	0	20	20	36	43	WK East A
WK East C	29	57	43	43	0		7	21	21	21	0	29	0	0	0	0	29	29	WK East C
WK	39	71	48	64	71	43		43	26	35	57	30	71	0	0	40	17	71	WK
HEB West 2a	48	14	30	64	64	43	52		0	9	52	0	79	10	0	50	4	43	HEB West 2a
HEB West 2b	48	36	30	57	71	57	43	13		22	52	0	71	0	0	60	13	43	HEB West 2b
HEB West 3	39	86	52	79	86	29	48	22	39		52	4	79	10	0	70	26	57	HEB West 3
HEB East	70	79	43	29	14	7	65	17	61	70		43	0	0	0	0	48	50	HEB East
PDK Trench IV	43	29	52	71	64	57	48	22	4	22	48		71	10	0	60	13	43	PDK Trench IV
BK	71	79	64	0	57	14	79	79	71	86	57	71		0	40	20	57	50	BK
TK Upper	20	40	20	40	0	0	20	30	30	40	20	50	20		20	0	0	20	TK Upper
TK Lower	0	0	20	20	20	0	10	0	20	0	60	10	40	20		0	0	20	TK Lower
SHK	60	60	40	20	20	20	80	80	60	80	0	60	20	20	40		40	40	SHK
EF-HR	30	36	17	57	57	29	43	22	22	30	61	26	57	20	20	50		21	EF-HR
MLK	57	43	29	71	57	29	79	50	43	64	57	43	57	20	20	60	50		MLK
Site	HK	TK Fish Gully	FLK Masek	PDK Trenches I-III	WK East A	WK East C	WK	HEB West 2a	HEB West 2b	HEB West 3	HEB East	PDK Trench IV	BK	TK Upper	TK Lower	SHK	EF-HR	MLK	Site

This treatment of the material is as used in the study of the Kalambo Falls Large Cutting Tools to which I have referred (Roe n.d.). It was always intended that the two sets of results should be directly comparable with each other and the data were gathered accordingly from the start. Other analytical systems could be used (e.g. Cahen and Martin 1972), but in some cases additional data might be needed.

Discussion of the results

The measurements just listed were applied to the eighteen Olduvai samples, and the various ratios calculated and the statistical comparisons carried out: full results are given in Tables 8.2–8.51 and the complete sets of handaxe and cleaver-shape diagrams are given in Figs 8.6–8.34. This formidable body of information is now commended to the reader's attention, and a brief commentary follows.

So far as the statistical comparisons are concerned, attention is drawn to the summarising tables (Tables 8.16, 8.31, 8.50 and 8.51) which, while they are no substitute for the ensemble of statistical comparison tables themselves, do indicate quite simply which pairs of sites show consistent or frequent differences and which pairs do not differ 'significantly' from each other on many occasions or even at all. Summary tables of this kind are given for the comparisons of the sets of handaxes (Table 8.31), the sets of cleavers (Table 8.50) and the complete sets of bifaces (Table 8.16), while finally in Table 8.51 there is a summary table in which all the positive results from all the statistical comparisons carried out are added up. Some individual results are undeniably impressive. For example, the general summary table (Table 8.51) shows that for fourteen comparisons between the Developed Oldowan B bifaces from BK in Bed II and those of the elegant Acheulean industry excavated by the late John Waechter at HEB Level 3 in Bed IV, the 'difference scores' out of a possible 100 each time are 79 (t-values) and 86 (estimates of significance). When the same BK sample is compared with two Bed IV Developed Oldowan C samples, admittedly small, from PDK Trenches I–III and WK East C, the results are: t-values, 0 and 0, and estimates of significance, 0 and 14, respectively, each score again being out of a possible 100. For fourteen comparisons between two of the Developed Oldowan sites in Bed IV, WK East A and WK East C, the difference scores for both t-values and estimates of significance were 0 out of 100.

These are extreme cases, deliberately chosen here, and of course all the figures need to be looked at carefully. Some results are predictable, some rather surprising. Thus the two PDK samples in Bed IV, one a Developed Oldowan C industry and one an Acheulean, could be expected to differ: fourteen comparisons between them produced scores in Table 8.51 of 64 (t-values) and 71 (estimates of significance), out of 100. One might have expected the two Acheulean samples from late-dating hill-wash deposits, HK and TK FG, to resemble each other closely (though in fact there is no particularly compelling reason why they should); the difference scores in fact are 50 out of 100 (t-values) and 71 out of 100 (estimates of significance), suggesting substantial differences between them. So far as TK FG is concerned, Table 8.51 in fact suggests a rather closer comparison between it and the one available sample from the Masek Beds, FLK Masek.

Naturally, all these summarised scores need to be checked in terms of their constituents, to see whether there is some particular bias in the cumulative difference between two samples – whether, for example, they differ only on the tests of size, which might be explainable by local factors in the availability of raw material. The frequency tables also need to be looked at carefully: has the mean value of a small sample been distorted by the presence of just one or two atypical pieces – one huge handaxe, perhaps, amongst a group of otherwise small implements? The standard deviation should reflect this, if so. Nor should the relative proportions of handaxes and cleavers in any sample be forgotten, when counts are being considered rather than percentages, or when one section of the assemblage is being considered. Table 8.1 is helpful here. Thus, in the two samples from HEB West, Levels 2a and 2b, separated by a marker horizon, there are twenty-eight handaxes each time, but they form respectively 80 per cent and 65 per cent of the bifaces. Or again, there are thirty-seven cleavers from HK, the second largest set of cleavers in absolute quantity, but they constitute only 9·6 per cent of the bifaces from that site, while the fifteen cleavers from HEB West Level 2b represent 34·9 per cent of their biface group. At WK, there are eighty-six handaxes, the second largest group of handaxes studied, but because this site is exceptionally rich in cleavers by Olduvai standards, the handaxes comprise only 58·1 per cent of the WK biface total.

Space does not permit an analysis here of the comparison of every pair of sites: the relevant information is contained in the tables. General conclusions about the biface morphology at Olduvai are likely to be of more interest.

The results of the present work suggest that there are indeed very real differences between the Developed Oldowan samples and the Acheulean ones, when it comes to a detailed study of their bifaces, quite apart from the more general differences reported elsewhere. There are two aspects to this. Individual Developed Oldowan samples differ from almost all individual Acheulean ones over a wide range of metrical tests. Also, the Developed Oldowan samples are very consistent amongst themselves, and differ as a group from the Acheulean group, which in fact shows much greater internal variability. As regards the comparisons between individual Acheulean samples and those of the Developed Oldowan, there is one important exception to the general rule that we have just noted of marked difference: HEB East in Bed IV, which maintains rather high scores in Table 8.51 for comparisons with the rest of the Acheulean samples, except TK Lower, and remarkably low ones for comparisons with all the Developed Oldowan occurrences. Morphologically, the HEB East bifaces therefore seem closer to Developed Oldowan than Acheulean, a point to which I shall return later. The situation regarding TK Lower is less clear: it is a very small sample, nine bifaces only, and primarily for this reason it produces very few significant differences over all the comparisons made. The industry was originally classified by Dr Leakey as Developed Oldowan, but this was amended to Acheulean in her paper on cultural patterns at Olduvai (1975a). The evidence from biface morphology in the small available sample does not offer clear evidence one way or the other.

The other major point about the Developed Oldowan biface sets is that the Bed II and Bed IV occurrences are impressively similar, generally speaking, as regards their morphology: this may be taken as firm support for Mary Leakey's (1975a) contention that the Developed Oldowan tradition continues after Bed II and is to be seen as a Developed Oldowan C in Bed IV. With Developed Oldowan A we are not here concerned, owing to the general absence of bifaces in its industries. Those who are not prepared to envisage separate but contemporary 'Acheulean' and 'Oldowan' populations in the Olduvai region will need to express the conclusion rather differently, perhaps by saying that whatever may have been the different ecological conditions or functional considerations that in Bed II times gave rise to variants as different as the EF-HR and BK industries, they must still have prevailed when Bed IV was being deposited, causing the inhabitants of the area to require stone tool-kits as different as those of WK East A and PDK Trenches I–III on the one hand and WK or HEB West 3 on the other. This is one version of a classic controversy in Palaeolithic studies, which could just as easily be fought out in one form or another on the foreshore at Clacton-on-Sea in England, in the Middle or Upper Palaeolithic levels of the caves of southwest France, or in south and east Asia, or indeed in other arenas. This controversy concerns the explanation of variability in lithic assemblages, and the evidence is perhaps not yet available anywhere to decide it finally. So far as Olduvai is concerned, I myself am very willing to accept for the present Dr Leakey's hypothesis that two distinct populations, Oldowan and Acheulean, were involved, though I would do so on a wider range of arguments than just those of biface morphology.

If we now turn from the Developed Oldowan to the industries which in Table 8.1 are designated Acheulean, it is perhaps the degree of variety that is impressive rather than the underlying similarities. If there were ever going to be clear time-trends in Acheulean handaxe morphology, should we not expect to see them here at Olduvai, with anything up to a million years of Acheulean time spread out before us? Perhaps we can indeed discern something of the sort, but only faintly if so. It is true that our control over the stratigraphic order of the industries concerned is far from perfect. It is also true that many different rock types are involved in the various samples, some of which may have had a considerable effect on biface morphology and technology (cf. Jones 1979, and in this volume), and we cannot be certain that each Acheulean population had equally free access to all the different raw materials. But even with these allowances, the morphological variety amongst the Acheulean biface samples is considerable, notably in Bed IV. Again, this need not surprise anyone who is prepared to accept that the bifaces were functional tools, made for immediate use rather than for the eventual employment of classifiers and metrical analysts.

When we consider the matter of time-trends, we may say that in the sense of broad stratigraphy there are three successive groups of Acheulean tool-kits at Olduvai: the EF-HR, MLK and TK Lower samples from Bed II; the Bed IV sites of PDK IV, WK and the HEB group; and the 'post-Bed IV' occurrences of HK, TK FG and FLK Masek. Considering these as groups for the moment, it would be fair to go this far in discerning differences that may be related to technological progress with the passing of time:

(a) *As a group*, the Bed II samples show heavier, thicker, less-standardised handaxes than those in either of the two later series of samples. Cleavers, also, are rare by comparison with the Bed IV industries.

(b) In the Bed IV samples, the *overall* regularity of the handaxe shapes and flatness of sections are improved, and cleavers are more frequent and often elegantly made. However, there are exceptions to these higher technological standards as reflected in handaxe morphology, notably the HEB East sample, if that is indeed Acheulean.

(c) The post-Bed IV samples, *taken as a group*, show the best control over raw material, in the sense of the makers' ability to repeat accurately what appear to be preferred biface shapes. That this was achieved so effectively at FLK Masek in the Naibor Soit quartzite, which is a very difficult material to work, is especially striking (Plates 20–21). There are some very finely shaped large handaxes both here and at TK FG. Cleavers, however, are not strongly represented in these three samples, for whatever reason.

These general comments will need to be considered in the light of Dr Callow's conclusions, which take far greater account of technology and raw material than does my own study. Meanwhile, if anyone wishes to say of Olduvai that Bed II contains rather crude Acheulean industries, Bed IV a varied range of developed Acheulean industries and the post-Bed IV levels a small number of Acheulean industries showing a high degree of technological competence, then the purely morphological evidence provided by my metrical and statistical analyses would be in sympathy with this. In view of the stratigraphic evidence, it would not be inappropriate to refer to the three groups of samples as Early, Middle and Late Acheulean stages respectively, *in a purely local sense*, though it would be quite another matter to say that such stages at Olduvai corresponded to similarly named divisions of the Acheulean elsewhere in Africa or beyond. In any case, it has been stressed that these comments are generalisations about groups of samples from broad stratigraphic units. It has also to be emphasised, as all the tables show clearly, that within each of these three groups of Acheulean biface samples there are very substantial differences between individual industries. To make this point, and as a final guide to the reader's study of the tables and figures, it seems appropriate to end with brief notes on the morphology of the bifaces from each of the Olduvai sites considered. These notes are not to be

METRICAL ANALYSIS OF HANDAXES AND CLEAVERS

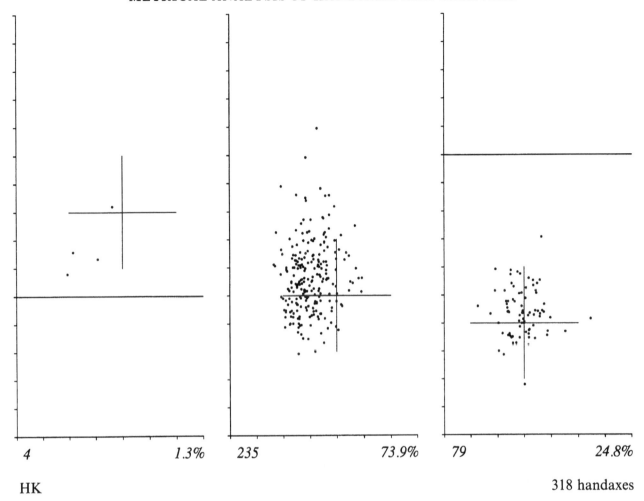

Fig. 8.6 Handaxe-shape diagrams: HK

regarded as comprehensive technical summaries of Acheulean industries; they are brief guides to the various tables and figures relating to the morphology of handaxes and cleavers. The sites are considered in their various Acheulean and Developed Oldowan groups and taken, as in the tables, in stratigraphic order, working from youngest to oldest.

Acheulean from deposits younger than Bed IV

HK (Figs 8.6, 8.24)
This is far and away the largest biface sample studied (Table 8.1), with 387 handaxes and thirty-seven cleavers. Although a few very large imple-ments occur, the average size and weight are relatively low; the cleavers are inclined to be larger and heavier than the handaxes (Tables 8.17, 8.18, 8.32, 8.33). The sections and tips of the implements are moderately flat, the handaxes being very slightly flatter than the cleavers on average. In shape, the cleavers tend to be broader than the handaxes. The handaxe-shape diagram (Fig. 8.6) shows a massive accumulation of ovates and pyriforms on the centre section: few of them are very pointed. On the right-hand section, this range is continued, with slightly lower positions of maximum breadth; a few broad squat shapes occur, but low B_1/B_2 values are again rare – i.e., acutely pointed forms are very scarce. The cleaver shape diagram (Fig. 8.24) shows 18.4 per cent

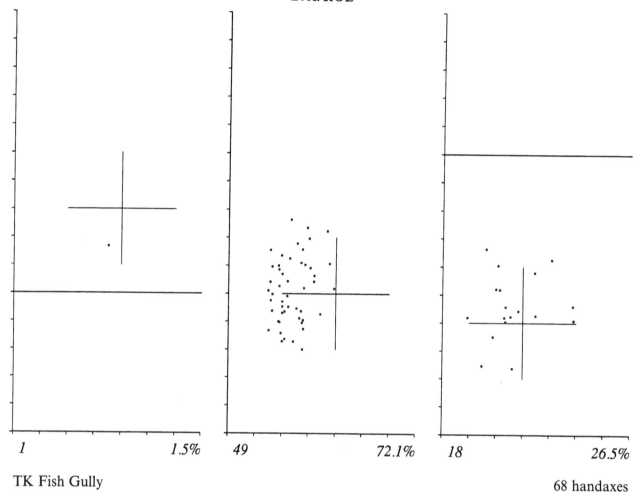

Fig. 8.7 Handaxe-shape diagram: TK FG

acutely angled cleavers, 26.3 per cent angled and 55.3 per cent transverse, and within this breakdown there is a wide range of different cleaver shapes, with only one of the nine sections left empty. All butt shapes occur, but U-shaped butts are easily the most frequent class. There is some suggestion of small concentrations – dare we perhaps regard them as groups of cleavers of specific 'types'? – though this is more clearly seen in the larger sample from WK, discussed below (Fig. 8.26). In the case of HK, look for example at the centre section of the transverse cleavers (Fig. 8.24). There is also a higher proportion of acutely angled cleavers at HK than at any other of the sites (discounting MLK, whose only cleaver was acutely angled).

TK FG (Fig. 8.7)

There are seventy-one bifaces, all handaxes. Their mean values for length and weight are high, but the presence of just a few exceptionally large and heavy handaxes has distorted them (Tables 8.17, 8.18): the main body of implements are only of medium or medium-large size by the local standards. The sections and tips are robust, but not exceptionally thick. There is a preference for narrow plan-forms, exceeded only by the PDK IV handaxes, which are really too small a sample to give a reliable picture (Table 8.21). In the handaxe-shape diagram for TK FG (Fig. 8.7), it is really only this preference for narrowness that prevents quite a close similarity with the HK plan-forms (Fig. 8.6), but over the whole range of

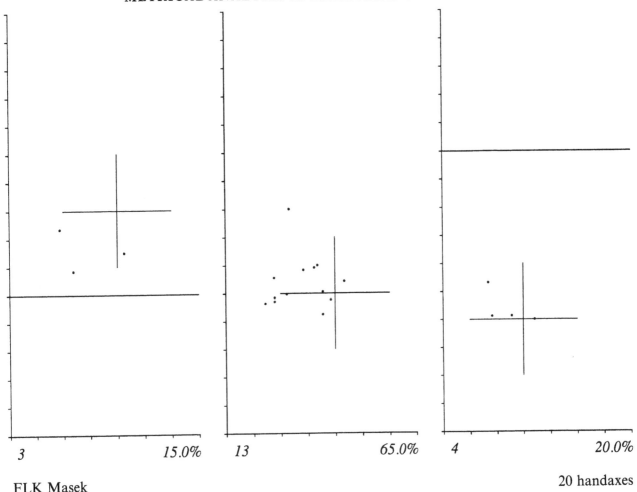

Fig. 8.8 Handaxe-shape diagram: FLK Masek

metrical analyses the two samples show various substantial differences, as the statistical comparison tables make clear. The TK FG handaxe shapes show, incidentally, a wider dispersal of the dots on the right-hand section than on the centre section of the shape diagram. It seems curious, and may well be significant (for whatever reason) that no cleavers at all were present at TK FG: this is certainly a basic difference between this site and HK.

FLK Masek (Figs 8.8, 8.25)
This is the only sample from the Masek Beds and the industry represents a remarkably skilful exploitation of the Naibor Soit white quartzite. There are two cleavers and twenty-one handaxes, and one wishes there were more to provide a clearer definitive picture of this very striking assemblage. As things are, size preference is in fact not strong, though the emphasis certainly falls at the large end of the scale and the biggest handaxes closely resemble each other across a whole range of measurements: one cannot but feel that a highly accomplished knapper, in full control of a difficult raw material, was aiming at a preferred size and shape, in those particular cases. It is to these dramatic objects (cf. Pl. 21) that we owe the fact that FLK Masek produces easily the highest mean values for length and weight (Tables 8.2, 8.3), but predictably the standard deviations are also large. The mean values for thickness of section and thickness of tip

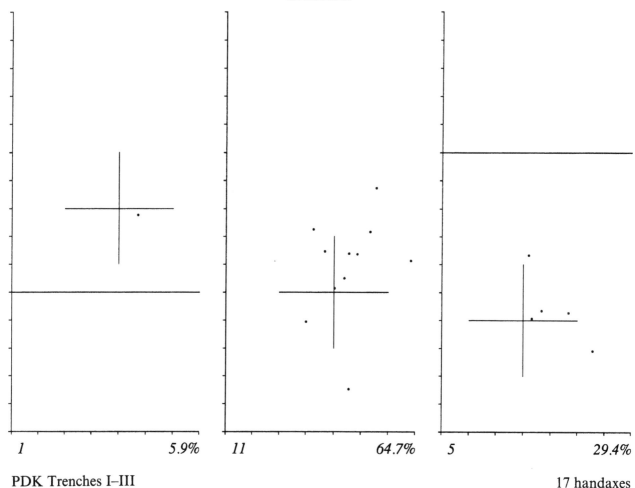

Fig. 8.9 Handaxe-shape diagram: PDK Trenches I–III

are also on the high side (Tables 8.4, 8.5), doubtless again by virtue of the raw material used. The mean value for B/L is a low to middle-range one (Table 8.6), which here reflects the presence of a range of shapes from narrow to moderately broad, rather than a strong preference for values close to the mean. On the other hand, the B_1/B_2 values in Tables 8.6 and 8.22 do have a much narrower range than usual, reflected in the lowest standard deviations for this ratio for any Olduvai site. The handaxe-shape diagram shows that there is indeed an overall similarity of planforms, regardless of the section on which the dots occur (Fig. 8.8). The vertical range of the distribution is narrow, controlled of course by the B_1/B_2 values, and no other of the Olduvai handaxe-shape diagrams compares really closely

with this one. The two cleavers are both angled forms with U-shaped butts and middle-range L_1/L values (Fig. 8.25).

Developed Oldowan C in Bed IV

PDK Trenches I–III (Fig. 8.9)

This is a small sample of eighteen bifaces, not all yielding a full set of measurements, in which no cleavers are present. The handaxes, as noted in earlier discussion, contrast sharply with those of the Bed IV Acheulean, but compare quite closely with those of the Developed Oldowan B of Bed II: they are small or sometimes tiny, in terms of both length and weight (Tables 8.17, 8.18); their cross-sections are often thick (Table 8.19), and

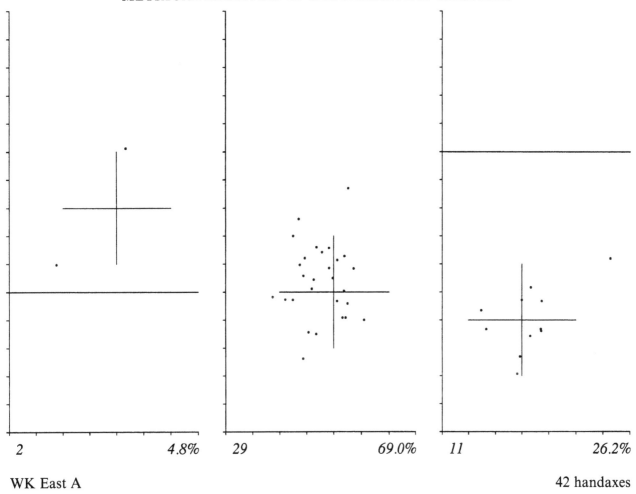

Fig. 8.10 Handaxe-shape diagram: WK East A

the mean value for the tip thickness measurement (T_1/L) is the highest of all at 0.343 (Table 8.20); so is the B/L mean value at 0.742 (Table 8.21). The B_1/B_2 mean value is also high, and is in fact identical with that of the BK Developed Oldowan B handaxe sample in Table 8.22. The handaxe-shape diagram (Fig. 8.9) predictably shows widely scattered dots, with a clear preference for broad, blunt-ended plan-forms, mostly of irregular ovate or ovoid shape.

WK East A (Fig. 8.10)

This is the strongest of the Developed Oldowan C samples numerically, with forty-two handaxes and no cleavers. The handaxes are generally similar to those just described from PDK Trenches I–III, but they cover a wider morphological range and the mean values are therefore less extreme. The handaxes are small and thick, with thick tips, though not as strikingly robust as those of PDK Trenches I–III. While their B/L mean is perhaps only moderate by Developed Oldowan standards, it still exceeds the reading for every Acheulean site except the atypical HEB East and the small sample from TK Lower in Bed II (Table 8.21). The shape diagram (Fig. 8.10) has a strong component of narrower handaxes than those of PDK Trenches I–III, and though there are again plenty of blunt-ended ovates and ovoids, there is also a rather higher proportion with lower B_1/B_2

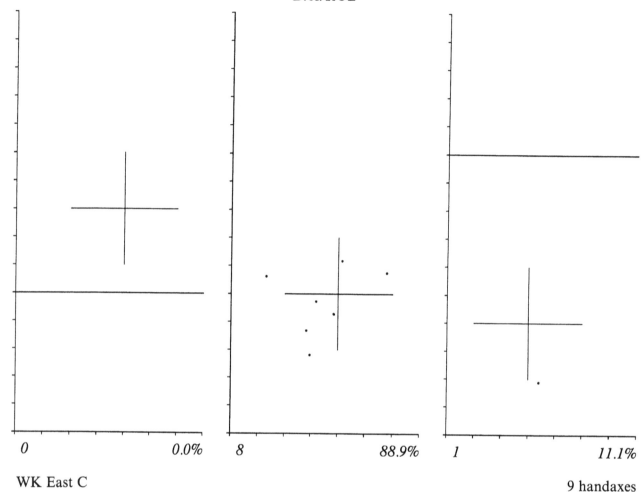

Fig. 8.11 Handaxe-shape diagram: WK East C

values, indicating a tendency towards pointedness. The diagram perhaps resembles that of BK (Fig. 8.18) most closely.

WK East C (Fig. 8.11)

This is a very small sample, with just nine handaxes and no cleavers. The implements are small, but less so than at the other two Developed Oldowan C sites. Their sections are as thick, however, and their tips are again pretty robust. The B/L mean is identical with that from WK East A (Table 8.6), but the B_1/B_2 mean is lower (Table 8.7). Of the small number of spots on the handaxe-shape diagram (Fig. 8.11), all except one fall on the centre section, near or below the middle, but there are so few of them that it is hardly appropriate to talk about shape preferences or to make firm comparisons with any of the other Developed Oldowan sites.

Acheulean in Bed IV

WK (Figs 8.12, 8.26).

There are sixty-two cleavers, constituting 41.9 per cent of the bifaces, the biggest proportion at any of the Olduvai sites studied. The eighty-six handaxes are a big enough sample to give reasonably smooth frequency curves. Implements are typically rather small, though not to the extent of those at HEB East or the Bed IV Developed Oldowan sites (Tables 8.2, 8.3); the cleavers are inclined to be larger than the handaxes (compare

METRICAL ANALYSIS OF HANDAXES AND CLEAVERS

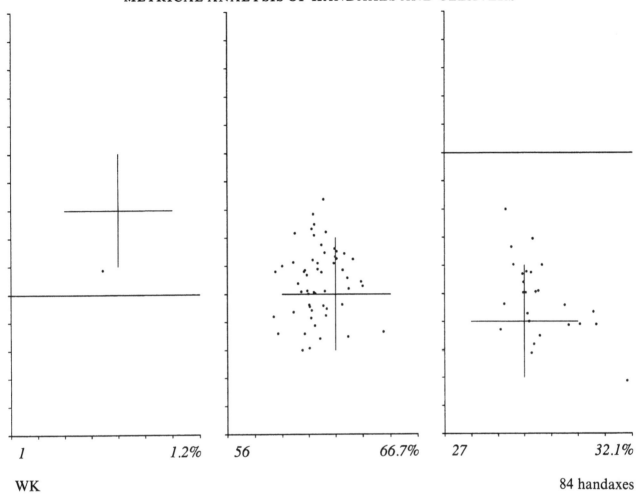

Fig. 8.12 Handaxe-shape diagram: WK

Tables 8.32 and 8.33 with Tables 8.17 and 8.18). The implements have fairly flat sections. Broader shapes are much commoner than at the HEB West sites to be considered next, notably on the right-hand section of the handaxe-shape diagram (Fig. 8.12); here, there is also a cluster near the top of the visual coordinate cross, in what is not usually a well-populated area in the Olduvai shape diagrams. The cleavers break down into 4.8 per cent acutely angled, 40.3 per cent angled and 54.8 per cent transverse. The shape diagram (Fig. 8.26) shows several good small clusters, featuring two or three implements of very similar shapes, often with the same butt symbol. There is a stronger preference for cleavers of broad proportions than in the HEB West samples. It is interesting to compare this cleaver-shape diagram with that for HK (Fig. 8.24), since they are the two largest series; there is much overlap, but each has some populated areas that are blank on the other. Overall, WK appears to have tighter concentrations, presumably representing genuine preferences for particular shapes: again, one might refer to 'cleaver types'.

HEB West Level 2a (Figs 8.13, 8.27)
There are seven cleavers (20 per cent) in a series of thirty-five bifaces. Metrically and morphologically, this series has much in common with HEB West Level 2b, discussed next, which underlies it at the same site, and the statistical comparisons reflect this by the scarcity of 'significant dif-

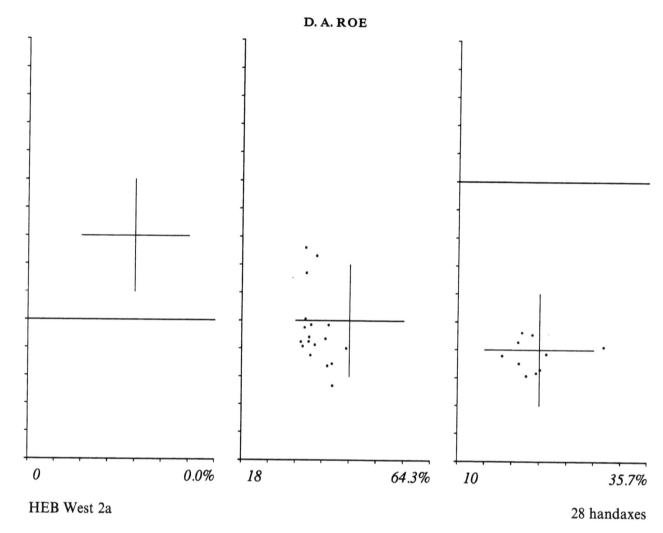

Fig. 8.13 Handaxe-shape diagram: HEB West Level 2a

ferences' between the two. The handaxes of Level 2a tend to be just a little larger and thicker, with slightly flatter tips (Tables 8.17, 8.18, 8.20); the B/L figures are pretty similar for the two sites (Table 8.21). The two sets of handaxe shapes fall into a very similar range of rather narrow ovates and pyriforms, but with a difference of emphasis: in this upper level there are rather more of the pyriforms with lower-placed maximum breadth on the right-hand section of the shape diagram (Fig. 8.13), and on the centre section there is a certain preference for lower B_1/B_2 values, i.e. a greater tendency towards pointedness within the pyriform shape. Angled cleavers are commonest, with one acutely angled and two transverse (Fig. 8.27). It is notable that all of the four angled cleavers fall on the centre section, and all have U-shaped butts – another close resemblance to the industry from the underlying Level 2b at HEB West.

HEB West Level 2b (Figs 8.14, 8.28)

The forty-three bifaces include fifteen cleavers (34.9 per cent). The metrical results place this industry, like the one just discussed, in the middle of most ranges. Thus, the handaxes are moderately large and heavy but not particularly thick in profile or at their tips. Their plan-forms tend towards narrowness. On the handaxe shape diagram (Fig. 8.14), the L_1/L values assign over 80 per cent of the implements to the centre section, most being rather narrow ovates or pyriforms.

METRICAL ANALYSIS OF HANDAXES AND CLEAVERS

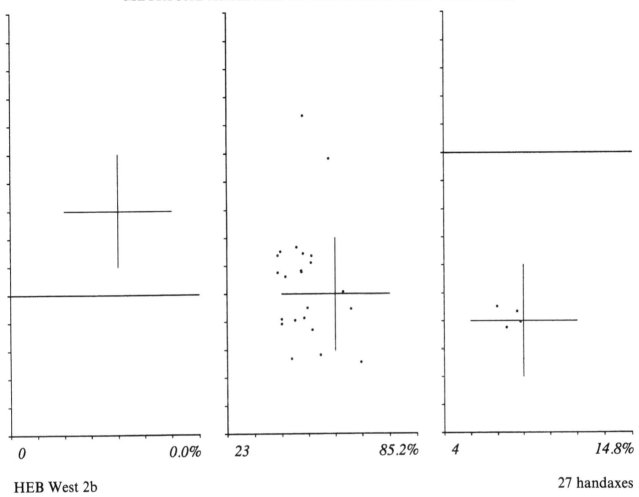

| 0 | 0.0% | 23 | 85.2% | 4 | 14.8% |

HEB West 2b

27 handaxes

Fig. 8.14 Handaxe-shape diagram: HEB West Level 2b

The only handaxes which appear on the right-hand section are closely concentrated and are also essentially pyriform shapes, with slightly lower position of maximum breadth. The handaxes from this site certainly give the impression of being one continuous range rather than a tool-kit composed of several distinct types. Amongst the cleavers, angled types are just commoner than transverse, and some clustering is apparent on the cleaver-shape diagram (Fig. 8.28): for example, on the centre section for angled cleavers, all six implements have U-shaped butts (cf. HEB West Level 2a). There is also one tight group of three cleavers on the centre section for transverse cleavers, two of these also with U-shaped butts.

The cleavers are of larger size than the handaxes on average (Tables 8.17, 8.18, 8.32, 8.33).

HEB West Level 3 (Figs 8.15, 8.29)

There are ten cleavers, constituting 16·7 per cent of the sixty bifaces. The bifaces are generally small by Olduvai Acheulean standards (Tables 8.2, 8.3) and notably flat in their cross sections and tip profiles (Tables 8.4, 8.5): these things probably reflect the extensive use for tool-making at this site of the green phonolite from Engelosen, though this is something discussed more fully by Dr Callow (chapter 9, this volume). Some of the handaxe shapes are also rather distinctive: on the shape diagram (Fig. 8.15), the right-hand section

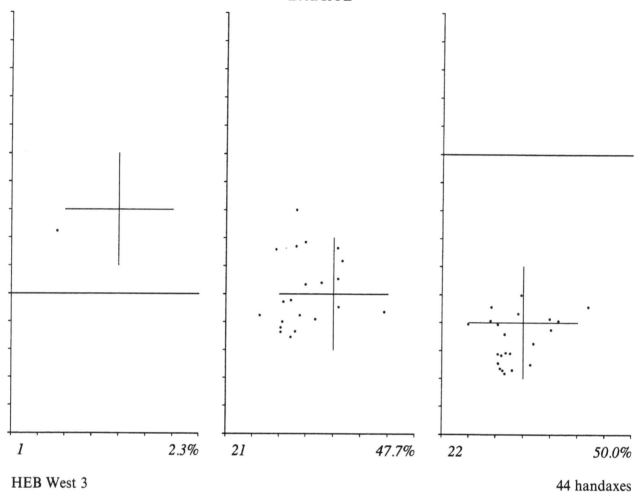

Fig. 8.15 Handaxe-shape diagram: HEB West Level 3

is better populated than usual (50 per cent of the handaxes), including some areas of it which remain virtually empty in all the other Olduvai Acheulean handaxe samples – there is a notable cluster of acutely pointed ovoid shapes, for example. The cleavers, often neatly made, are widely scattered on their shape diagram (Fig. 8.29); transverse types account for five out of ten. Butt shapes are also varied. The cleavers do not differ much in size from the handaxes.

HEB East (Figs 8.16, 8.30)

There are two cleavers only, in a set of forty-four bifaces that stand out as different from the rest of the Olduvai Acheulean samples in several ways, as can be seen from almost any of the metrical and statistical tables and from the handaxe-shape diagram (Fig. 8.16). In terms of mean values, for example, the HEB East bifaces are by some way the smallest and lightest Acheulean series (Tables 8.2, 8.3) and have the thickest tips (Table 8.5) and some of the thickest sections (Table 8.4), exceeded only by EF-HR, a Middle Bed II site, and FLK Masek, a somewhat specialised industry. Their plan view proportions are broadest (highest Acheulean B/L mean value in Table 8.6) and they even have the highest mean value for L_1/L (position of maximum breadth) of all sites (Table 8.8). The handaxe-shape diagram is quite unlike the other Acheulean ones, with 90 per cent of the forty-two handaxes falling on the centre section, where there is a notable component of broad,

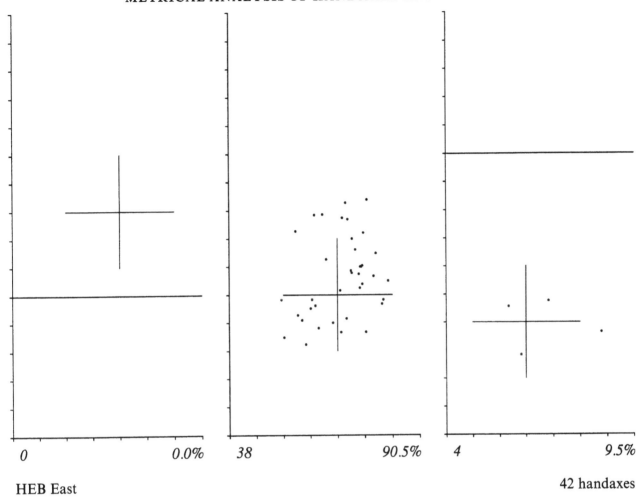

Fig. 8.16 Handaxe-shape diagram: HEB East

rather blunt-ended plan-forms (Fig. 8.16). The two cleavers are both small, one being angled and one transverse (Fig. 8.30). It is not surprising, then, that the statistical comparisons show numerous significant differences when HEB East is compared with the other Acheulean samples and few in comparison with those of the Developed Oldowan. It is hardly for me to assert on the evidence of biface morphology alone that HEB East should be regarded as a Developed Oldowan site, but it is quite clear to me that the biface morphology has much more in common with that of the Developed Oldowan of Olduvai than with the local Acheulean.

PDK Trench IV (Figs 8.17, 8.31)

This sample consists of seven cleavers and eleven handaxes: one wishes it were larger, because the industry is elegant and probably distinctive. Implement sizes are moderately small, the cleavers tending to be larger and heavier than the handaxes (Tables 8.17, 8.18, 8.32, 8.33). The means for handaxe section and tip thickness (Th/B and T_1/L, Tables 8.19 and 8.20) have the lowest value for any Olduvai site, and the cleaver tips are also the flattest (Table 8.35). The B/L means are also the lowest (i.e., there is a preference for narrower shapes), both for handaxes and cleavers and indeed for all biface samples

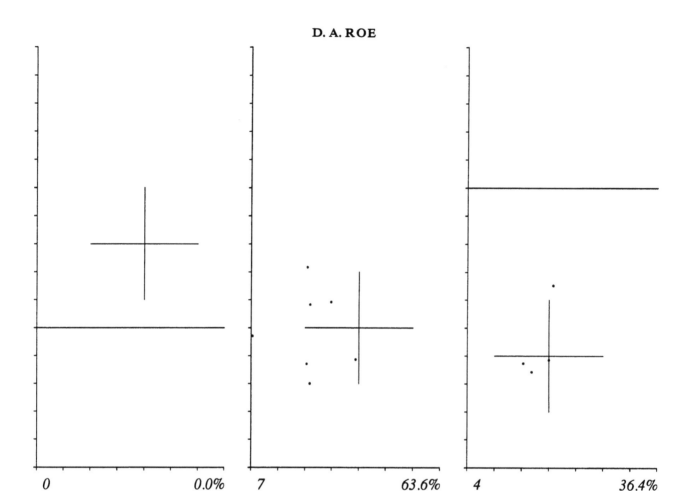

PDK Trench IV   11 handaxes

Fig. 8.17 Handaxe-shape diagram: PDK Trench IV

(Tables 8.6, 8.21, 8.36). The scatter of handaxe shapes (Fig. 8.17) is really too thin to assess, but it looks as if it might have resembled those from HEB West Levels 2a and 2b as closely as any. The cleaver sample is also too small to be definitive: two angled and four transverse were measurable for the purposes of Fig. 8.31. Two of the PDK Trench IV cleavers are in fact splayed forms, very rare at Olduvai. The PDK Trench IV cleavers are mostly distinctly narrow in their general proportions.

Developed Oldowan B in Bed II

BK (Figs 8.18, 8.32)
There is only one cleaver, out of sixty-two bifaces.

The implements vary in size, but many are small and some are tiny. They have notably thick sections and tips, the mean values for Th/B and T_1/L each being the second highest for all the sites studied (Tables 8.4, 8.5). There is a preference for broad plan-forms and on the handaxe-shape diagram (Fig. 8.18) the dots are widely scattered, indicating a general lack of strong shape preferences or perhaps inability to repeat shapes accurately; these are quite different things. Various squat oval shapes seem to be the most popular. Numerically, the BK sample is the best of those studied from the Developed Oldowan of Upper Bed II.

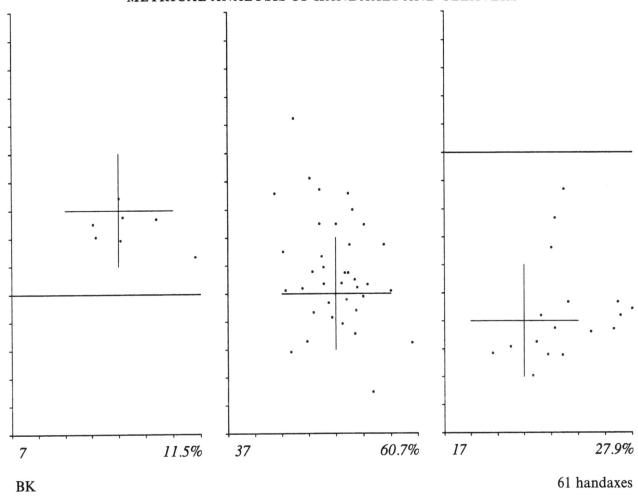

Fig. 8.18 Handaxe-shape diagram: BK

SHK (Figs 8.21, 8.33)

There is one cleaver, out of forty-seven bifaces. The implements are variable in size; there is a rather higher proportion of large handaxes than there was at BK, though some tiny ones are again present. The sections are typically thick (Table 8.4); it is unfortunate that no data for tip thickness are available. Plan-forms are frequently broad, with various irregular outlines; the handaxe shapes (Fig. 8.21) are not quite so scattered as at BK, since there is a preference for rather blunt-ended forms, this sample having the highest mean value of all for the ratio B_1/B_2 applied to handaxes (Table 8.22).

TK Upper Level (Fig. 8.19)

This is a small sample with only fifteen bifaces and is hard to assess. There are no cleavers. Biface sizes are very variable and no really tiny specimens are present (Table 8.2). The implements have thick sections; again, no tip thickness data are available. The handaxe-shape diagram (Fig. 8.19) shows a diffuse scatter. Broad shapes appear less frequent than at BK and SHK but with only fifteen specimens to plot the reality of this observation must remain open to doubt.

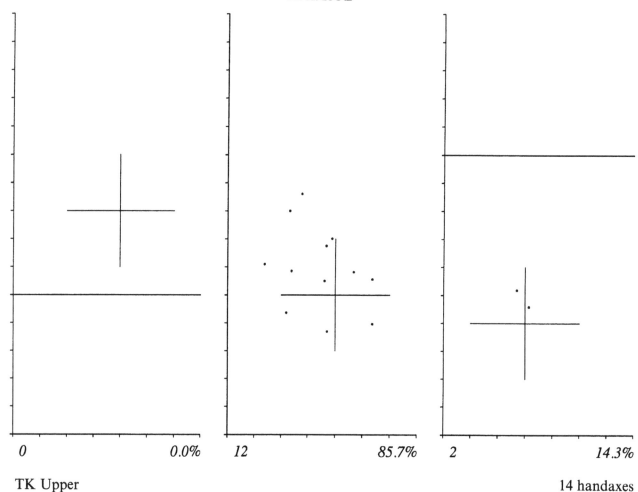

Fig. 8.19 Handaxe-shape diagram: TK Upper Level

Acheulean in Bed II

TK Lower Level (Fig. 8.20)

This is unfortunately a very small sample of only nine handaxes – no cleavers are present – but Bed II Acheulean occurrences are rare and of considerable interest, so it seemed worth including. Some of the handaxes are large, and one very large indeed; there are no very small ones at all. The length mean is actually almost double that of the BK Developed Oldowan B handaxes, though the largest handaxe has had a disproportionate effect on it because the sample is so small (Table 8.17). None of the handaxes has a very thick section, and the Th/B mean is actually the lowest for all sites except PDK Trench IV in Table 8.19, but again the sample is too small for this to carry much weight. No data are available on tip thickness. The handaxe shapes are widely scattered (Fig. 8.20), and some notably broad forms are included. Accordingly, the shape diagram does not much resemble those of the other Olduvai Acheulean samples, but the comparison is hardly a valid one.

EF-HR (Figs 8.22, 8.34)

There are six cleavers according to Roe, or two according to M. D. Leakey, and thirty-one handaxes. Moderately large handaxes, with thick sections and tips, are characteristic: the EF-HR

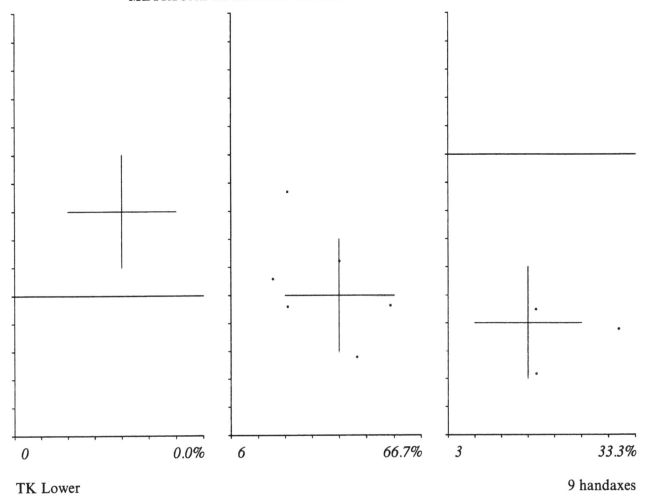

Fig. 8.20 Handaxe-shape diagram: TK Lower Level

handaxe Th/B mean is higher than its counterpart for all the Bed IV and post-Bed IV Acheulean samples (Table 8.19) and the handaxe T_1/L mean is only exceeded by that from the rather unusual HEB East assemblage amongst the Acheulean samples younger than Bed II (Table 8.20). The handaxe plan-forms are mostly narrow – various irregular narrow oval or pyriform shapes are commonest, giving a quite different distribution on the shape diagram from that typical of the Developed Oldowan series. The cleavers are all different (Fig. 8.34), but transverse types are commonest if Roe's figure is accepted. At many sites, as we have seen, the weight and length means are perceptibly higher for cleavers than handaxes, but this is not the case at EF-HR (Tables 8.17, 8.18, 8.32, 8.33).

MLK (Figs 8.23, 8.35)

There is one cleaver out of twenty-nine bifaces. The industry has a substantial proportion of heavy handaxes: the mean length and weight figures (Tables 8.17, 8.18) are exceeded amongst the sites studied only by those from FLK Masek, which is a rather specialised industry arguably the best part of a million years younger. A few small bifaces are in fact present at MLK. The implements have notably thick sections (highest Th/B mean value for all sites in Table 8.4) and thick tips (the T_1/L mean is marginally exceeded only by

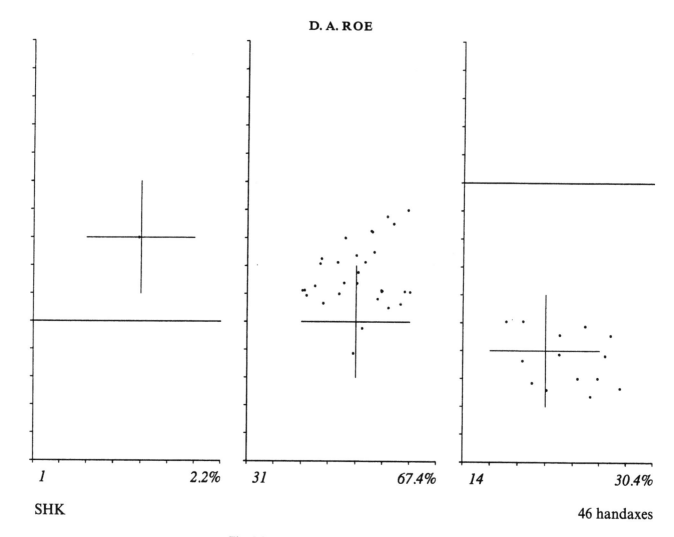

Fig. 8.21 Handaxe-shape diagram: SHK

HEB East amongst the Acheulean samples, and we have already noted more than once the somewhat atypical status of HEB East). The MLK handaxe shapes are typically narrow and are distributed much like those of EF-HR on the shape diagram (Fig. 8.23). Ardent typologists might think the physical appearance of the MLK assemblage a classic Early Acheulean one; from the stratigraphic point of view, it certainly seems to be the oldest of all the Olduvai Acheulean occurrences.

Conclusion

The meat of this report necessarily comes in the dauntingly numerous tables and figures, which need to be studied carefully with due regard to such things as sample sizes and levels of probability. Different views are perfectly possible on various points, but my own overall assessment suggests to me the following conclusions:

1. The differences between the Developed Oldowan and Acheulean series of biface samples are substantial and genuine, however one may choose to interpret the significance of this situation.

2. The Developed Oldowan B of Bed II and the Developed Oldowan C of Bed IV have a great deal in common, so far as biface morphology is concerned. All the Developed Oldowan biface sets are pretty similar to each other and all are different from the classic Acheulean samples.

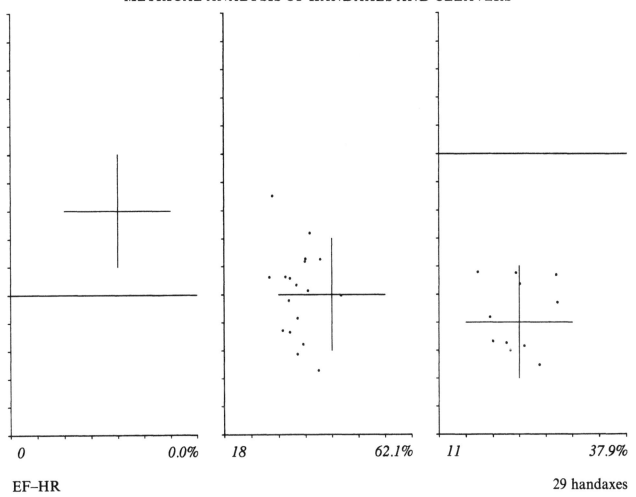

Fig. 8.22 Handaxe-shape diagram: EF–HR

3. Within the Acheulean, it is not unreasonable to refer to time trends in the long period represented by the deposits from Lower Bed II to the Masek Beds and the post-Masek hill washes, but such time trends are not strong and they concern general rather than detailed aspects of biface morphology.

4. There is much variability amongst the Acheulean industries, especially within Bed IV, many of them having original features. HEB East looks rather an odd site out in the Acheulean list, and the bifaces there have much in common with those of the Developed Oldowan; perhaps the whole status of this assemblage would be worth reexamination, on a broader basis than that of biface morphology, to see whether its present attribution to the Acheulean is correct. The two Acheulean samples that have most in common with each other in terms of the present study are Levels 2a and 2b at HEB West.

5. In general terms and with due caution, the Acheulean industries of Bed II, Bed IV and the post-Bed IV levels can reasonably be called Early, Middle and Late Acheulean respectively, in a purely local sense.

It must always be remembered that these conclusions are based entirely on a study of biface morphology, and not on full analyses of complete lithic industries. I am perfectly aware of this: that was the task that was assigned to me, and it is for

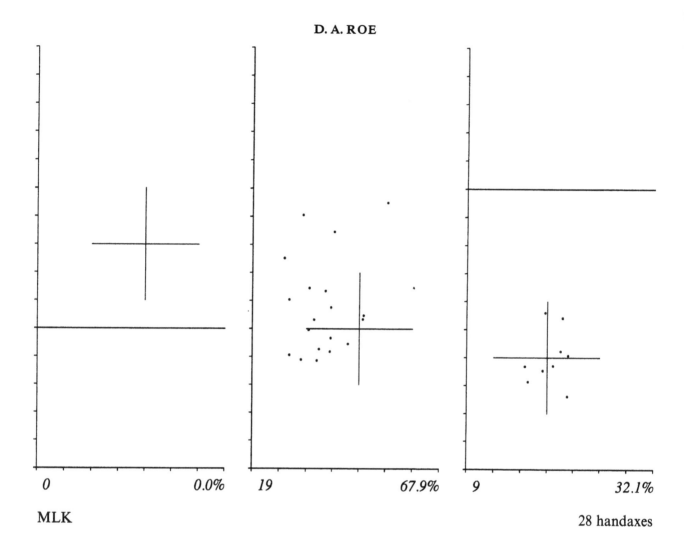

Fig. 8.23 Handaxe-shape diagram: MLK

others to broaden the study in whatever directions they may wish. Broader approaches will indeed be found elsewhere in the present volume, and plenty more research possibilities offered by the Olduvai bifaces remain to be followed up.

METRICAL ANALYSIS OF HANDAXES AND CLEAVERS

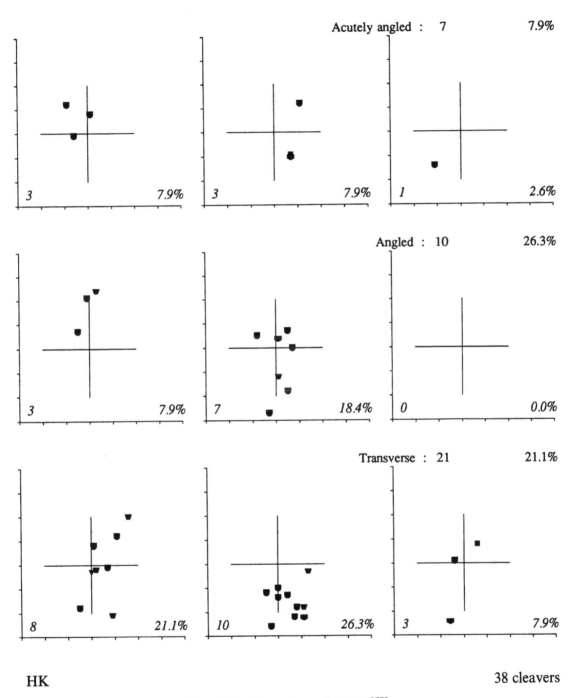

Fig. 8.24 Cleaver-shape diagram: HK

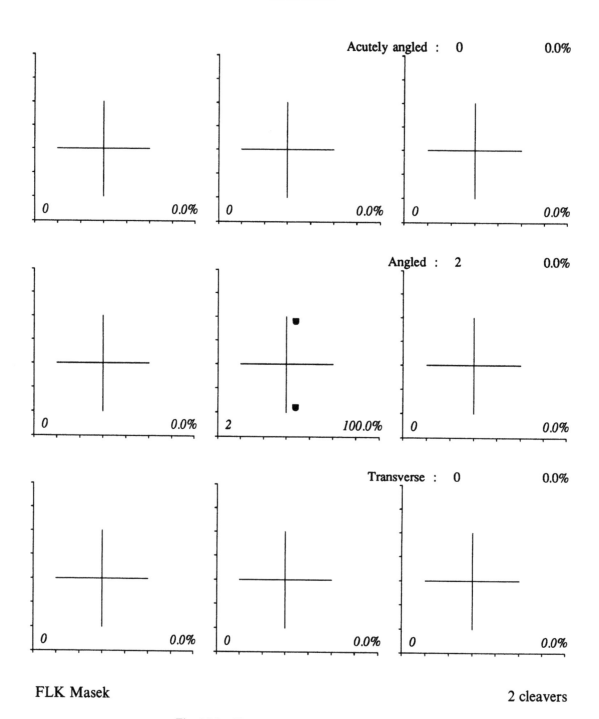

Fig. 8.25 Cleaver-shape diagram: FLK Masek

METRICAL ANALYSIS OF HANDAXES AND CLEAVERS

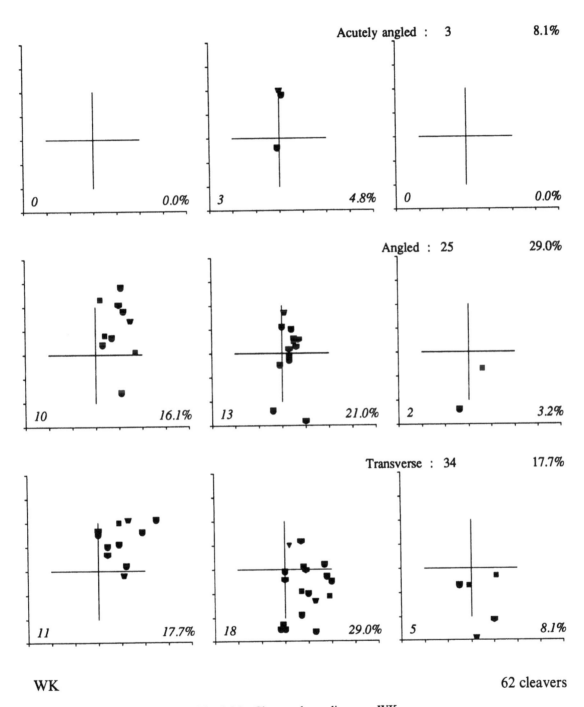

WK
62 cleavers

Fig. 8.26 Cleaver-shape diagram: WK

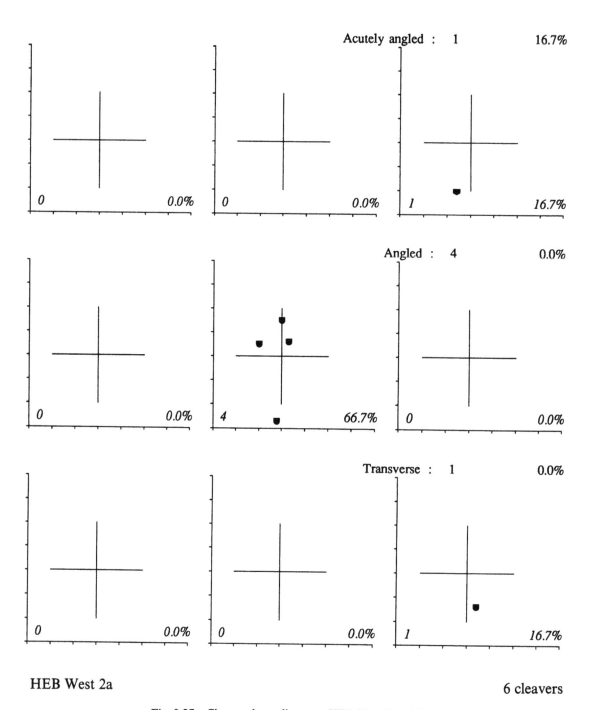

Fig. 8.27 Cleaver-shape diagram: HEB West Level 2a

METRICAL ANALYSIS OF HANDAXES AND CLEAVERS

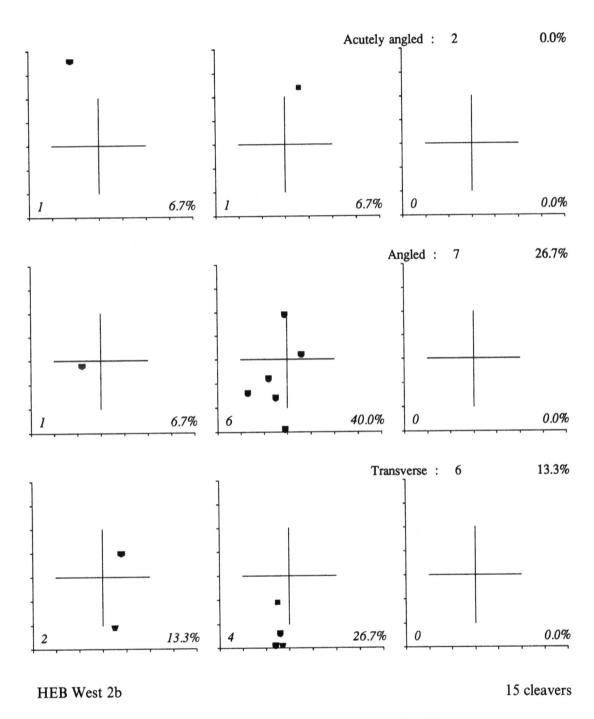

Fig. 8.28 Cleaver-shape diagram: HEB West Level 2b

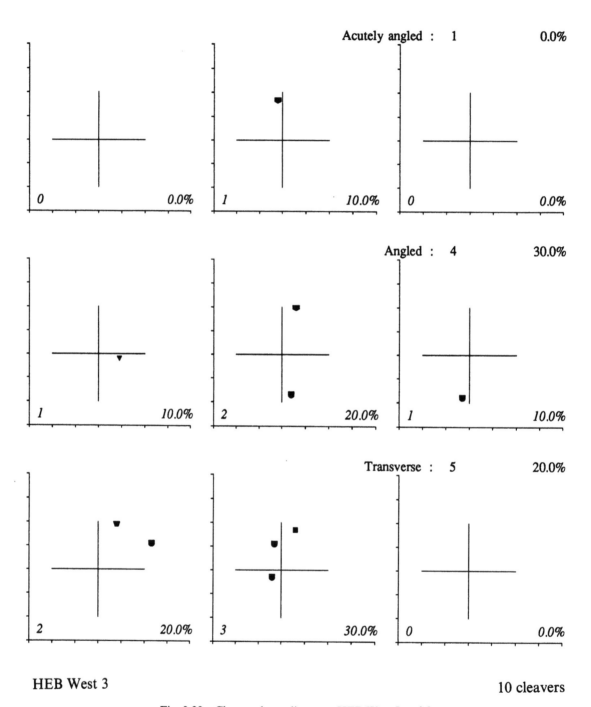

Fig. 8.29 Cleaver-shape diagram: HEB West Level 3

METRICAL ANALYSIS OF HANDAXES AND CLEAVERS

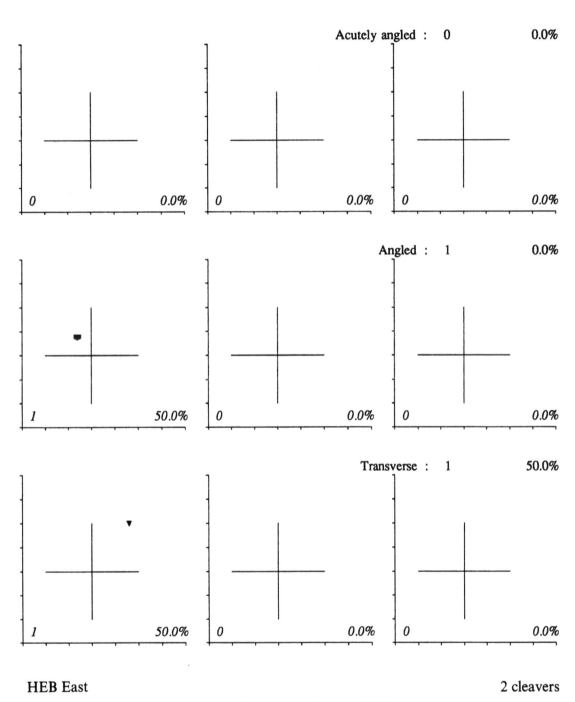

HEB East

Fig. 8.30 Cleaver-shape diagram: HEB East

2 cleavers

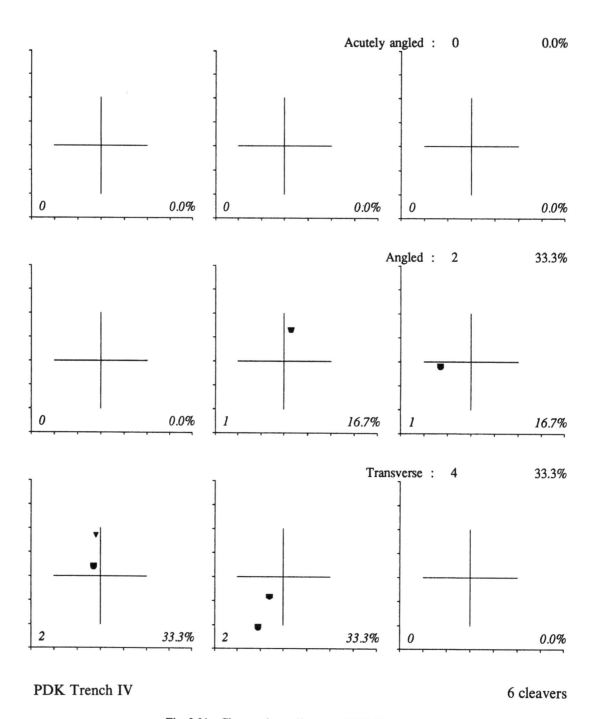

Fig. 8.31 Cleaver-shape diagram: PDK Trench IV

METRICAL ANALYSIS OF HANDAXES AND CLEAVERS

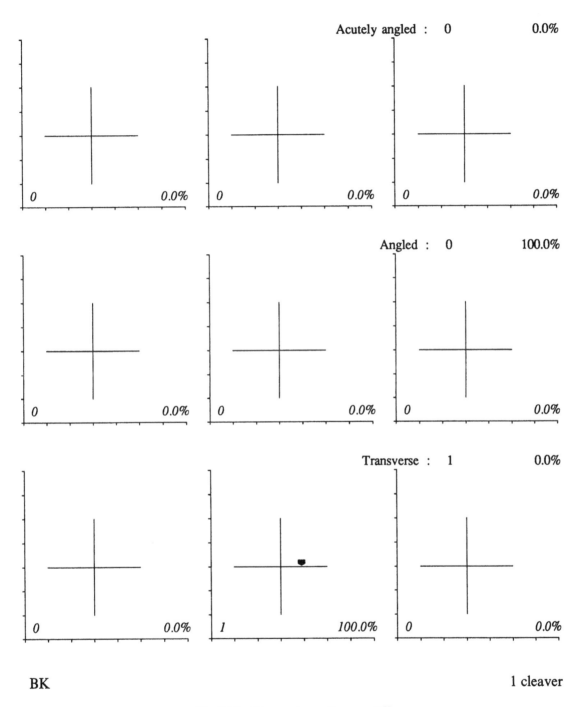

Fig. 8.32 Cleaver-shape diagram: BK

D. A. ROE

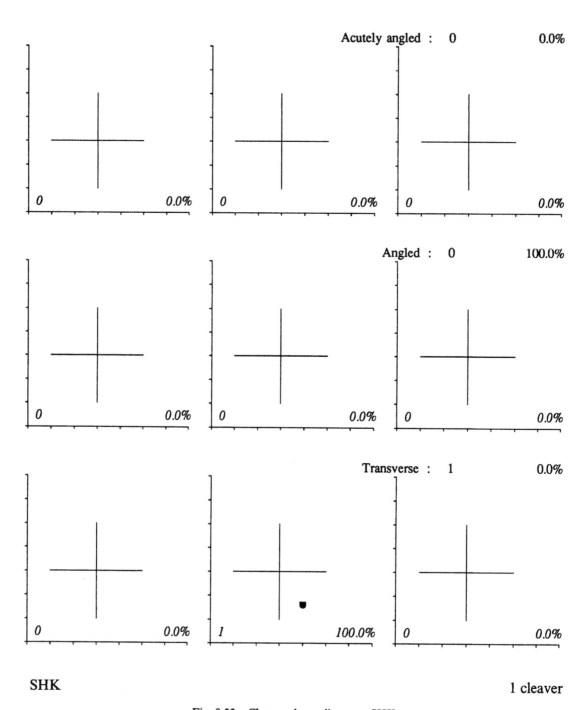

SHK

Fig. 8.33 Cleaver-shape diagram: SHK

METRICAL ANALYSIS OF HANDAXES AND CLEAVERS

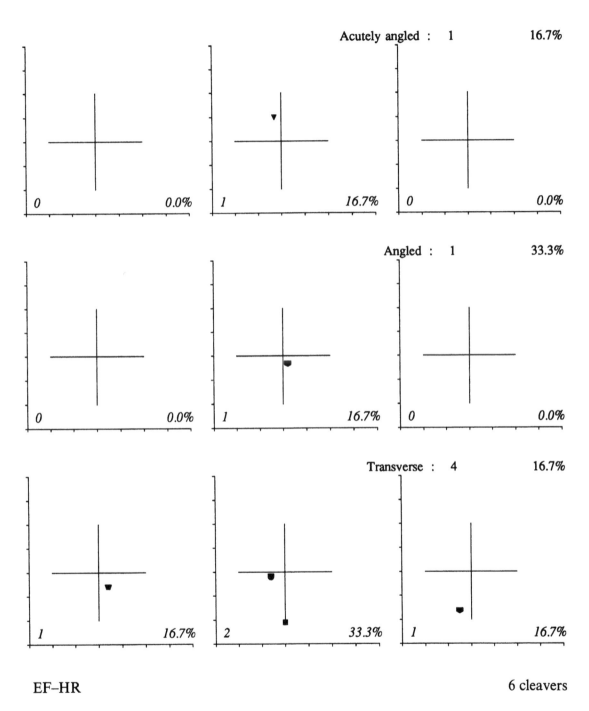

EF–HR   6 cleavers

Fig. 8.34 Cleaver-shape diagram: EF–HR

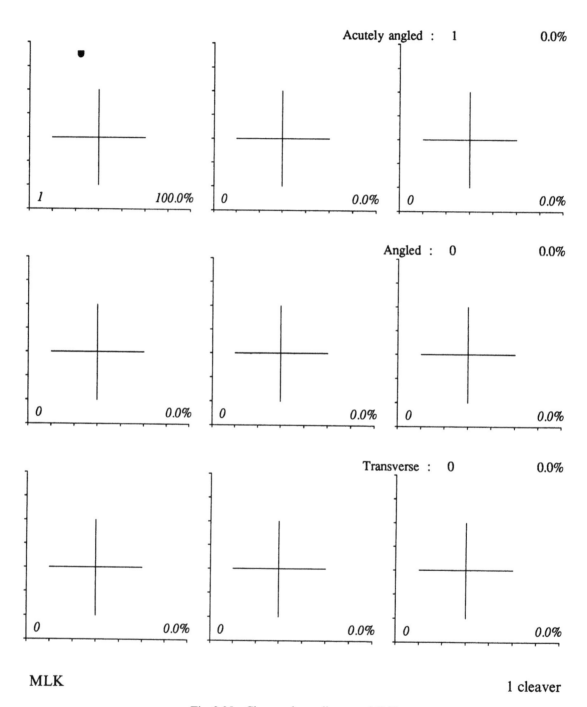

Fig. 8.35 Cleaver-shape diagram: MLK

9

THE OLDUVAI BIFACES: TECHNOLOGY AND RAW MATERIALS

P. CALLOW

Introduction

Metrical characteristics of the Olduvai bifaces have been discussed in the previous chapter by Dr Derek Roe. To complement his work, Dr Mary Leakey asked me to undertake a further analysis of these pieces with particular emphasis on the technological information provided by scar counts, and by the presence of primary flake surfaces and cleavage planes. It was expected that these would reflect not only the limitations of the various raw materials but perhaps also cultural preferences. A key objective of the investigation was simply to document the variation between assemblages, taking note of raw material, stratigraphic provenance and the type of industry (Acheulean or Developed Oldowan). Bifaces from throughout the Olduvai sequence are included, though particular attention is paid to those from Bed IV. It also covers both handaxes and cleavers.

As Dr Roe describes, the dispersal of the artefacts among several institutions was a major source of difficulty, in that bifaces were unexpectedly 'rediscovered' from time to time over a period of nearly ten years (and all the calculations had to be repeated for the enlarged sample). These delays yielded one important benefit, though. The experimental knapping carried out by Peter Jones in the late 1970s provided fresh insights into the flaking qualities of the local rocks (Jones 1979, for example), and suggested possible explanations for some features of the archaeological material.

The data

The bifaces studied are from sites ranging in stratigraphic position from the middle of Bed II to post-Masek hill wash. The numbers of pieces from different sites vary greatly, as Tables 9.1–9.2 show. This has two particular implications. Some samples are really too small for comparisons to be very useful; this is a problem not infrequently encountered when working with bifaces from sites in primary or near-primary context. Also, when the biface series are grouped into major stratigraphic units some series will tend to predominate – notably the large one from WK (one solution would be to apply weights based on the reciprocal of the sample size, but in this case the very small series would create too much 'noise', anyway, because of the importance given to individual pieces from them).

Only 'finished' pieces were included: in all, 114 cleavers and 520 handaxes. The breakdown into these classes follows Dr Leakey rather than Dr Roe; inevitably, there are some differences in the way artefacts have been classified by these two colleagues. In our sample there are 108 pieces for which Dr Roe has recorded a cleaver edge breadth measurement; 12 of these are classed by Dr Leakey as handaxes.

Sources

The data are based on information provided by Dr Leakey (rather than on this author's own direct observations on the artefacts). She supplied drawings at natural size of the biface outlines

Table 9.1 *Handaxes: frequency of occurrence of the various raw materials, by site (HEB East is marked 'Acheulean?', since its attribution is questioned here)*

Site	Industry	Basalt/tr. n	Basalt/tr. %	Trachyte n	Trachyte %	Phonolite n	Phonolite %	Nephelinite n	Nephelinite %	Quartzite n	Quartzite %	Gneiss n	Gneiss %	Total n
LATER BEDS														
HK	Acheulean	0	0.0	0	0.0	0	0.0	0	0.0	34	100.0	0	0.0	34
TK Fish Gully	Acheulean	3	10.3	1	3.4	3	10.3	0	0.0	22	75.9	0	0.0	29
FLK Masek	Acheulean	1	5.0	0	0.0	0	0.0	0	0.0	19	95.0	0	0.0	20
BED IV														
PDK Trenches I–III	Dev. Old.	7	43.8	1	6.3	5	31.3	1	6.3	2	12.5	0	0.0	16
WK East A	Dev. Old.	10	24.4	6	14.6	15	36.6	0	0.0	10	24.4	0	0.0	41
WK East C	Dev. Old.	2	22.2	3	33.3	3	33.3	1	11.1	0	0.0	0	0.0	9
WK	Acheulean	51	59.3	2	2.3	10	11.6	0	0.0	23	26.7	0	0.0	86
HEB Layer 2a	Acheulean	23	85.2	3	11.1	1	3.7	0	0.0	0	0.0	0	0.0	27
HEB Layer 2b	Acheulean	15	57.7	5	19.2	0	0.0	0	0.0	6	23.1	0	0.0	26
HEB Layer 3	Acheulean	3	6.8	1	2.3	21	47.7	0	0.0	19	43.2	0	0.0	44
HEB East	Acheulean?	16	40.0	3	7.5	9	22.5	0	0.0	12	30.0	0	0.0	40
PDK Trench IV	Acheulean	7	77.8	0	0.0	0	0.0	1	11.1	1	11.1	0	0.0	9
BED II														
BK	Dev. Old.	3	20.0	0	0.0	0	0.0	0	0.0	12	80.0	0	0.0	15
TK Upper	Dev. Old.	4	26.7	3	20.0	0	0.0	0	0.0	8	53.3	0	0.0	15
TK Lower	Acheulean	1	11.1	1	11.1	0	0.0	0	0.0	7	77.8	0	0.0	9
SHK	Dev. Old.	18	40.0	5	11.1	0	0.0	0	0.0	21	46.7	1	2.2	45
EF-HR	Acheulean	22	73.3	1	3.3	0	0.0	1	3.3	6	20.0	0	0.0	30
MLK	Acheulean	3	12.0	0	0.0	0	0.0	0	0.0	22	88.0	0	0.0	25

Table 9.2 *Cleavers: frequency of occurrence of the various raw materials, by site*

Site	Industry	Basalt/tr. n	Basalt/tr. %	Trachyte n	Trachyte %	Phonolite n	Phonolite %	Nephelinite n	Nephelinite %	Quartzite n	Quartzite %	Total n	Percentage of bifaces %
LATER BEDS													
HK	Acheulean	1	33.3	0	0.0	0	0.0	0	0.0	2	66.7	3	8.1
TK Fish Gully	Acheulean	0	0.0	0	0.0	0	0.0	0	0.0	1	100.0	1	3.3
BED IV													
WK East A	Dev. Old.	0	0.0	1	100.0	0	0.0	0	0.0	0	0.0	1	2.4
WK	Acheulean	41	67.2	0	0.0	5	8.2	0	0.0	15	24.6	61	41.5
HEB Layer 2a	Acheulean	6	85.7	1	14.3	0	0.0	0	0.0	0	0.0	7	20.6
HEB Layer 2b	Acheulean	11	68.8	2	12.5	0	0.0	1	6.3	2	12.5	16	38.1
HEB Layer 3	Acheulean	0	0.0	0	0.0	4	66.7	0	0.0	2	33.3	6	12.0
HEB East	Acheulean?	1	50.0	0	0.0	0	0.0	0	0.0	1	50.0	2	4.8
PDK Trench IV	Acheulean	3	33.3	0	0.0	0	0.0	5	55.6	1	11.1	9	50.0
BED II													
SHK	Dev. Old.	1	100.0	0	0.0	0	0.0	0	0.0	0	0.0	1	2.2
EF-HR	Acheulean	3	50.0	0	0.0	0	0.0	1	16.7	2	33.3	6	16.7
MLK	Acheulean	0	0.0	0	0.0	0	0.0	0	0.0	1	100.0	1	3.8

(obtained by running a pencil round the piece); on these were marked any areas lacking secondary flaking, with indications as to their nature (primary flake surface, cleavage plane, etc.). Also noted were scar counts for each face, the raw material, and an attribution as a handaxe or cleaver. As a further source of data, Dr Roe kindly gave free access to his own observations.

As a first step the outline drawings were digitised. Two points, at the tip and butt of the biface, were recorded for orientation purposes; the outline was then traced. This resulted in a set of about thirty coordinates representing the shape of the piece (initial experiments had suggested that increasing the number of points beyond this gave negligible improvement in the accuracy of area calculations). If natural or primary flake surfaces were present, these were likewise digitised. If any such features on the other face had to be recorded it was not necessary to take care in aligning the drawing, nor to re-record the outline. Instead, the locations of the same reference points were obtained so that the computer would be able to perform any rotation needed and then superimpose the data for the two faces for inspection purposes, after mirror-imaging that for the second face. The equipment used was a D-MAC digitiser, as described by Doran and Hodson (1975: 325). The paper tape it produced – remember that this work was being done in the 1970s – was processed on the University of Cambridge's IBM 370 mainframe computer, and its successors.

To check the accuracy of the recording, the outlines were plotted out for visual inspection. After any errors had been corrected, for each piece its total area and those of any unflaked surfaces were calculated and then merged with other information, including Dr Roe's measurements, which had been keypunched separately. The data was then analysed using the SPSS program package (Nie *et al.* 1975, and more recent publications of SPSS Inc.).

Because of the circumstances under which data were obtained, as described above, there are some differences between the totals (and hence some of the statistics) obtained by Dr Roe and by this author. The sample used here for HK is very incomplete, for example, as no data were provided by Dr Leakey for the majority of its pieces, stored at Dar-es-Salaam. On the other hand, a few bifaces from HEB West Layer 3 came to light after Dr Roe's analysis had been completed.

The variables employed

The following quantitative variables were calculated from the digitised information for each biface (for the two faces separately, and for the whole piece), to be used both directly and to derive further indices:

area (area of the biface outline, in sq mm)
count (total secondary scar count)
scar size (mean secondary scar area, in sq mm – i.e. the ratio area/count)
% scars (percentage area covered by secondary flaking)
% cortex (percentage area covered by cortex, etc.)
% primary (percentage area covered by residual primary, i.e. positive, flake scar surface)

For the analyses described below, length rather than area was taken as the preferred measure of biface size, so area is not listed in the tables. This was partly for historical reasons (the use of length is deeply embedded in the traditions of lithic metrology), but it also avoids the very high degree of mutual redundancy existing between the area of the piece, the scar count and the mean scar size.

Two presence/absence variables were derived from Dr Leakey's outline drawings, for primary flake scars and 'cortex'. For the latter, all natural surfaces (including cleavage planes) and old scars have been treated as a single category.

In addition, three simple ratios were derived in order to compare the secondary working of the two faces. They were computed in such a way as to give values between zero and one (the latter implies full symmetry), by always choosing the larger of the two observations as the denominator:

Table 9.3 *Handaxes and cleavers: frequency of cortex and primary scars, by biface type and raw material*

| | Handaxes | | | | Cleavers | | | |
| | Primary scar | | Cortex | | Primary scar | | Cortex | |
Raw Material	n	%	n	%	n	%	n	%
LATER BEDS								
Basalt/tr.	0	0.0	0	0.0	1	100.0	1	100.0
Trachyte	0	0.0	1	100.0	0	0.0	0	0.0
Phonolite	0	0.0	2	66.7	0	0.0	0	0.0
Quartzite	13	17.3	36	48.0	1	33.3	2	66.7
BED IV								
Basalt/tr.	62	46.3	100	74.6	51	82.3	53	85.5
Trachyte	4	16.7	14	58.3	3	75.0	0	0.0
Phonolite	12	18.8	39	60.9	7	77.8	8	88.9
Nephelinite	1	33.3	3	100.0	4	66.7	5	83.3
Quartzite	21	28.8	54	74.0	7	33.3	20	95.2
BED II								
Basalt/tr.	21	41.2	43	84.3	3	75.0	3	75.0
Trachyte	4	40.0	9	90.0	0	0.0	0	0.0
Nephelinite	1	100.0	1	100.0	1	100.0	1	100.0
Quartzite	17	22.4	57	75.0	1	33.3	3	100.0
Gneiss	0	0.0	1	100.0	0	0.0	0	0.0
ALL BEDS								
Basalt/tr.	83	43.9	143	75.7	55	82.1	57	85.1
Trachyte	8	22.9	24	68.6	3	75.0	0	0.0
Phonolite	12	17.9	41	61.2	7	77.8	8	88.9
Nephelinite	2	50.0	4	100.0	5	71.4	6	85.7
Quartzite	51	22.8	147	65.6	9	33.3	25	92.6
Gneiss	0	0.0	1	100.0	0	0.0	0	0.0

scar count ratio (ratio of the scar counts for the two faces

scar area ratio (ratio of the areas of secondary flaking on the two faces)

scar size ratio (ratio of the mean scar sizes on the two faces)

Raw materials

Basalt and trachyandesite have been treated as a single class; also included in this is lava of indeterminate character, likely to be one or other of these materials. (In the interests of brevity, this group is referred to as 'bas/tr' in the text which follows.) Other categories used are trachyte, phonolite, nephelinite, quartzite and gneiss; this last is represented by one piece only, and nephelinite is also extremely scarce. The raw material frequencies are set out in Tables 9.1 and 9.2.

The sources and flaking properties of the rocks used by the Olduvai hominids were discussed by Hay (1976) and more recently by Jones (1979, for instance, and in this volume). The quartzite was evidently obtained as exfoliated fragments or was flaked off larger slabs or the outcrops themselves. Phonolite took the form of slabs and blocks, often with marked cleavage planes; as Jones has remarked, although it lends itself to careful shaping, in use the edges are fragile and need frequent resharpening. Most of the lavas are obtainable as water-rounded cobbles.

THE BIFACES: TECHNOLOGY AND RAW MATERIALS

Table 9.4 *Handaxes and cleavers: occurrence of different combinations of cortex and primary scars on the two faces. Key: C cortex; P primary scar; - neither cortex nor primary scar, i.e. entirely secondarily flaked. Note that the combinations -/- and C/- give no indication of the technique used in the manufacture of the piece because cortical or primary surface may have been removed.*

		Handaxes						Cleavers				
Site	Industry	n	-/- %	C/- %	C/C %	C/P %	P/- %	n	-/- %	C/- %	C/C %	C/P %	P/- %
LATER BEDS													
HK	Acheulean	34	41.2	20.6	11.8	14.7	11.8	3	33.0	0.0	0.0	66.7	0.0
TK Fish Gully	Acheulean	29	41.4	20.7	34.5	3.4	0.0	1	0.0	100.0	0.0	0.0	0.0
FLK Masek	Acheulean	20	60.0	20.0	5.0	5.0	10.0	0	0.0	0.0	0.0	0.0	0.0
BED IV													
PDK Trenches I–III	Dev. Old.	16	25.0	25.0	43.8	6.3	0.0	0	0.0	0.0	0.0	0.0	0.0
WK East A	Dev. Old.	41	29.3	34.1	29.3	4.9	2.4	1	0.0	0.0	0.0	0.0	100.0
WK East C	Dev. Old.	9	22.2	44.4	33.3	0.0	0.0	0	0.0	0.0	0.0	0.0	0.0
WK	Acheulean	86	9.3	15.1	18.6	50.0	7.0	61	0.0	4.9	24.6	59.0	11.5
HEB Layer 2a	Acheulean	27	22.2	14.8	22.2	29.6	11.1	7	0.0	14.3	14.3	57.1	14.3
HEB Layer 2b	Acheulean	26	23.1	23.1	23.1	26.9	3.8	16	6.3	12.5	31.3	25.0	25.0
HEB Layer 3	Acheulean	44	20.5	25.0	22.7	18.2	13.6	6	0.0	0.0	16.7	66.7	16.7
HEB East	Acheulean?	40	52.5	35.0	12.5	0.0	0.0	2	0.0	50.0	50.0	0.0	0.0
PDK IV	Acheulean	9	0.0	44.4	11.1	11.1	33.3	9	0.0	22.2	33.3	33.3	11.1
BED II													
BK	Dev. Old.	15	26.7	40.0	20.0	0.0	13.3	0	0.0	0.0	0.0	0.0	0.0
TK Upper	Dev. Old.	15	40.0	13.3	33.3	13.3	0.0	0	0.0	0.0	0.0	0.0	0.0
TK Lower	Acheulean	9	11.1	11.1	33.3	44.4	0.0	0	0.0	0.0	0.0	0.0	0.0
SHK	Dev. Old.	45	26.7	31.1	20.0	22.2	0.0	1	100.0	0.0	0.0	0.0	0.0
EF-HR	Acheulean	30	3.3	6.7	36.7	50.0	3.3	6	0.0	0.0	16.7	83.3	0.0
MLK	Acheulean	25	0.0	40.0	56.0	0.0	4.0	1	0.0	0.0	100.0	0.0	0.0

Table 9.5 *Handaxes and cleavers: occurrence of different combinations of cortex and primary scars on the two faces, by bed and industry type. For an explanation of the column heading see Table 9.4. In this and similar tables, the set of figures in italics shows the effect of reclassifying HEB East as Developed Oldowan.*

	Handaxes						Cleavers					
Industry	n	-/- %	C/- %	C/C %	C/P %	P/- %	n	-/- %	C/- %	C/C %	C/P %	P/- %
LATER BEDS												
Acheulean	83	45.8	20.5	18.1	8.4	7.2	4	25.0	25.0	0.0	50.0	0.0
BED IV												
Combined	298	22.8	24.8	22.1	23.5	6.7	102	1.0	8.8	25.5	50.0	14.7
Acheulean	232	21.6	22.4	19.0	28.9	8.2	101	1.0	8.9	25.7	50.5	13.9
Dev. Old.	66	27.3	33.3	33.3	4.5	1.5	1	0.0	0.0	0.0	0.0	100.0
Acheulean	*192*	*15.1*	*19.8*	*20.3*	*34.9*	*9.9*	*99*	*1.0*	*8.1*	*25.3*	*51.5*	*14.1*
Dev. Old.	*106*	*36.8*	*34.0*	*25.5*	*2.8*	*0.9*	*3*	*0.0*	*33.3*	*33.3*	*0.0*	*33.3*
BED II												
Combined	139	17.3	25.2	32.4	22.3	2.9	8	12.5	0.0	25.0	62.5	0.0
Acheulean	64	3.1	20.3	43.8	29.7	3.1	7	0.0	0.0	28.6	71.4	0.0
Dev. Old.	75	29.3	29.3	22.7	16.0	2.7	1	100.0	0.0	0.0	0.0	0.0

Variability in the Olduvai bifaces: general remarks

Cortical and primary flake surfaces

The frequencies of relict surfaces of various kinds ('cortex') and of primary flake scars, and their presence in different combinations, are listed in Tables 9.3–9.6. (A minor caveat: it is to be expected that in some cases very small patches of cortex or primary scar will have been overlooked, so the figures given will be underestimates.) The three late series are characterised by a very high percentage of pieces which have been entirely secondarily worked. Beds II and IV differ chiefly in respect of bifaces with cortex on both faces (common in Bed II) and of those made on flakes with complete removal of dorsal cortex. Furthermore, few of the bifaces from series attributed to the Developed Oldowan in Bed IV preserve evidence of having been made on flakes, whereas this is quite frequent in Bed II.

Both cortical and primary surfaces are most strongly represented on bas/tr and nephelinite, which also exhibit a very high frequency of the combination primary/cortex; they are more common on cleavers than on handaxes. There are important differences in the treatment of bifaces attributed to the Acheulean and Developed Oldowan: thus for phonolite the combination primary/cortex occurs frequently on the former but is absent on the latter; a similar pattern exists for bas/tr, but for quartzite the industrial differences chiefly concern the amount of cortex.

Metrical data

Means and standard deviations for some of the observations are listed in Tables 9.7–9.10. The mean percentage area formed of each type of surface is represented graphically in Figs 9.1 and 9.2. However, in tabulating the percentage areas and the three scar ratios in Tables 9.11–9.14, distribution-free statistics (the median and interquartile range) have been preferred to the mean and standard deviation. This is because the distribution of these variables is not always Gaussian, as confirmed by Kolmogorov-Smirnov tests on stratigraphic and raw material subsets – thus where a large proportion of bifaces entirely lack cortex and primary scars the distribution may be heavily skewed.

There are considerable differences between the bifaces when grouped according to bed, raw material or industry. Thus the predominantly quartzite handaxes of the three late sites are larger and more intensively worked than are those of earlier assemblages. Again, bas/tr bifaces, being often made from flakes, tend to have relatively few removals compared to those of phonolite and, to a lesser extent, trachyte and quartzite. The scars on the phonolite pieces are relatively small (as are the bifaces themselves). On the other hand, as Dr Leakey has observed, the contrasts between assemblages termed Acheulean and Developed Oldowan include not only length but all of the variables except perhaps percentage of cortex and scar size ratio (the most extreme being for percentage of primary).

The variation in Dr Roe's biface shape indices with respect to raw material is shown for Bed IV in Tables 9.9 and 9.10. The raw material proves to have a very slight influence on B/L in the Acheulean, but a more marked one in the Developed Oldowan (contrasting phonolite and quartzite). L_1/L is also quite stable, though in the Acheulean the phonolite pieces tend to be pointed (see also B_1/B_2). Trachyte bifaces, and also the few made on nephelinite, are relatively thin (Th/B).

The relationship between raw material and other variables

Handaxes and cleavers

As indicated earlier, the classification of bifaces as either handaxes or cleavers is an inexact process. Being essentially morphological and technical, moreover, it begs the functional questions raised by their considerable size range and the differences of form within either class. But once the categorisation has been performed, however arbitrarily, it both increases the number

THE BIFACES: TECHNOLOGY AND RAW MATERIALS

Table 9.6 *Handaxes and cleavers: occurrence of different combinations of cortex and primary scars on the two faces, by raw material and industry type (for an explanation of the column headings see Table 9.4)*

		Handaxes						Cleavers				
Material	Industry	n	-/- %	C/- %	C/C %	C/P %	P/- %	n	-/- %	C/- %	C/C %	C/P %	P/- %
LATER BEDS													
Basalt/tr.	Acheulean	4	100.0	0.0	0.0	0.0	0.0	1	0.0	0.0	0.0	100.0	0.0
Trachyte	Acheulean	1	0.0	0.0	100.0	0.0	0.0	0	0.0	0.0	0.0	0.0	0.0
Phonolite	Acheulean	3	33.3	33.3	33.3	0.0	0.0	0	0.0	0.0	0.0	0.0	0.0
Quartzite	Acheulean	75	44.0	21.3	17.3	9.3	8.0	3	33.3	33.3	0.0	33.3	0.0
BED IV													
Basalt/tr.	Combined	134	17.9	23.1	15.7	35.8	7.5	62	0.0	9.7	17.7	58.1	14.5
	Acheulean	115	15.7	21.7	14.8	40.0	7.8	62	0.0	9.7	17.7	58.1	14.5
	Dev. Old.	19	31.6	31.6	21.1	10.5	5.3	0	0.0	0.0	0.0	0.0	0.0
	Acheulean	*99*	*9.1*	*19.2*	*16.2*	*46.5*	*9.1*	*61*	*0.0*	*9.8*	*16.4*	*59.0*	*14.8*
	Dev. Old.	*35*	*42.9*	*34.3*	*14.3*	*5.7*	*2.9*	*1*	*0.0*	*0.0*	*100.0*	*0.0*	*0.0*
Trachyte	Combined	24	33.3	16.7	37.5	4.2	8.3	4	25.0	0.0	0.0	0.0	75.0
	Acheulean	14	35.7	21.4	21.4	7.1	14.3	3	33.3	0.0	0.0	0.0	66.7
	Dev. Old.	10	30.0	10.0	60.0	0.0	0.0	1	0.0	0.0	0.0	0.0	100.0
	Acheulean	*11*	*36.4*	*9.1*	*27.3*	*9.1*	*18.2*	*3*	*33.3*	*0.0*	*0.0*	*0.0*	*66.7*
	Dev. Old.	*13*	*30.8*	*23.1*	*46.2*	*0.0*	*0.0*	*1*	*0.0*	*0.0*	*0.0*	*0.0*	*100.0*
Phonolite	Combined	64	34.4	35.9	15.6	9.4	4.7	9	0.0	11.1	11.1	66.7	11.1
	Acheulean	41	39.0	26.8	12.2	14.6	7.3	9	0.0	11.1	11.1	66.7	11.1
	Dev. Old.	23	26.1	52.2	21.7	0.0	0.0	0	0.0	0.0	0.0	0.0	0.0
	Acheulean	*32*	*37.5*	*21.9*	*12.5*	*18.8*	*9.4*	*9*	*0.0*	*11.1*	*11.1*	*66.7*	*11.1*
	Dev. Old.	*32*	*31.3*	*50.0*	*18.8*	*0.0*	*0.0*	*0*	*0.0*	*0.0*	*0.0*	*0.0*	*0.0*
Nephelinite	Combined	3	0.0	33.3	33.3	33.3	0.0	6	0.0	16.7	16.7	50.0	16.7
	Acheulean	1	0.0	0.0	0.0	100.0	0.0	6	0.0	16.7	16.7	50.0	16.7
	Dev. Old.	2	0.0	50.0	50.0	0.0	0.0	0	0.0	0.0	0.0	0.0	0.0
Quartzite	Combined	73	19.2	20.5	34.2	19.2	6.8	21	0.0	4.8	61.9	28.6	4.8
	Acheulean	61	18.0	21.3	31.1	21.3	8.2	21	0.0	4.8	61.9	28.6	4.8
	Dev. Old.	12	25.0	16.7	50.0	8.3	0.0	0	0.0	0.0	0.0	0.0	0.0
	Acheulean	*49*	*8.2*	*22.4*	*32.7*	*26.5*	*10.2*	*20*	*0.0*	*0.0*	*65.0*	*30.0*	*5.0*
	Dev. Old.	*24*	*41.7*	*16.7*	*37.5*	*4.2*	*0.0*	*1*	*0.0*	*100.0*	*0.0*	*0.0*	*0.0*
BED II													
Basalt/tr.	Combined	51	11.8	27.5	27.5	29.4	3.9	4	25.0	0.0	0.0	75.0	0.0
	Acheulean	26	7.7	7.7	30.8	50.0	3.8	3	0.0	0.0	0.0	100.0	0.0
	Dev. Old.	25	16.0	48.0	24.0	8.0	4.0	1	100.0	0.0	0.0	0.0	0.0
Trachyte	Combined	10	10.0	20.0	40.0	30.0	0.0	0	0.0	0.0	0.0	0.0	0.0
	Acheulean	2	0.0	0.0	50.0	50.0	0.0	0	0.0	0.0	0.0	0.0	0.0
	Dev. Old.	8	12.5	25.0	37.5	25.0	0.0	0	0.0	0.0	0.0	0.0	0.0
Nephelinite	Acheulean	1	0.0	0.0	0.0	100.0	0.0	1	0.0	0.0	0.0	100.0	0.0
Quartzite	Combined	76	22.4	23.7	35.5	15.8	2.6	3	0.0	0.0	66.7	33.3	0.0
	Acheulean	35	0.0	31.4	54.3	11.4	2.9	3	0.0	0.0	66.7	33.3	0.0
	Dev. Old.	41	41.5	17.1	19.5	19.5	19.5	0	0.0	0.0	0.0	0.0	0.0
Gneiss	Dev. Old.	1	0.0	100.0	0.0	0.0	0.0	0	0.0	0.0	0.0	0.0	0.0

of possible queries and reduces the size of samples to the point at which some queries become unanswerable.

Table 9.15 shows the extent to which each raw material was used to make either type, in Bed IV. It indicates a very high degree of selectivity, with a Chi-Square of 20.9 (P = .00033). Thus although nephelinite is rare, when available it tended to be used for cleavers rather than handaxes, as did bas/tr (far more abundant); phonolite, on the

Table 9.7 *Handaxes: means and standard deviations for quantitative attributes, by site*

Site	Industry	Length (mm)			Scar count			Average scar size (sq mm)			B/L (ratio)			T/B (ratio)			B_1/B_2 (ratio)			L_1/L (ratio)		
		mean	sd	n	mean	sd	n	mean	sd	n	mean	sd	n	mean	sd	n	mean	sd	n	mean	sd	n
LATER BEDS																						
HK	Acheulean	123.9	21.1	34	20.0	5.1	34	597	228	34	0.617	0.060	34	0.527	0.110	34	0.785	0.166	34	0.411	0.049	34
TK Fish Gully	Acheulean	159.4	37.8	29	27.5	9.2	29	642	276	29	0.533	0.071	29	0.593	0.141	29	0.681	0.116	29	0.387	0.076	29
FLK Masek	Acheulean	185.9	57.5	20	26.6	9.7	20	1006	350	20	0.566	0.086	20	0.619	0.079	20	0.744	0.112	20	0.426	0.105	20
BED IV																						
PDK Trenches I-III	Dev. Old.	73.3	17.7	16	12.7	4.5	16	400	302	16	0.727	0.085	16	0.602	0.108	8	0.748	0.187	16	0.405	0.095	16
WK East A	Dev. Old.	91.2	25.7	41	14.6	4.6	41	404	181	41	0.648	0.103	41	0.595	0.146	41	0.711	0.160	41	0.417	0.095	41
WK East C	Dev. Old.	107.0	29.2	9	14.3	3.2	9	568	254	9	0.648	0.120	9	0.601	0.177	9	0.635	0.141	9	0.411	0.053	9
WK	Acheulean	129.4	32.0	84	14.3	7.1	86	596	278	86	0.643	0.093	84	0.535	0.094	73	0.729	0.132	84	0.387	0.069	84
HEB Layer 2a	Acheulean	154.6	24.3	27	17.8	4.9	27	770	284	27	0.569	0.048	27	0.565	0.107	27	0.637	0.119	27	0.380	0.070	27
HEB Layer 2b	Acheulean	150.0	30.3	26	17.2	6.6	26	724	287	26	0.589	0.081	26	0.525	0.095	26	0.727	0.206	26	0.424	0.063	26
HEB Layer 3	Acheulean	138.5	26.2	44	23.5	10.1	44	540	222	44	0.570	0.083	44	0.525	0.090	38	0.646	0.156	44	0.362	0.074	44
HEB East	Acheulean?	94.0	32.1	40	16.8	5.1	40	470	237	40	0.709	0.109	40	0.607	0.116	40	0.741	0.145	40	0.423	0.065	40
PDK Trench IV	Acheulean	157.9	24.1	9	16.9	5.3	9	742	122	9	0.527	0.104	9	0.484	0.102	9	0.664	0.137	9	0.397	0.075	9
BED II																						
BK	Dev. Old.	62.5	24.4	15	11.7	5.1	15	285	107	15	0.761	0.105	15	0.730	0.077	15	0.774	0.219	15	0.409	0.088	15
TK Upper	Dev. Old.	110.7	59.5	15	14.3	4.7	15	455	243	15	0.624	0.125	15	0.611	0.138	13	0.763	0.151	15	0.412	0.049	15
TK Lower	Acheulean	151.0	57.2	9	15.7	7.6	9	882	449	9	0.670	0.169	9	0.512	0.078	9	0.677	0.193	9	0.357	0.075	9
SHK	Dev. Old.	103.4	40.6	45	16.4	5.4	45	462	199	45	0.706	0.120	45	0.619	0.115	44	0.784	0.174	45	0.403	0.087	45
EF-HR	Acheulean	146.0	29.3	30	13.0	6.0	29	652	391	29	0.584	0.077	30	0.634	0.143	30	0.689	0.164	30	0.386	0.089	30
MLK	Acheulean	169.3	31.9	25	17.2	7.4	25	954	376	25	0.573	0.080	25	0.689	0.099	24	0.710	0.152	25	0.395	0.074	25

Table 9.8 Cleavers: means and standard deviations for quantitative attributes, by site

Site	Industry	Length (mm) mean	sd	n	Scar count mean	sd	n	Average scar size (sq mm) mean	sd	n	B/L (ratio) mean	sd	n	T/B (ratio) mean	sd	n	B_1/B_2 (ratio) mean	sd	n	L_1/L (ratio) mean	sd	n
LATER BEDS																						
HK	Acheulean	139.7	11.1	3	13.7	2.5	3	1130	604	3	0.782	0.182	3	0.388	0.083	3	0.844	0.293	3	0.479	0.076	3
TK Fish Gully	Acheulean	205.0	—	1	26.0	0.0	1	1026	—	1	0.507	—	1	0.490	—	1	0.805	—	1	0.346	—	1
BED IV																						
WK East A	Dev. Old.	120.0	—	1	17.0	—	1	501	—	1	0.683	—	1	0.402	—	1	0.786	—	1	0.492	—	1
WK	Acheulean	138.8	14.3	61	11.8	3.9	61	741	300	61	0.667	0.066	61	0.520	0.080	59	1.061	0.236	61	0.506	0.161	61
HEB Layer 2a	Acheulean	156.1	26.6	7	15.3	4.8	7	773	439	7	0.580	0.050	7	0.560	0.087	7	0.929	0.169	7	0.413	0.090	7
HEB Layer 2b	Acheulean	166.6	24.8	16	17.0	6.4	16	881	374	16	0.557	0.059	16	0.524	0.129	16	0.933	0.177	16	0.453	0.093	16
HEB Layer 3	Acheulean	136.0	12.0	6	18.7	5.2	6	537	258	6	0.604	0.051	6	0.501	0.063	6	1.062	0.186	6	0.504	0.174	6
HEB East	Acheulean	97.0	11.3	2	13.5	3.5	2	488	344	2	0.676	0.188	2	0.424	0.029	2	1.200	0.228	2	0.581	0.107	2
PDK Trench IV	Acheulean	158.9	25.3	9	16.3	8.4	9	695	218	9	0.551	0.053	9	0.478	0.077	9	1.040	0.321	9	0.446	0.155	8
BED II																						
SHK	Dev. Old.	118.0	—	1	26.0	—	1	581	—	1	0.703	—	1	0.699	—	1	0.940	—	1	0.441	—	1
EF-HR	Acheulean	150.6	7.7	5	9.8	4.6	6	904	348	6	0.575	0.036	6	0.549	0.110	6	1.009	0.156	6	0.428	0.073	5
MLK	Acheulean	252.0	—	1	9.0	—	1	793	—	1	0.516	—	1	0.638	—	1	1.071	—	1	0.687	—	1

Table 9.9 Handaxes from Bed IV only: means and standard deviations for quantitative attributes, by raw material and industry type

Material	Industry	Length (mm) mean	Length (mm) sd	Length (mm) n	Scar count mean	Scar count sd	Scar count n	Average scar size (sq mm) mean	Average scar size (sq mm) sd	Average scar size (sq mm) n	B/L (ratio) mean	B/L (ratio) sd	B/L (ratio) n	T/B (ratio) mean	T/B (ratio) sd	T/B (ratio) n	B_1/B_2 (ratio) mean	B_1/B_2 (ratio) sd	B_1/B_2 (ratio) n	L_1/L (ratio) mean	L_1/L (ratio) sd	L_1/L (ratio) n
Basalt/tr.	Combined	131.7	34.2	132	16.0	6.5	134	635	283	134	0.627	0.096	132	0.546	0.098	124	0.700	0.148	132	0.400	0.076	132
	Acheulean	137.9	31.8	113	16.4	6.7	115	653	279	115	0.618	0.093	113	0.543	0.096	108	0.697	0.149	113	0.398	0.070	113
	Dev. Old.	94.5	22.1	19	13.4	4.3	19	522	289	19	0.685	0.097	19	0.570	0.116	16	0.717	0.144	19	0.414	0.107	19
	Acheulean	*145.8*	*23.6*	*97*	*16.5*	*6.9*	*99*	*683*	*271*	*99*	*0.601*	*0.079*	*97*	*0.535*	*0.093*	*92*	*0.689*	*0.147*	*97*	*0.396*	*0.069*	*97*
	Dev. Old.	*92.4*	*27.5*	*35*	*14.5*	*4.7*	*35*	*499*	*274*	*35*	*0.701*	*0.104*	*35*	*0.579*	*0.107*	*32*	*0.728*	*0.149*	*35*	*0.413*	*0.092*	*35*
Trachyte	Combined	132.6	37.9	24	18.4	7.1	24	633	295	24	0.615	0.101	24	0.550	0.100	22	0.731	0.176	24	0.407	0.080	24
	Acheulean	142.1	40.3	14	20.6	7.0	14	658	330	14	0.599	0.110	14	0.560	0.104	12	0.750	0.139	14	0.397	0.054	14
	Dev. Old.	119.2	31.2	10	15.4	6.4	10	599	251	10	0.638	0.088	10	0.539	0.100	10	0.703	0.224	10	0.422	0.108	10
	Acheulean	*148.0*	*40.7*	*11*	*20.4*	*7.4*	*11*	*657*	*344*	*11*	*0.554*	*0.059*	*11*	*0.556*	*0.069*	*9*	*0.744*	*0.127*	*11*	*0.388*	*0.053*	*11*
	Dev. Old.	*119.5*	*31.3*	*13*	*16.8*	*6.7*	*13*	*613*	*259*	*13*	*0.667*	*0.102*	*13*	*0.547*	*0.120*	*13*	*0.719*	*0.214*	*13*	*0.424*	*0.096*	*13*
Phonolite	Combined	107.1	39.4	64	20.0	9.0	64	410	184	64	0.613	0.102	64	0.571	0.152	57	0.653	0.156	64	0.384	0.070	64
	Acheulean	123.7	38.9	41	23.2	9.4	41	453	204	41	0.598	0.090	41	0.533	0.112	36	0.640	0.145	41	0.373	0.064	41
	Dev. Old.	77.6	16.4	23	14.2	3.8	23	331	105	23	0.641	0.118	23	0.637	0.189	21	0.676	0.174	23	0.403	0.077	23
	Acheulean	*130.9*	*39.8*	*32*	*24.2*	*10.3*	*32*	*467*	*217*	*32*	*0.584*	*0.082*	*32*	*0.490*	*0.067*	*27*	*0.627*	*0.142*	*32*	*0.353*	*0.051*	*32*
	Dev. Old.	*83.4*	*20.2*	*32*	*15.7*	*4.4*	*32*	*352*	*123*	*32*	*0.643*	*0.114*	*32*	*0.644*	*0.170*	*30*	*0.680*	*0.167*	*32*	*0.414*	*0.073*	*32*
Nephelinite	Combined	110.3	32.6	3	11.3	4.2	3	351	160	3	0.634	0.046	3	0.423	0.071	2	0.751	0.109	3	0.359	0.090	3
	Acheulean	144.0	—	1	8.0	—	1	526	—	1	0.597	—	1	0.372	—	1	0.792	—	1	0.368	—	1
	Dev. Old.	93.5	20.5	2	13.0	4.2	2	263	71	2	0.653	0.046	2	0.473	—	1	0.730	0.146	2	0.355	0.126	2
Quartzite	Combined	114.3	37.9	73	14.3	6.3	73	557	266	73	0.660	0.122	73	0.574	0.111	66	0.743	0.151	73	0.401	0.081	73
	Acheulean	121.8	35.6	61	14.4	6.6	61	600	262	61	0.649	0.123	61	0.565	0.113	56	0.738	0.154	61	0.395	0.082	61
	Dev. Old.	76.3	24.4	12	14.0	4.6	12	334	153	12	0.714	0.103	12	0.625	0.089	10	0.765	0.139	12	0.434	0.066	12
	Acheulean	*129.7*	*31.2*	*49*	*14.2*	*7.0*	*49*	*632*	*261*	*49*	*0.629*	*0.120*	*49*	*0.556*	*0.113*	*44*	*0.731*	*0.164*	*49*	*0.389*	*0.085*	*49*
	Dev. Old.	*83.0*	*30.6*	*24*	*14.4*	*4.5*	*24*	*403*	*206*	*24*	*0.721*	*0.102*	*24*	*0.612*	*0.101*	*22*	*0.768*	*0.121*	*24*	*0.427*	*0.065*	*24*

Table 9.10 Cleavers from Bed IV only: means and standard deviations for quantitative attributes by raw material and industry type

Site	Industry	Length (mm)			Scar count			Average scar size (sq mm)			B/L (ratio)			T/B (ratio)			B_1/B_2 (ratio)			L_1/L (ratio)		
		mean	sd	n	mean	sd	n	mean	sd	n	mean	sd	n	mean	sd	n	mean	sd	n	mean	sd	n
Basalt/tr.	Combined	144.6	23.2	62	12.9	4.6	62	759	325	62	0.638	0.083	62	0.515	0.087	61	1.044	0.255	61	0.494	0.145	61
	Acheulean	144.6	23.2	62	12.9	4.6	62	759	325	62	0.638	0.083	62	0.515	0.087	61	1.044	0.255	61	0.494	0.145	61
	Acheulean	145.2	22.8	61	12.8	4.6	61	768	321	61	0.639	0.082	61	0.517	0.087	60	1.039	0.254	60	0.491	0.145	60
	Dev. Old.	105.0	—	1	16.0	—	1	245	—	1	0.543	—	1	0.404	—	1	1.361	—	1	0.657	—	1
Trachyte	Combined	153.0	22.6	4	21.5	4.7	4	706	218	4	0.587	0.081	4	0.534	0.109	4	0.970	0.221	4	0.479	0.046	4
	Acheulean	164.0	6.6	3	23.0	4.4	3	775	209	3	0.554	0.060	3	0.577	0.080	3	1.032	0.224	3	0.475	0.055	3
	Dev. Old.	120.0	—	1	17.0	—	1	501	—	1	0.683	—	1	0.402	—	1	0.786	—	1	0.492	—	1
	Acheulean	164.0	6.6	3	23.0	4.4	3	775	209	3	0.554	0.060	3	0.577	0.080	3	1.032	0.224	3	0.475	0.055	3
	Dev. Old.	120.0	—	1	17.0	—	1	501	—	1	0.683	—	1	0.402	—	1	0.786	—	1	0.492	—	1
Phonolite	Acheulean	147.7	12.2	9	18.4	4.0	9	569	255	9	0.624	0.052	9	0.486	0.062	8	1.077	0.158	9	0.547	0.204	9
Nephelinite	Acheulean	158.2	26.4	6	16.7	9.4	6	600	218	6	0.537	0.058	6	0.518	0.177	6	1.058	0.146	6	0.461	0.098	6
Quartzite	Combined	139.6	20.8	21	12.1	5.0	21	811	346	21	0.644	0.075	21	0.524	0.075	21	0.974	0.207	21	0.454	0.154	21
	Acheulean	139.6	20.8	21	12.1	5.0	21	811	346	21	0.644	0.075	21	0.524	0.075	21	0.974	0.207	21	0.454	0.154	21
	Acheulean	142.2	17.7	20	12.2	5.1	20	815	355	20	0.636	0.066	20	0.528	0.075	20	0.971	0.212	20	0.451	0.157	20
	Dev. Old.	89.0	—	1	11.0	—	1	732	—	1	0.809	—	1	0.444	—	1	1.038	—	1	0.506	—	1

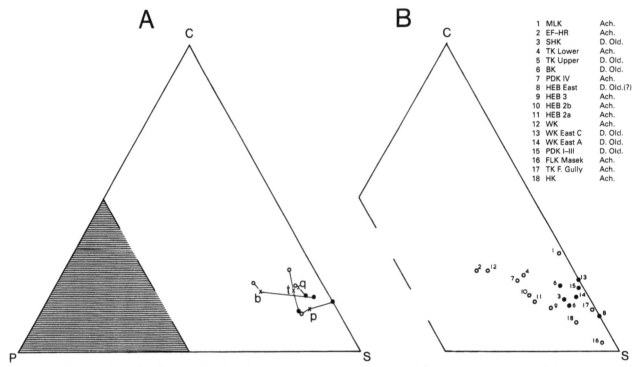

Fig. 9.1 Bifaces: triangular diagram showing the mean percentage area occupied by cortex (C), primary flake scar (P) and secondary flaking (S). Each apex of the diagram represents 100 per cent for the appropriate surface type (and 0 per cent for the other two). The shading marks the zone corresponding to more than 50 per cent primary flake surface (i.e. extending over more than the area of one face), which in practice can be ignored. The Acheulean is indicated by open circles, the Developed Oldowan by filled circles. As explained in the text, HEB East has been included as *Developed Oldowan*.
(A) The principle raw material types (the grand mean is indicated by a cross). *Key:* (b) basalt and trachyandesite; (p) phonolite; (q) quartzite; (t) trachyte.
(B) The individual series.

other hand, was under-used for this purpose and was particularly favoured for handaxes.

To draw contrasts between handaxes and cleavers runs some risk of circularity, but Table 9.3 records that cleavers tend to possess cortex more often than do handaxes. It also shows that the vast majority retain primary scar surface, except on quartzite.

The handaxes from Bed IV

There are enough handaxes from Bed IV to permit a summary description of the Acheulean and Developed Oldowan pieces made on different raw materials. In general, the Developed Oldowan handaxes are appreciably smaller in size and relatively broader than the Acheulean ones and on average have a lower scar count which is more uniform across raw materials. Primary scars are present if scarce on bas/tr, but do not occur on other materials, whereas they are more generally abundant on Acheulean pieces. The following accounts take note of these observations; comparative terms are therefore used to refer *within* each industry type. Italics denote extreme characteristics (shown by the Scheffé multiple comparison test to be significantly different from their equivalent for at least one of the other materials).

Acheulean handaxes:

Bas/tr: large handaxes with *few and very large scars*, often mostly on one face. The various scar ratios are all below 0.7, emphasising the contrasting treatment of the faces. Primary scar surface and cortex occur

THE BIFACES: TECHNOLOGY AND RAW MATERIALS

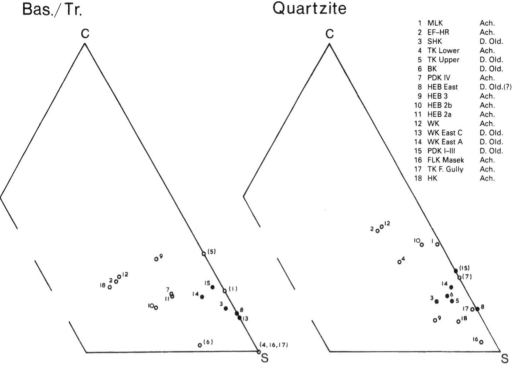

Fig. 9.2 Bifaces: as Fig. 9.1B, but for the two most common raw materials

very frequently (only 9 per cent of pieces are entirely worked).

Trachyte: large, relatively narrow but broad-tipped handaxes, with fairly large and numerous scars, and occasionally primary scar surface or cortex.

Phonolite: small, relatively thin, pointed handaxes with *numerous small flake scars covering most of the surface. Cortex* and primary surfaces are occasionally present.

Quartzite: rather small but quite broad handaxes, relatively broad-tipped, and with *few* and large secondary scars. Cortex is very abundant, primary surfaces rather abundant (only 8 per cent of the pieces are entirely worked).

Developed Oldowan handaxes:

Bas/tr: relatively broad handaxes, *moderate in size*, thickness and scar size. Often cortical, and more rarely with primary scar surface (43 per cent are entirely worked, strongly contrasting with the Acheulean).

Trachyte: *exceptionally large* (for this industry) and rather thin handaxes, *with very large scars*. Cortex is very abundant.

Phonolite: *very small*, very thick handaxes, with *small scars* and distinctly pointed in shape. Cortex is very abundant.

Quartzite: *very small* handaxes, rather less thick than those of phonolite (also broader, and broad-tipped). Cortex is abundant; 42 per cent are entirely worked, again in contrast to the Acheulean quartzite handaxes.

Table 9.11 *Handaxes: medians and interquartile ranges for types of surface and for secondary scar ratios, by site*

Site	Industry	Scars (%)			Cortex (%)			Primary (%)			Scar count (ratio)			Total scar area (ratio)				Average scar size (ratio)		
		med	iqr	n	med	iqr	n	med	iqr	n	med	iqr	n	mean	iqr	n	med	iqr	n
LATER BEDS																				
HK	Acheulean	86.5	35.2	34	0.0	18.0	34	0.0	22.3	34	0.862	0.317	34	0.897	0.366	34	0.869	0.219	34
TK Fish Gully	Acheulean	91.8	27.1	29	8.2	26.7	29	0.0	0.0	29	0.900	0.100	27	0.958	0.201	29	0.853	0.140	27
FLK Masek	Acheulean	100.0	12.3	20	0.0	4.8	20	0.0	0.0	20	0.833	0.203	20	1.000	0.201	20	0.811	0.203	20
BED IV																				
PDK Trenches I–III	Dev. Old.	80.1	29.9	16	19.1	25.0	16	0.0	0.0	16	0.800	0.107	13	0.773	0.386	16	0.670	0.292	13
WK East A	Dev. Old.	80.6	38.0	41	14.4	35.3	41	0.0	0.0	41	0.778	0.286	33	0.770	0.402	41	0.723	0.403	31
WK East C	Dev. Old.	86.2	36.3	9	13.8	36.3	9	0.0	0.0	9	0.750	0.319	7	0.899	0.263	9	0.622	0.352	7
WK	Acheulean	50.1	44.6	86	20.1	22.8	86	24.6	40.7	86	0.667	0.500	85	0.543	0.543	86	0.701	0.324	70
HEB Layer 2a	Acheulean	77.9	35.3	27	10.9	22.1	27	0.0	27.8	27	0.636	0.333	27	0.774	0.256	27	0.693	0.259	26
HEB Layer 2b	Acheulean	66.3	41.5	26	16.9	30.6	26	0.0	34.1	26	0.766	0.420	26	0.607	0.575	26	0.750	0.389	22
HEB Layer 3	Acheulean	74.1	31.7	44	6.5	22.5	44	0.0	22.0	44	0.769	0.252	44	0.830	0.392	44	0.716	0.306	42
HEB East	Acheulean?	100.0	15.9	40	0.0	15.9	40	0.0	0.0	40	0.833	0.195	40	1.000	0.268	40	0.802	0.322	40
PDK Trench IV	Acheulean	67.0	25.9	9	15.5	25.0	9	0.0	37.7	9	0.733	0.515	9	0.612	0.349	9	0.680	0.283	9
BED II																				
BK	Dev. Old.	81.3	35.5	15	6.0	28.1	15	0.0	0.0	15	0.714	0.625	15	0.875	0.433	15	0.856	0.255	14
TK Upper	Dev. Old.	88.1	59.1	15	11.9	34.1	15	0.0	0.0	15	0.833	0.375	15	0.903	0.302	15	0.770	0.206	14
TK Lower	Acheulean	60.6	16.6	9	23.5	28.7	9	0.0	33.3	9	0.714	0.554	9	0.543	0.397	9	0.593	0.307	8
SHK	Dev. Old.	80.3	35.6	45	17.2	27.9	45	0.0	0.0	45	0.727	0.232	45	0.785	0.352	45	0.734	0.302	42
EF-HR	Acheulean	39.6	38.6	30	24.5	24.1	30	30.4	35.5	30	0.472	0.627	30	0.516	0.801	30	0.484	0.337	23
MLK	Acheulean	71.9	28.1	25	25.7	29.4	25	0.0	0.0	25	0.833	0.247	24	0.707	0.350	25	0.743	0.246	22

Table 9.12 *Cleavers: medians and interquartile ranges for types of surface and for secondary scar ratios, by site*

Site	Industry	Scars (%)			Cortex (%)			Primary (%)			Scar count (ratio)			Total scar area (ratio)			Average scar size (ratio)		
		med	iqr	n	med	iqr	n	med	iqr	n	med	iqr	n	mean	iqr	n	med	iqr	n
LATER BEDS																			
HK	Acheulean	60.0	53.1	3	16.7	20.4	3	23.3	32.7	3	0.589	0.378	2	0.802	0.414	3	0.827	0.289	2
TK Fish Gully	Acheulean	86.5	0.0	1	13.5	0.0	1	0.0	0.0	1	0.857	0.000	1	0.730	0.000	1	0.851	0.000	1
BED IV																			
WK East A	Dev. Old.	55.4	0.0	1	0.0	0.0	1	44.6	0.0	1	0.214	0.000	1	0.108	0.000	1	0.505	0.000	1
WK	Acheulean	37.5	26.8	61	28.1	19.4	61	33.6	23.7	61	0.600	0.500	61	0.446	0.402	61	0.685	0.340	51
HEB Layer 2a	Acheulean	60.7	49.3	7	22.8	24.2	7	24.6	19.3	7	0.700	0.361	7	0.742	0.289	7	0.808	0.506	7
HEB Layer 2b	Acheulean	56.1	18.1	16	16.9	31.5	16	30.0	40.6	16	0.464	0.516	16	0.566	0.564	16	0.689	0.303	15
HEB Layer 3	Acheulean	60.1	40.7	6	24.9	35.6	6	29.6	19.7	6	0.586	0.167	6	0.410	0.379	6	0.549	0.249	6
HEB East	Acheulean?	65.4	43.3	2	34.5	43.3	2	0.0	0.0	2	0.806	0.056	2	0.639	0.206	2	0.654	0.071	2
PDK Trench IV	Acheulean	49.1	36.4	9	31.0	35.0	9	0.0	37.7	9	0.500	0.485	9	0.488	0.443	9	0.687	0.146	9
BED II																			
SHK	Dev. Old.	100.0	0.0	1	0.0	0.0	1	0.0	0.0	1	0.529	0.000	1	1.000	0.000	1	0.529	0.000	1
EF-HR	Acheulean	39.7	44.6	6	23.4	16.9	6	45.6	30.0	6	0.182	0.500	6	0.139	0.837	6	0.554	0.677	3
MLK	Acheulean	16.0	0.0	1	84.0	0.0	1	0.0	0.0	1	0.000	0.000	1	0.000	0.000	1	—	—	0

Table 9.13 *Handaxes from BED IV only: medians and interquartile ranges for types of surface and for secondary scar ratios, by raw material and industry type*

Material	Industry	Scars (%)			Cortex (%)			Primary (%)			Scar count (ratio)			Total scar area (ratio)				Average scar size (ratio)		
		med	iqr	n	med	iqr	n	med	iqr	n	med	iqr	n	mean	iqr	n	med	iqr	n
Basalt/tr.	Combined	67.2	42.7	134	14.1	26.7	134	0.0	35.2	134	0.700	0.379	129	0.651	0.469	134	0.703	0.315	116
	Acheulean	63.4	42.2	115	14.9	23.3	115	17.5	37.4	115	0.700	0.396	114	0.625	0.508	115	0.713	0.298	102
	Dev. Old.	86.2	38.0	19	11.5	26.7	19	0.0	0.0	19	0.714	0.247	15	0.770	0.428	19	0.619	0.422	14
	Acheulean	*58.8*	*37.2*	*99*	*16.2*	*21.6*	*99*	*23.0*	*39.9*	*99*	*0.667*	*0.402*	*98*	*0.601*	*0.521*	*99*	*0.697*	*0.285*	*86*
	Dev. Old.	*92.1*	*24.8*	*35*	*7.9*	*22.7*	*35*	*0.0*	*0.0*	*35*	*0.800*	*0.317*	*31*	*0.843*	*0.388*	*35*	*0.766*	*0.418*	*30*
Trachyte	Combined	85.4	42.6	24	11.7	34.3	24	0.0	0.0	24	0.817	0.337	22	0.899	0.330	24	0.833	0.296	21
	Acheulean	89.6	23.5	14	2.8	23.5	14	0.0	11.1	14	0.833	0.289	14	0.848	0.321	14	0.875	0.140	14
	Dev. Old.	65.6	43.9	10	34.3	43.9	10	0.0	0.0	10	0.775	0.461	8	0.927	0.339	10	0.800	0.431	7
	Acheulean	*88.9*	*57.8*	*11*	*0.0*	*24.3*	*11*	*0.0*	*15.4*	*11*	*0.875*	*0.317*	*11*	*0.888*	*0.321*	*11*	*0.882*	*0.140*	*11*
	Dev. Old.	*79.4*	*41.4*	*13*	*20.6*	*41.4*	*13*	*0.0*	*0.0*	*13*	*0.750*	*0.494*	*11*	*0.911*	*0.339*	*13*	*0.810*	*0.431*	*10*
Phonolite	Combined	87.1	28.7	64	7.4	17.8	64	0.0	0.0	64	0.800	0.232	59	0.856	0.339	64	0.742	0.297	57
	Acheulean	87.8	29.7	41	4.2	13.6	41	0.0	14.1	41	0.833	0.246	41	0.880	0.303	41	0.760	0.308	39
	Dev. Old.	85.6	21.5	23	14.4	21.5	23	0.0	0.0	23	0.800	0.130	18	0.744	0.388	23	0.635	0.248	18
	Acheulean	*89.2*	*37.0*	*32*	*4.1*	*15.9*	*32*	*0.0*	*20.0*	*32*	*0.838*	*0.277*	*32*	*0.891*	*0.389*	*32*	*0.764*	*0.304*	*30*
	Dev. Old.	*87.1*	*20.9*	*32*	*12.9*	*20.9*	*32*	*0.0*	*0.0*	*32*	*0.800*	*0.161*	*27*	*0.779*	*0.303*	*32*	*0.649*	*0.322*	*27*
Nephelinite	Combined	23.2	72.2	3	35.6	72.2	3	0.0	41.2	3	0.429	0.444	3	0.837	0.375	3	0.545	0.231	3
	Acheulean	23.2	0.0	1	35.6	0.0	1	41.2	0.0	1	0.333	0.000	1	0.612	0.000	1	0.545	0.000	1
	Dev. Old.	55.7	72.2	2	44.2	72.2	2	0.0	0.0	2	0.603	0.349	2	0.912	0.150	2	0.537	0.231	2
Quartzite	Combined	66.6	39.4	73	20.1	35.1	73	0.0	26.3	73	0.778	0.275	71	0.725	0.461	73	0.670	0.292	63
	Acheulean	65.5	43.2	61	20.4	34.7	61	0.0	29.4	61	0.750	0.274	61	0.689	0.524	61	0.636	0.283	53
	Dev. Old.	73.2	28.6	12	20.1	32.2	12	0.0	0.0	12	0.875	0.182	10	0.804	0.295	12	0.801	0.298	10
	Acheulean	*62.2*	*37.7*	*49*	*23.4*	*27.8*	*49*	*0.0*	*33.4*	*49*	*0.714*	*0.413*	*49*	*0.559*	*0.482*	*49*	*0.630*	*0.290*	*41*
	Dev. Old.	*80.5*	*35.3*	*24*	*16.9*	*34.3*	*24*	*0.0*	*0.0*	*24*	*0.840*	*0.111*	*22*	*0.911*	*0.287*	*24*	*0.818*	*0.309*	*22*

Table 9.14 *Cleavers from Bed IV only: medians and interquartile ranges for types of surface and for secondary scar ratios, by raw material and industry type*

Material	Industry	Scars (%)			Cortex (%)			Primary (%)			Scar count (ratio)			Total scar area (ratio)			Average scar size (ratio)		
		med	iqr	n	med	iqr	n	med	iqr	n	med	iqr	n	mean	iar	n	med	iqr	n
Basalt/tr.	Combined	43.2	29.4	62	24.4	20.1	62	33.1	19.3	62	0.600	0.514	62	0.445	0.510	62	0.685	0.367	55
	Acheulean	43.2	29.4	62	24.4	20.1	62	33.1	19.3	62	0.600	0.514	62	0.445	0.510	62	0.685	0.367	55
	Acheulean	*42.7*	*29.4*	*61*	*24.1*	*19.9*	*61*	*33.6*	*17.8*	*61*	*0.600*	*0.514*	*61*	*0.443*	*0.510*	*61*	*0.681*	*0.367*	*54*
	Dev. Old.	*43.8*	*0.0*	*1*	*56.2*	*0.0*	*1*	*0.0*	*0.0*	*1*	*0.778*	*0.000*	*1*	*0.536*	*0.000*	*1*	*0.689*	*0.000*	*1*
Trachyte	Combined	71.7	30.6	4	0.0	0.0	4	28.3	30.6	4	0.513	0.445	4	0.434	0.612	4	0.773	0.303	4
	Acheulean	80.0	36.6	3	0.0	0.0	3	20.0	36.6	3	0.636	0.468	3	0.599	0.732	3	0.857	0.253	3
	Dev. Old.	55.4	0.0	1	0.0	0.0	1	44.6	0.0	1	0.214	0.000	1	0.108	0.000	1	0.505	0.000	1
	Acheulean	*80.0*	*36.6*	*3*	*0.0*	*0.0*	*3*	*20.0*	*36.6*	*3*	*0.636*	*0.468*	*3*	*0.599*	*0.732*	*3*	*0.857*	*0.253*	*3*
	Dev. Old.	*55.4*	*0.0*	*1*	*0.0*	*0.0*	*1*	*44.6*	*0.0*	*1*	*0.214*	*0.000*	*1*	*0.108*	*0.000*	*1*	*0.505*	*0.000*	*1*
Phonolite	Acheulean	41.2	26.0	9	27.0	19.4	9	31.8	5.8	9	0.571	0.167	9	0.417	0.414	9	0.642	0.285	9
Nephelinite	Acheulean	39.5	40.1	6	24.7	35.6	6	31.5	37.7	6	0.450	0.485	6	0.554	0.358	6	0.638	0.173	6
Quartzite	Combined	44.5	40.2	21	33.7	46.8	21	0.0	25.4	21	0.600	0.433	21	0.637	0.312	21	0.704	0.339	17
	Acheulean	44.5	40.2	21	33.7	46.8	21	0.0	25.4	21	0.600	0.433	21	0.637	0.312	21	0.704	0.339	17
	Acheulean	*44.3*	*38.2*	*20*	*36.8*	*42.8*	*20*	*0.0*	*27.6*	*20*	*0.600*	*0.439*	*20*	*0.631*	*0.329*	*20*	*0.711*	*0.377*	*16*
	Dev. Old.	*87.1*	*0.0*	*1*	*12.9*	*0.0*	*1*	*0.0*	*0.0*	*1*	*0.833*	*0.000*	*1*	*0.742*	*0.000*	*1*	*0.618*	*0.000*	*1*

Table 9.15 *Handaxes and cleavers from Bed IV only: typological frequencies for each raw material. Large values of the standardized residual indicate a frequency higher or lower than expected for the type in question, depending on the sign.*

	Handaxes			Cleavers			Total
Material	n	%	st. resid.	n	%	st. resid.	n
Basalt/tr.	134	68.4	−1.0	62	31.6	1.7	196
Trachyte	24	85.7	0.7	4	14.3	−1.2	28
Phonolite	64	87.7	1.3	9	12.3	−2.2	73
Nephelinite	3	33.3	−1.4	6	66.7	2.4	9
Quartzite	73	77.7	0.4	21	22.3	−0.6	94
Total	298	74.5		102	25.5		400

In general, then, there is a lot of variation within the industrial groupings (contrasting the different materials) but also substantial differences between them, often with technological implications. The trachyte bifaces go against the general trend, however, as the Acheulean and Developed Oldowan examples are very much alike apart from a scale factor.

Discrimination between groups

Simple quotation of means and standard deviations gives only a poor idea of the extent of overlap between the groups. What is the best separation that can be obtained by appropriately weighting the original variables, then? An indication is provided by the percentage of pieces which can be correctly assigned to their original group by means of the classification functions produced during discriminant analysis (Blackith and Reyment 1971).

Good results were obtained when discriminating between the two industries, leading to a success rate of *c*. 80 per cent (several slightly different selections of variables were used). Strictly speaking, this was a somewhat circular exercise because the initial classification is likely to be based on essentially the same criteria as those we are using, but it may be regarded as a test of consistency. A more modest 50–60 per cent was obtained for discrimination between the three major stratigraphic units. But percentages of only 40–50 per cent were typical for the raw materials when all of these (except gneiss) were employed for the initial partition; thus at least half of the pieces fell within a 'common' area of variation. Attempts to discriminate between any two of the materials were much more effective, generally yielding classifications which were at least 70 per cent and sometimes over 90 per cent correct. Since the loadings of variables on the discriminant functions tended to be in broad agreement with the comments made above (with primary emphasis on length, scar count, etc.), they need not be discussed here.

The industrial discriminant scores assigned at least three-quarters of the HEB East pieces to the Developed Oldowan. It is therefore suggested that, to judge from the bifaces at least, the original designation of this series as Acheulean may have been erroneous – a conclusion reached independently by Dr Roe in the previous chapter – and accordingly an alternative set of descriptive statistics is provided in the tables.

The Acheulean and the Developed Oldowan

One of the most important questions to be borne in mind during the work was whether this proposed industrial dichotomy has any validity. A study based on just a single artefact class is not likely to resolve this. Nevertheless, it is fair to enquire whether the characteristics of Developed Oldowan bifaces suggest that these represent anything more than one end of a size/morphology spectrum, with Acheulean ones forming the other. Unfortunately, none of the raw materials is represented by enough handaxes, over a restricted stratigraphic range, to provide an unequivocal, bimodal frequency distribution curve (if appropriate) to prove the point.

Developed Oldowan handaxes are not scaled-down replicas of the Acheulean ones. The contrast between the handaxes of the two industries (cleavers are irrelevant) is not merely one of size, though this is the most conspicuous difference; technological differences are also apparent. A case in point is the occurrence of pieces with primary flake scars as against those with cortex on both faces. Though the very broad definition of cortex employed here precludes the assumption that all of the latter were made on cobbles, it is clear that in the Acheulean the use of flakes as blanks was a common practice for all the lavas, *even including phonolite* (Table 9.6). On the other hand there is little evidence for such a practice in the Developed Oldowan; most lava pieces whose technology is discernible were manufactured as core tools. The case of quartzite is a little different; here the Acheulean possesses a higher percentage of bifaces with cortex on both sides, the result of rather marginal shaping of flat slabs.

An obvious question concerns a possible causal, mechanical, relationship which might result in the technical features observed. Are the various differences between the Developed Oldowan and Acheulean bifaces imposed by the size or form of the raw material? The variation in the quartzite pieces is certainly explicable in these terms – to attempt to make a small biface from a slab that is too thick calls for the removal of much or all of the cortex. But the other rocks allow and require a choice to be made as to the type of blank to be used.

At the beginning of this project the prevailing ideas about assemblage formation were rather simplistic. Assemblage variation was almost always interpreted in cultural terms, or otherwise in terms of quite straightforward 'mental templates', etc.; quite simple statistical models appeared appropriate, requiring modest amounts of data. This was long before the concept of the dynamic, complex 'operational chain' was developed, in effect moving the goalposts. Consider the fourth of the models proposed by Jones in this volume (based on a continuum of sharpening and use). If, as seems likely, the preferred technological trajectories for resharpening bifaces differ from one material to another, the model will require not only more handaxes than are in fact available, but more detailed observations on them and on the rest of their assemblages.

To conclude, the differences between the two proposed industrial types are certainly very strong regarding biface technology, as well as size. Though the sample for the Developed Oldowan is rather small, on the present evidence the non-stylistic arguments of raw material and size – albeit important in imposing technical and morphological constraints – appear insufficient to explain the contrast on the basis of a very simple model. However, more complex ones may prove appropriate.

10

RESULTS OF EXPERIMENTAL WORK IN RELATION TO THE STONE INDUSTRIES OF OLDUVAI GORGE

P. R. JONES

General introduction

This chapter deals mostly with a set of topics that come under the heading of experimental archaeology. The experiments described were carried out over a period of seven years (1976 to 1983) and originally were a series of unrelated experiments intended to solve particular research problems as they arose. The chapter draws together much of that experimental work which relates to the assemblages and artefact types of Beds III and IV. It was never intended that I should specifically cover all the individual assemblages found in these beds, nor all the various tool types defined by Dr Leakey. Instead, I consider some of the major factors that must have influenced the assemblages, both Acheulean and Developed Oldowan, and some of the major tool types that characterise them.

Many of my initial experiments were designed to establish and document the probable methods of flaking needed to arrive at particular tool types; to assess the complexity of the various processes of tool manufacture, the amounts of time involved, and the materials selected. Further work was then done to relate these factors to observed tool morphology and to possible function. The whole programme should, therefore, be seen as an attempt to go beyond the traditional archaeological methods of the definition and description of artefacts and assemblages. While detailed descriptions and the juggling of various percentages and measurements of tools or assemblages will tell you what you have, it is purely descriptive and does not tell you what they might mean. Experimental archaeology can break through this barrier and, when used properly, can provide unique insights helpful in solving the problem of what artefacts or assemblages actually mean in a given situation.

The major raw materials used at Olduvai for tool making

When working with the artefact assemblages at Olduvai we possess the unusual advantage of knowing the exact sources of all the major types of lithic raw materials that were used by early tool makers throughout most of the archaeological sequence. From Bed I until the top of the Ndutu Beds, the artefact assemblages are dominated by quartzite, basalt, trachyandesite and phonolite (in rough order of importance), and to a lesser degree nephelinite, trachyte and chert. Fig. 10.1 shows the positions of the major sources of these rocks in relation to the Gorge and to the archaeological sites.

A knowledge of the location and nature of a source of raw material in relation to an archaeological assemblage can provide much information to the archaeologist and may be of fundamental importance for the interpretation of the site. The distance over which raw materials were transported by hominids can provide evidence for the possible ranges of early hominid groups. Broadly speaking, the evidence from Olduvai

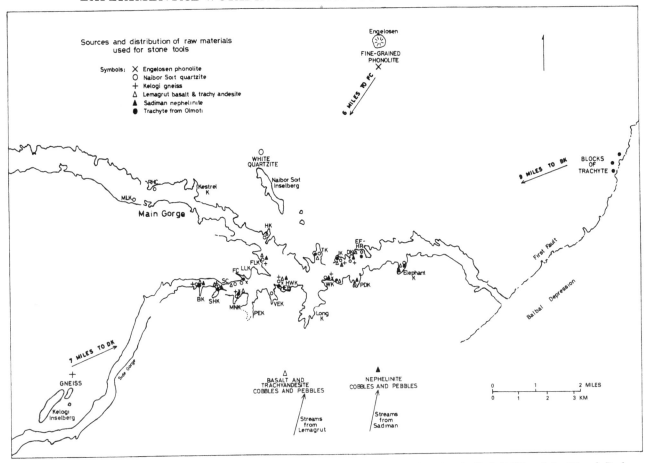

Fig. 10.1 Sources and distribution of raw materials used for stone tools found at sites in Beds I–IV and the Masek Beds

shows that during Bed I times, while people were living close to the palaeo-lake shore, the favoured stones used in tool manufacture were small river cobbles, usually of nephelinite and basalt, that were readily available close at hand in the beds of streams entering the lake. In the slightly younger industries of the Developed Oldowan and Acheulean tool makers from Upper Middle Bed II through to the Masek Beds, regular use was made of materials whose sources were up to 5 km from the sites, and sometimes material was transported up to 13 km. By roughly 20,000 years ago, in a Late Stone Age industry contemporary with the Naisiusiu Beds, we find artefacts in obsidian, a material that was most probably transported several hundred kilometres.* While this does not prove that individual human movement had increased to that extent, it certainly suggests that inter-group contact had.

The raw materials are also important because their differing qualities, such as the size and shape of the units in which they occur and their mechanical properties, often have a great effect on the type and morphology of the tools that are made from them (Jones 1979, Clark 1980). A material that splits cleanly along cleavage planes from the parent rock, producing slabs an inch thick, is necessarily approached by the tool maker in a different manner from a material that occurs in huge irregular boulders. A fine-grained material will permit subtle flaking techniques whilst a coarse-grained material will not. The physical properties of a rock are also important when it comes to tool use. In some cases, the primary flake edge may be strong and durable, while in

* *Note by M.D.L.*: The sources of the obsidians used for artefacts found in the Olduvai region are now known to be in the Naivasha basin of Kenya (Merrick and Brown 1984).

others it will be brittle and delicate. When it comes to using flakes for different cutting tasks, one made from a brittle material may need regular edge retouch in order to maintain a serviceable tool while one from a more durable material can be used for much longer periods without such retouch. These differences will be reflected in the general character of the industry at a site and in the detailed morphology of artefacts.

It is thus clear that the location and study of the sources and types of raw material can be of great importance. For these reasons the following sections describe in some detail the different types of raw materials with which we are concerned at Olduvai, the location of their sources, the sizes and shapes in which each stone normally occurs and the different sets of mechanical or flaking qualities as I have determined them through experimental tool manufacture.

Quartzite

The major source of quartzite in the Olduvai palaeo-lake basin was Naibor Soit, a Precambrian inselberg, which is about 3·5 km north of the confluence of the Main and Side Gorges. This inselberg is a clearly visible feature for many kilometres, and the outcrops of the rock show up very clearly. Another known source also used by early tool makers is Naisiusiu Hill, a Precambrian outcrop 12 km from the confluence of the gorges. The texture and general appearance of the Naibor Soit quartzite, together with its regular cleavage pattern along micaceous layers, make it easily recognisable, and the vast majority of quartzite tools found at the Olduvai sites are undoubtedly made from this material. The lithologic description given by Hay (1976) is: 'Quartzite, white or colourless, rarely pale brown or green; generally very coarse-grained; micaceous layers are foliated and lineated' (1976, p. 11). The basic mineral composition is given as quartz and muscovite mica.

At Naibor Soit (Ki Maasai for 'white stone') quartzite is available in several forms. It can be found on the main outcrop (total area less than 1 sq km) as massive blocks which are *in situ* and immovable. Large chunks which have fallen from the main outcrop also occur on the sides of the hill. However, the major occurrence as far as most of the tool makers were concerned is undoubtedly the scree slopes of the inselberg which consist of a mass of small chunks and platey exfoliated slabs up to 5 kg in weight. Blocks of this material can also be found today scattered on the surface of the grass plains up to 400 m from the slopes of the inselberg.

This source of quartzite, being within 5 km of most of the major sites at Olduvai, was undoubtedly of great economic importance to early hominids in the Olduvai basin. The source was exposed at all times but might not always have been readily accessible owing to fluctuations in lake level. The material which it yielded was extensively and almost exclusively used for the manufacture of small light-duty tools such as *outils écaillés* and scrapers at all sites of all periods in the gorge, though on occasion larger tools were also made from it. The majority of debitage from most sites is also made up of quartzite.

Flaking qualities

The Naibor Soit quartzite is tough and coarse grained. Flakes need to be detached by means of swinging, follow-through blows. Retouch is possible on an acute edge, but retouch flakes tend to crush rather than step fracture if the edge is more obtuse. Small secondary retouching needs to be carried out with the aid of much finger pressure on the dorsal surface of the intended flake to aid flake 'carry'. Soft hammer flaking is possible, although blows should not be directed too much into the material.

Hammerstone flaking requires experience as blows too close to the edge will only crush the platform, while blows directed too far in from the edge can split the entire block in two. Forceful flaking regularly causes flakes to split longitudinally from the point of impact down their length. Regular hard hammer flaking is most easily carried out at angles of 45°–85°.

As regards tool use, the quartzite edges are very efficient because of the serrated effect caused by

Phonolite

This material is a fine-grained lava, usually green, that was often chosen by tool makers for biface manufacture. The phonolite occurs at a steep-sided volcanic neck called Engelosen (Ki Maasai for 'the hill that stands alone') which forms a prominent landmark about 7 km north of the nearest part of the Main Gorge. The lithology is described by Hay (1976) as: 'Flow-banded nepheline phonolite lava; slightly porphyritic, dark greenish grey' (1976, p. 12).

Today the hill is 500 m in diameter and 145 m high, and is highly eroded, suggesting a Pliocene or older date. Engelosen is largely mantled by what Hay describes as 'talus breccia cemented by calcrete'. This mantling is most probably quite recent, and Engelosen as a rock source would have had a quite different aspect during Beds III and IV times. Whereas today phonolite only outcrops in a few places on the breccia-covered slopes giving exposures of some 200 to 300 sq m in all, in the past it would probably have been far more abundant and easily obtainable over most or all of the 0·25 sq km area taken up by the hill.

The phonolite can be collected in the form of slabs which have exfoliated along flow lines, and occasionally also as large blocks at the base of the hill. Outcrops of *in situ* material are only exposed in a few erosion gullies on the slopes. The slabs are quite often the best-quality material and are found in convenient sizes both for carrying and for biface manufacture.

Engelosen is one of the remoter rock sources of the Olduvai palaeobasin, being situated at its northernmost edge. Interestingly, it was not used or transported to any of the Olduvai sites until early Bed II times, though the source must have been exposed throughout the period of Bed I.

Many of the phonolite tools found in Bed II sites are between 9 and 10 km from Engelosen. The most notable occurrence of phonolite tools is in the Bed IV site of HEB Level 3 with forty-eight bifaces and much debitage.

the coarse-grained crystalline structure of the material.

Flaking qualities

While the phonolite is a very fine-grained material, the flaking of it is greatly affected by flow banding, which gives the stone a grain. Flaking must be directed either with or directly across this grain to ensure the best results. It is interesting to note that the fine phonolite bifaces from HEB Level 3 (which are not obviously affected by this grain, are made of a type of phonolite whose source on Engelosen is no longer exposed today. The phonolite is quite brittle and easily accepts many sorts of subtle flaking techniques of the kinds that are regularly applied to English flint. Fine retouch is possible, and finger pressure can be used a great deal to direct flakes. This is in marked contrast to any of the other materials used at Olduvai, with the exception of chert (see below).

Phonolite flakes on the whole produce very sharp edges, but as the material is brittle these edges are highly susceptible to damage or blunting. Secondary retouch is easily carried out and need not reduce the tool greatly in size. If soft-hammer flaking is to be employed, platform preparation is required, and thinning flakes can then be removed almost as readily as with flint. Hard-hammer flaking can often leave traces which one would normally attribute to soft-hammer work, particularly in terms of the diffuse bulb and flake feathering at the edges. It was found that a quartzite hammerstone could produce results very similar to that of a soft hammer.

Basalt and trachyandesite

These materials are from the volcano Lemagrut which borders the Olduvai basin to the south. The lithology of the Lemagrut basalt and trachyandesite is described by Hay as follows: 'Clasts are dominantly grey trachyandesite, and soda trachyandesite, fine to coarse grained, and non porphyritic to very porphyritic; common is olivine basalt and rare is trachyte' (1976, p. 12).

For my experimental work I gathered samples of these materials from the river channels which flow seasonally from the volcano to the plains, as these were the most convenient sources. Throughout the Pleistocene these materials occurred in and were evidently obtained from broadly similar palaeochannels that ran from these same volcanoes but in a more northerly direction and for greater distances than those of today. Several such channels are exposed in Bed IV deposits, and can be seen to contain boulders of the same materials as those commonly used for biface manufacture. My modern source of material for experimental work is probably much closer to the volcanic highlands than it was necessary for early tool makers to go.

In the modern river channels, these materials are found in all sizes from pebbles up to boulders of over 1·5 m in diameter, though, as one would expect, the larger boulders have not travelled so far down the streams from their source. Such boulders and cobbles of selected sizes were commonly used for biface manufacture and for polyhedrons and choppers. WK in Upper Bed IV and EF-HR in Bed II are the sites where such raw material was most extensively used.

Flaking qualities

Basalt and trachyandesite are tough and coarse grained and there can be considerable variation from boulder to boulder in the quality of the material. Some of the finer olivine basalt flakes extremely well, while other cobbles will hardly show any conchoidal features at all when a flake is removed. On the whole, flaking has to be carried out by means of large, swinging, follow-through blows. These are applied with the arm being turned at the elbow, and the wrist stiff, thus allowing considerable force to be applied. Alternate flakes are easily struck using natural platforms. Platform preparation is rarely possible and in fact is rarely needed. Small retouch or the use of any subtle flaking techniques generally results in the loss of the platform altogether. I have found that finger pressure helps very little with flake carry, except with the finest-grained boulders. The cushioning needs to be light and the flakes should be allowed to fall free. Secondary flake scars are generally deep. Careful use of natural platforms does allow control of the material and edges. While the edges of the retouched tools are not always very sharp they are tough and jagged, features that can be as important as actual sharpness for certain tasks. Primary flake edges, however, are strong and durable.

Nephelinite

This material comes from the volcano Sadiman, near Lemagrut, to the south of the Olduvai basin. The lithology of the Sadiman nephelinite is described by Hay as follows: 'Clasts are chiefly porphyritic nephelinite and nepheline phonolite, dark greenish grey; few are ijolite' (*op cit*, p. 13).

As with the rock types just described which came from Lemagrut, the nephelinite occurs in river channels running in this case down from the parent volcano towards the Ol Balbal depression to the east of Olduvai. In Pleistocene times the drainage took a more northerly direction and brought this material directly into the Olduvai basin. As with basalt and trachyandesite, the tool makers would have had to go upstream for a distance to find useful blocks for large flake manufacture.

Cobbles and pebbles of this material were most commonly used for chopper manufacture by tool makers in Bed I. The most specialised use of the nephelinite, however, occurs in Bed IV at PDK Trench IV where several bifaces are found made on large flakes.

Flaking qualities

This is an extremely tough and coarse-grained material. While choppers are easily made and general secondary flaking easily carried out on carefully selected natural blanks, large flake removal from big blocks is difficult. The success of the latter depends very much on the selection of an appropriate boulder beforehand, since nephelinite boulders tend to occur in rather more spherical shapes than do basalt or trachyandesite cobbles and it is therefore hard to begin flaking. All that has been said with reference to the coarse blocks of basalt and trachyandesite applies to this material. Large swinging follow-through blows are used for secondary flaking, with no platform preparation, and such flaking has to be kept as simple and direct as possible to avoid ruining the edge. This material was not often used by the hominids and my experiments with it have therefore been less frequent than those on the other materials.

Nephelinite produces a very good edge, stronger and sharper than that of basalt. One may guess that were it not for its occurrence in such unmanageable blocks it would probably have been utilised much more than the present evidence from Olduvai suggests.

Trachyte

Trachyte occurred as a source of raw material in the form of large boulders and cobbles in the Bed II Eastern Fluvial Deposits (Hay 1976, pp. 94–6). These are exposed today only along the scarp edge of the first fault near the Ol Balbal depression. Many of the currently exposed boulders of this material are not usable for experimental tool-making owing to extensive weathering, and given their mode of occurrence there is no specific source as such; block size, quality and chemical composition vary considerably from area to area. In the past trachyte boulders would probably have been available from an exclusive area to the east of the present confluence of the gorges.

For these reasons experimental flaking of this material was only briefly carried out. The unweathered blocks were found to be very tough and hard to flake because of their large size and because they were rarely angular enough to offer an easy starting point. In general, trachyte was found to be similar to nephelinite in flaking quality and, like nephelinite, occurs in Bed IV predominantly as bifaces made on large flakes.

Chert

Chert may be briefly mentioned here even though it is not found in any of the Beds III–IV assemblages. The Olduvai chert consists of microcrystalline quartz in the form of small irregular nodules (Hay 1976, p. 184–5). Several horizons of chert nodules are known from Beds I and II, and as a raw material it was extensively used in Bed II for the manufacture of flakes and light-duty tools. This material flakes in a very similar way to good-quality English flint, producing very sharp durable edges. The only technological drawback to its use, and most probably the reason why it was only used for light-duty tools at Olduvai, is the small size and irregularity of the nodules in which it occurs (rarely larger than 15 cm at Olduvai). The complete lack of chert artefacts in Beds III–IV after its extensive use in Bed II presumably indicates that the source horizons were not exposed during these later periods.

Experimental tool manufacture

For my experimental tool manufacture using the Olduvai raw materials as described above, I collected stone from each of the sources mentioned. Both surface scree slabs and large chunks and boulders of each raw material were collected. These were then transported by Land Rover or truck back to camp at Olduvai. All of the experimental work was carried out within a small defined area within the Olduvai camp compound, where there was no chance of modern artefacts or debitage ever contaminating any area that might contain original stone artefacts. No flaking was carried out at any of the sources of stone or in the gorge, since this could affect a possible future study of surface material or exfoliation patterns in these areas.

Most of the hammerstones that I used were collected from the modern seasonal river beds that flow from Lemagrut and Sadiman, and were of basalt and trachyandesite. Most flaking was carried out by means of direct freehand percussion. Although some experimental flaking was done with soft hammers (for example, bone and wood), I discovered that I was able to duplicate almost all the features seen on the Olduvai stone artefacts with direct hammerstone percussion alone.

Prior to my arrival at Olduvai in 1976, I had had six years of flint-knapping experience in Europe and was very familiar with several types of European flint and also English Bunter quartzite as raw materials for making bifaces, choppers, flake tools, etc., as well as a wide range of Upper Palaeolithic types of tool (that is, worked from blade blanks). I soon found that very few of the techniques that had been appropriate for flint were useful for the Olduvai materials. This was due to the coarse grain and different internal structures of the local rocks. Hence, most of my early experiments were aimed at learning the basic mechanical properties of each of the main stone types previously described.

During this initial learning period, much raw material was used up and wasted as block after block was entirely flaked in order to discern the subtly different effects of different blow types and of variations in finger pressure on flake release and type. This is a particularly good method of learning the qualities of a new material, and it was important to study the mechanical properties of the material as well as basic flaking techniques before I attempted to replicate any of the archaeological specimens.

When starting with a new raw material, or when learning to flake for the first time, a knapper will tend to use much larger amounts of force than is strictly necessary to achieve a particular goal. As an interesting example, I found that while experimenting with the quartzite, each blow which removed a flake would also often split it longitudinally from the point of impact down its length and sometimes the core split as well. Each resulting piece had a triangular shape accordingly which in fact compares well with much of the debitage from BK in Upper Bed II. In due course I learned to flake the quartzite without these side effects.

Over a period of time I found that the force used to remove a flake became gradually reduced as technique and experience took over. This is clearly seen in the experiments leading to the large basalt flake removal technique: in this case, I was able to refine flake removal from a process initially requiring a 13 kg hammerstone to one that used a 1·5 kg stone. The development of this technique took well over a month of hard work, but was rewarding since it showed that with skill a 4·5 kg flake can easily be removed from a boulder by a 1·5 kg hammerstone.

The following sections for the most part do not attempt to describe the long and tedious learning process and concern themselves only with the methods by which many of the Olduvai artefacts can actually be made. This may not necessarily be exactly how all the originals were manufactured, but my methods have a particular value because they were the first to be arrived at during experimentation and are thus likely to be the simplest.

The artefact groups discussed in the following sections are: bifaces; hammerstones; polyhedrons, spheroids and subspheroids; and *outils écaillés*, punches and pitted anvils. In all cases I have worked with the terminology and basic specimen classifications as defined by M. D. Leakey. With a knowledge of the archaeological specimens, I would then design a series of experiments to replicate them and hence to understand the technological steps involved in their manufacture. I may offer a final interpretation of the tool type that suggests it to be, for instance, a debitage feature or to be composed of several technologically separate components, but this does not alter the basic validity of the groupings. My experimental flaking and the observations of how different tool types can be made accidentally has made me despair of developing any tool typology that can reasonably cope with such a variable thing as stone technology. There are many ways of grouping artefacts, and although it may be possible to quibble over the assignment of some individual specimens in any typology, the broad categories of artefact as they have been defined by Dr Leakey undeniably exist and I have stayed with her definitions throughout her work. In any case, it is with the broad subdivisions of assemblages, be they consistently shaped stone tools or mere types of by-product, that I am concerned and only rarely the individual specimens.

It may be interesting for others to understand the methods by which some of these ideas and interpretations arose. The initial flaking experiments were carried out with the twofold aim of understanding the flaking qualities of the raw materials and also to establish what were the particular skills and techniques required for several of the most common tool types to be made. These experiments resulted in a considerable amount of debitage on my flaking floor. This waste material in itself provided a very valuable store of information about the various rocks. Later, when examining the collections and noticing certain problems or anomalies, I was able to refer back to this collection of thousands of 'random' flakes and fracture patterns to compare with the archaeological situation. The benefits of this can be clearly seen, for example, in the hammerstone study where a simple comparison could be made between the archaeological sample and my own experimental one. This type of comparative study is valuable since at the time of flaking I was not concerned with hammerstones as such but with tool manufacture. Thus, the hammerstones on the flaking floor represent a combination of presumably inherent qualities, including the ability to do the job and my preference for selecting certain weights, shapes and sizes.

It should be noted that my results and interpretations apply only to the specimens and raw materials that I have studied. My conclusions, for instance, on subspheroids relate to the sample that Dr Leakey has identified from Beds II and IV and are constrained by the fact that the specimens are predominantly made in quartzite. These results may have little bearing, for example, on a collection of basalt tools that some other worker has also called subspheroids.

Bifaces

The occurrence and distribution of bifaces, handaxes and cleavers are discussed in detail in other sections of this volume. To follow the format of my treatment of other artefact types, however, I will briefly outline their occurrence throughout the archaeological sequence at Olduvai. The original definitions of these artefact types at Olduvai are as follows.

Proto-biface: This type of tool is intermediate between a biface and a chopper. They are generally bifacially flaked along both lateral edges as well as at the tip. The butts are thick and are often formed by the cortex surface of a cobblestone. Some specimens are high backed with a flat under-surface, and others are biconvex or lenticular in cross section. The edges are jagged as in choppers, and are often utilised (Leakey 1971, p. 5).

Handaxe: Handaxe has been dropped in favour of the more non-committal term biface since this can be applied to specimens of any size, including the diminutive examples found in the later stages of the Developed Oldowan (Leakey 1971, p. 3).

Biface: The bifaces from sites in Middle and Upper Bed II (apart from the Lower Acheulean site of EF-HR) are generally crude and there is such a degree of individual variation that it has often been necessary to describe each specimen separately (Leakey 1971).

No bifaces occur in any of the Oldowan assemblages and proto-bifaces are only found in sites between Upper Bed I and the lowest horizon of Middle Bed II: MNK Main Occupation Level, FC West, SHK, TK Upper Level and BK. Site EF-HR in Upper Middle Bed II is the first occurrence of the Acheulean at Olduvai with large bifaces making up 53·8 per cent of the total artefact assemblage. The TK Lower Occupation Floor has also been called Acheulean on the basis of biface size and morphology.

The Developed Oldowan B assemblages in Middle and Upper Bed II also contain bifaces, however, these differ from the Acheulean bifaces by being very variable in morphology within each site, and on average substantially smaller at all sites. The Developed Oldowan B bifaces make up between 2 per cent and 8·7 per cent of their artefact assemblages.

Bed IV seems a continuation of this basic pattern with assemblages being categorised as either Acheulean or Developed Oldowan C largely on the basis of biface occurrence and size. The Acheulean bifaces are mostly made on large flakes of stone and are quite consistent in basic morphology within each site. The Developed

Oldowan assemblages contain smaller numbers of bifaces which are of varying morphology. No Developed Oldowan sites have yet been found in post-Bed IV deposits though the Acheulean continues at FLK in the Masek Beds and HK and TK Fish Gully which are in post-Masek deposits.

The Olduvai bifaces, both Acheulean and Developed Oldowan, are made in a variety of raw materials which vary in their proportions from site to site. Within all the assemblages one finds bifaces occurring in several different materials: sometimes the biface assemblage will be dominated by one material and sometimes it will be split evenly between two or even three. This is in marked contrast to some of the other artefact categories where a definite preference for certain materials can be seen.

Experimental manufacture

The following sections describe briefly the various methods by which the bifaces can be made in the raw materials available in the Olduvai basin, and briefly describe some of the important features of the archaeological collections of bifaces found in each material.

The biface samples have already been studied in several different ways, and are discussed in several sections of this volume. Dr Leakey discusses their occurrence and describes the sample from each site for Beds III through to the post-Masek sites. Drs Roe and Callow have undertaken a metrical and shape analysis of all these samples including those of Bed II. My own study takes in only the samples from Bed III through to the post-Masek sites, and focuses on three main aspects of this sample that I consider influenced the tool makers' approach to biface manufacture, as well as relating to the basic functions of these tools. These three aspects are (1) the blanks from which bifaces were made, (2) the weight of the tool, and (3) the relative occurrences of primary flake edge and retouched edges and total edge lengths. For this last point, edges are measured to the nearest centimetre by tape measure. Thus, a handaxe made on a large flake may have a total edge length of 30 cm, of which 10 per cent consists of primary, unretouched flake edge, and 90 per cent therefore is an edge made by secondary retouch.

Basalt and trachyandesite

Since these materials occur in the form of water-rounded cobbles and boulders there are only two approaches available for tool manufacture. Small cobbles of stone can be selected from which a small core tool can be made; or the large boulders can be flaked in an organised way to produce suitable pieces for subsequent tool manufacture. The first approach restricts the tool maker greatly in the types of tools that can be made and their morphology. This is due to the general difficulty of carrying out extensive, controlled secondary flaking from rounded cobble surfaces, and to the fact that cobbles are generally thick in relation to their length and breadth. Thus, unless a great deal of time and effort is spent shaping and reducing a large cobble one is restricted to flaking an edge around the perimeter of a small cobble.

The second approach, that of breaking into the large boulders to produce angular pieces from which to make tools, is evident at most of the Acheulean sites at Olduvai where bifaces are generally made on large flakes. This method vastly increases the gross amount of raw material available to tool makers and there is considerably more control of the tool blank size and shape and final tool morphology. Without this technique of extracting large flakes from boulders, it would be hard to produce consistently large sharp-edged tools in these materials.

In my experimental manufacture I was first faced with the task of developing from scratch a method of striking large flakes from basalt boulders. I began my experimentation by using great amounts of force in order to ensure the removal of at least some flakes, and with the hope that I would eventually be able to refine the process.

The first experiments began by smashing one large boulder of about 15 kg from a height of 1 m on to the edge of a second boulder. Flakes were indeed occasionally removed in this manner but it was difficult to remove more than two flakes from any one boulder. This was due to the difficulty of accurately directing the blow to the intended platform, the large area that was hit which diffused the force, and a difficulty of holding the core steady when hit. The flakes that were produced in this manner were rarely more than ten centimetres long, often shattered on impact and had very flat ventral surfaces with no bulb of percussion.

Working on the observed inadequacies of this method, but having learned the best place to strike on a core, I began using a smaller hammerstone of about 4 kg, which was cylindrical and held in both hands at one end, in order to strike the core with the other end. In this way I was able to remove many flakes from a boulder, although these were still generally small, about ten by four centimetres at the most, and were not of a type from which bifaces could be made. I also found that after eight or ten flakes had been removed I could remove no more due to the lack of suitable striking platforms. I realised that I needed to deliver more force to points farther in from the edge of the core in order to remove larger flakes. Although this phase of experimentation was not very productive, I was learning much that was useful about the flaking qualities of basalt and trachyandesite.

The next phase of experimentation involved changing the cushioning of the core boulder from sandy soil to a platform that was composed of other boulders. This helped to keep the core in place when it was struck. I was also able to start using a smaller hammerstone of only 2 kg. This combination allowed me to produce consistently the correct types of large flake for biface manufacture, where the ventral surface is almost entirely taken up by a large diffuse bulb of percussion. Continued experimentation led me to the stage where I could regularly produce five to ten large usable flakes each of over 0·5 kg in weight from a 13 kg core. These were of the basic type on which most of the basalt and trachyandesite bifaces at Olduvai had been made.

Basalt and trachyandesite boulders vary considerably in quality and shape, and I soon found that flaking oval boulders with particular surface textures and colour produced better results. Interestingly, the cushioning arrangement that was eventually arrived at copies neatly a river-bed setting where boulders occur piled up in a similar manner. It would naturally be in the river beds of the Olduvai basin that the large flakes would have been struck. Indeed, only one large flake core has been found on an archaeological site (WK), but a survey of a Bed IV river conglomerate nearby revealed four large cores and many flakes showing varying degrees of river abrasion.

Bifaces are made from these large flakes by means of simple secondary flaking of the parts of the flake that are not already useful. This can be a very simple process since most flakes, even those produced without a very standardised method of manufacture or technique, preserve long primary flake edges that can be incorporated into the final tool. Thus, minimal retouch can be easily carried out, and alternate flaking can then extend the existing primary edges on the tool. I have found that bifaces can be made from basalt flakes from between fifty seconds to three minutes at the most. This gives an average manufacture time of about $1\frac{1}{2}$ minutes per biface. Making bifaces from selected cobbles of stone is much harder since the blank perimeters are smoothly rounded rather than angular and already sharp.

The largest sample of basalt and trachyandesite bifaces is from WK Upper Channel where they make up 63·2 per cent of the bifaces. The manufacture times quoted above relate specifically to this sample but also can apply to other sites; 88·7 per cent of these bifaces are clearly made on large flakes of stone, and 85·4 per cent of these preserve areas of primary flake edge which make up between 10 per cent and 100 per cent of the total tool edge. The average edge length is 34.1 cm and the average tool weight is 527.5 g. Forty-nine per cent of the WK East A bifaces are made in these same materials, but 43 per cent of this sample are made on cobbles and only 21 per cent (5) can be seen to have been made on flakes. The remaining 36 per cent are so extensively flaked that their blank type cannot be determined. Figs 10.2 and 10.3 show the weight ranges and

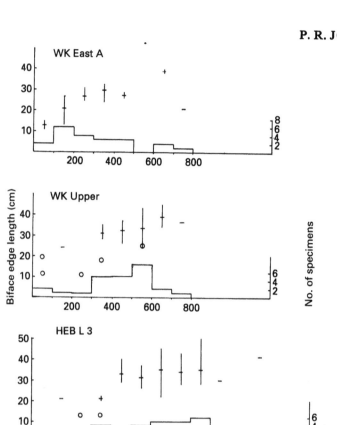

Fig. 10.2 Lava bifaces: charts showing tool frequency within weight classes and the range of edge length preserved on them. The small cross bars indicate average edge length per weight class

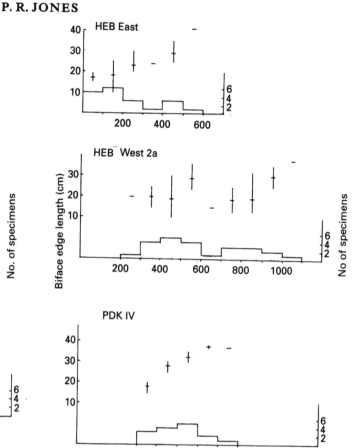

Fig. 10.3 Lava bifaces: charts showing tool frequency in various weight categories and the range of edge lengths preserved on them. The small cross bars indicate average edge lengths for the sample

edge lengths of basalt and trachyandesite bifaces from the main Bed IV sites.

Phonolite (Fig. 10.4)

Phonolite bifaces can be made either on the natural slabs of stone that occur at Engelosen, or from large flakes of stone struck from rock outcrops. In either case, the flaking of these blanks into tools is dictated to a great extent by the grain of the piece. A few well-placed and large roughing-out blows are necessary to shape the blank and determine the grain. More normal secondary flaking shapes the tool, and the final stage of fine flaking, which is generally missing from other bifaces, straightens the edge. Direct freehand percussion easily detaches flakes that will carry well over the surface and many techniques that can be applied to English flint can also be used with phonolite. The tool edges can be made by the removal of many small alternate flakes from a prepared edge platform.

I have found that the phonolite bifaces are most easily made from large flakes, although these themselves are quite difficult to remove from the parent rock, and that they require much less retouch than do the phonolite bifaces made on slabs. This is shown by the times required for manufacture in which bifaces made on slabs require generally between five and ten minutes of work, and bifaces made on flakes only require between two and five minutes. Quartzite chunks

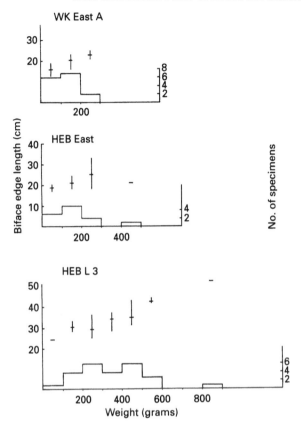

Fig. 10.4 Phonolite bifaces: charts showing tool frequency in various weight categories and the range of edge lengths preserved on them. The small cross bars indicate average edge lengths for the sample

were found to be particularly successful as hammerstones for flaking this material since their crystalline structure crushed easily on impact giving a soft hammerstone effect. Even without quartzite hammerstones much phonolite debitage resulting from biface manufacture shows soft-hammer features, notably diffuse bulbs and thin, feathered edges.

As a material, the phonolite allows more subtlety of flaking technique, and these qualities can be seen to have been used to their limit particularly by the Acheulean tool makers at HEB Level 3, the largest single collection of phonolite bifaces, though it is also seen in the phonolite bifaces at several other sites where only one or two specimens have been found. I found that considerably more time is needed to copy such finely made bifaces as those from HEB than is required to make bifaces in any other material type at Olduvai.

A total of 55·6 per cent of the bifaces at HEB Level 3 are made in phonolite, 64 per cent of which were undoubtedly made on flakes. None was definitely made on a slab of stone, and of the 36 per cent which are extensively flaked and mask the original blank, a majority were probably made on flakes on the basis of their morphology. Only half of the bifaces preserve areas of primary flake edge, and many of these are cleavers. The handaxes preserve an average of 27 per cent primary flake edge per tool. With the exception of a few bifaces from TK Fish Gully this high degree of retouch for the large tool size makes them quite different from any other collection of bifaces from Bed IV.

WK East A and HEB East also preserve interesting, if small, collections of phonolite bifaces which make up 32 per cent and 22 per cent of their assemblages respectively. Only two specimens from each sample can be seen to have been made on flakes and in all of these cases the flakes were small and only a little larger than the final tool. The WK East A sample has an average weight of 125 g and an average edge length of 19 cm. For HEB East these figures are 185 g and 21 cm respectively. This is in marked contrast to the Acheulean sample from HEB Level 3 with an average biface weight of 340 g and an average edge length of 32·7 cm (Fig. 10.4).

The highest amount of secondary retouch which is seen on most phonolite bifaces can be attributed to several factors and is most probably the result of a combination of them. The material is very fine grained and allows much small, delicate retouch. Because of the fine grain, each small flake scar is clearly visible in contrast with, for example, quartzite, where scar boundaries are hard to see and, hence, uncountable. Also, the use of phonolite tools has shown the edges to be sharp but brittle, and secondary retouch is a means of strengthening the edge as well as resharpening it.

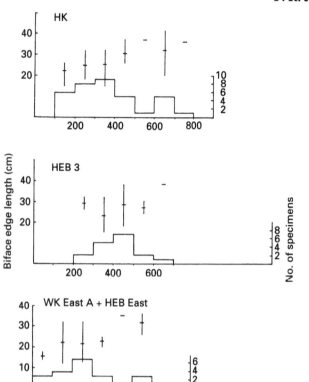

Fig. 10.5 Quartzite bifaces: charts showing tool frequency in various weight categories and the range of edge lengths preserved on them. The small cross bars indicate average edge lengths for the sample

Quartzite (Fig. 10.5)

Bifaces can be made from selected slabs or from large flakes, as with phonolite. My large flake experiments dealt only with large boulders that had detached naturally from the outcrop at Naibor Soit, but flakes could also probably be struck directly from the immovable outcrop of parent rock. I found that both block-on-block and direct percussion methods could successfully produce flakes of a suitably large size, although the structure of the material limited the size of flakes since thin flakes (relative to length and breadth) were crushed with the force that was needed to remove them, and thick flakes are not useful for biface manufacture. The direct percussion method rarely resulted in large unusable cores as occurred with the coarser and tougher lavas, and the whole block could generally be flaked with the final piece usable for biface manufacture or resulting in a large polyhedron/subspheroid piece.

As with the large flakes in other materials, very little retouch is necessary to produce Acheulean bifaces that incorporate suitable primary flake edges in the final tool The secondary retouch is simply carried out and edges are easily flaked. Small retouch tends to destroy the edge and platform by crushing.

When selected slabs are used as blanks more flaking has to be carried out, since an edge usually has to be flaked around most of the perimeter of the piece. Alternate flaking is commonly used and this results in edges central to the thickness of the tool. Edges are also found that have only been flaked unifacially. Cleavers are easily worked on slabs of stone with a tranchet blow across one end to make an edge.

At the WK Upper Channel site, 55·5 per cent of the quartzite bifaces are made on flakes, 93 per cent of which preserve primary flake edges which made up 10 per cent to 55 per cent of the total edge. Only 29·6 per cent can be identified with certainty as having been made on slabs, and most of these tools are cleavers. HEB Level 3, where more than 37 per cent of the biface sample is in quartzite, shows a preference for flake blanks (88·2 per cent), 48 per cent of which preserve primary flake edges to 10 to 50 per cent with an average of 25·7 per cent. Only one specimen was made on a slab. FLK Masek, where 86·3 per cent of the biface sample are in quartzite, shows quite a different pattern though with a smaller sample size. Flake and slab blanks make up 18·1 and 9 per cent respectively. The average weight of the three flake bifaces is 498 g, with an average edge length of 46·5 cm. The average weight of the four slab-blank tools is 954·7 g, with an average edge length of 36·5 cm. Since most of the remaining specimens from the site weigh well over 1,500 g each they were most probably made on slabs of stone. The tools weighing more than 1 kg (45·4 per cent of the total) have an average edge length of 52 cm, whereas the average for the total sample is 41·7 cm (Fig. 10.5).

The post-Masek site of HK shows a slightly different usage of quartzite again. The average tool weight of the measured sample of 35 is 370 gr with an average edge length of 27 cm. Of these tools, 40 per cent made made on flakes and a further 40 per cent are so extensively flaked that the blank type cannot be determined. Only two specimens preserve any primary flake edge.

It is hard to draw any conclusions from this sequence of collections. In the Bed IV assemblages more bifaces are made in quartzite than any other material, though basalt and then phonolite come a close second and third in frequency of occurrence. At the Developed Oldowan sites though, the quartzite bifaces made up an average of only 20 per cent of the total sample. Later than Bed IV, however, the biface assemblages are all dominated by this material. In terms of edges per unit stone the FLK Masek collections are the least efficient and more or less any two HK bifaces will have greater edge length and less weight than a Masek biface.

Trachyte and nephelinite

The toughness of these two materials and the manner in which they occur require the manufacture of large flakes rather than the use of slabs for virtually all biface manufacture. Where nephelinite cobbles were used for chopper manufacture in Bed I, few were used for biface manufacture. Large flake manufacture is made difficult by the commonly spherical shapes of the nephelinite boulders, and the massive size of the trachyte ones. Careful boulder selection and the use of broadly similar techniques to those used for basalt, however, did result in the production of large flakes. But I found that without consciously attempting to do so, I generally produced side flakes in these materials. The flakes produced tended also to be thinner than those of basalt and trachyandesite, and with quite different ventral characteristics. The ventral surfaces tended to be flatter with diffuse but distinct bulbs of percussion.

Biface manufacture from these thin flakes is easily carried out in the same basic manner as with basalt. The nephelinite flakes, being thinner, can be fragile and hence bifaces commonly include large areas of primary flake edge in the final tool. This is not commonly the case with the trachyte bifaces, however, which are often more heavily worked than nephelinite ones.

At PDK Trench IV, five large nephelinite bifaces make up 29·4 per cent of the biface collection, and all these are made on large flakes. The average primary flake edge length makes up 39·7 per cent of the total tool edge which averages 39·6 cm per specimen with a maximum of 48 cm. Since these tools have an average weight of 506.5 g, they have relatively long edges. Nephelinite bifaces do occur in small numbers at other sites, but are usually more heavily flaked, and do not seem anywhere to make up a group as the PDK Trench IV collection does.

An interesting occurrence of trachyte bifaces is seen in the HEB sequence of sites although they only make up small percentages of the biface group: HEB West 2B has 9 per cent, HEB West 2A 12 per cent and HEB East 7 per cent. All these bifaces are made on flakes and all have remarkably similar colour and surface textures.

Technological considerations

This section deals with a series of broader technological issues which were noted during my experimental tool manufacture and relate to bifaces generally and to the interpretations of the archaeological material. Several of these points may not relate directly to the Olduvai sample but to the technology of bifaces in general.

Flake blanks versus slab blanks for biface manufacture

The type of blank chosen for biface manufacture has a great effect on the morphology of the final tool and its edge qualities (Jones 1979). Previous sections have shown how the quartzite and phonolite materials occur naturally as slabs from which bifaces could easily have been made, and indeed for much of my own experimental work I found it easy to make them this way. The archaeological evidence shows that at Acheulean sites the surface slabs of phonolite were rarely, if ever, used. Some quartzite slabs were used for biface manufacture (particularly at WK Upper Channel), but make up only a small percentage of the total sample of Bed IV quartzite bifaces. Most of the bifaces in both these materials were definitely made on large flakes struck from boulders or rock outcrops, a seemingly more difficult approach to the manufacture of simple bifaces. Perhaps, however, the answer is simply that I am exploiting these rock sources 10,000 years after the last intensive use of the source by tool makers, and I have at my disposal only the accumulated weathered slabs from this period.

Experimental manufacture and use of these tools have shown several important differences between the tools made on flakes and those made from slabs. These relate to (1) the overall time of tool manufacture, (2) the qualities of the tool edge, and (3) the weight of the tool in relation to its edge length.

Only two sites provide a comparison of bifaces made on these two blank types: WK Upper Channel and FLK Masek. Fig. 10.6 illustrates the differences in changing mass per unit edge as biface size is increased for the WK sample. While for the smaller sizes of any blank type there is little difference, for sizes over 500 g there is an important difference, with tools gaining considerable weight at a faster rate than that of the edge increases. The FLK Masek sample in a larger size range shows a greater difference between bifaces of the two types:

	Average weight (g)	Average edge length (cm)	Mass weight per unit edge
Flake blank	498·1	46·5	10·7
Slab blank	636·6	35·0	18·2

This means that about twice the amount of edge is found per unit flake tool than is found per unit slab tool. This archaeological data supports my own observation that the core tools will generally be heavier. This type of comparative material

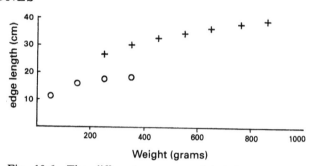

Fig. 10.6 The different average weight/edge-length relationships of basalt bifaces from WK made on cores (circles) and on large flakes (crosses)

does not exist for the phonolite bifaces, but I would predict a similar situation.

The differences between basalt or trachyandesite core and flake bifaces are even more pronounced. In these materials there is no real choice between blank type if large tools and long edges are needed. The Bed IV material suggests that cobble core tools rarely produce edge lengths of more than 18 cm and these core tools commonly exceed 20 g per 1 cm edge. If larger tools with longer edges are needed the only solution is to break down the larger cobbles and boulders and make tools on the flakes produced from them. Basalt bifaces made in this way commonly have edge lengths of up to 40 cm and average weight per unit edge of about 15 g. Fig. 10.6 illustrates the differences between basalt bifaces made on these two blank types.

Different edge qualities also result from the different blank types. At the most basic level, core/slab tools will necessarily have secondary flaked edges because of the need to remove the cortex, and flake bifaces will preserve significant amounts of primary flake edge. More important, however, are the edge angles produced and the degree to which they can easily be re-sharpened. Fig. 10.7 illustrates this with cross-section diagrams of core/slab and flake bifaces. A large flake with thin tapering edge submits easily to secondary retouch and can be flaked to produce edges of an acute angle for most of its perimeter. The core/slab tool, having greater edge angles to begin with, is flaked at greater angles (approaching 90°). Much flaking technique is concerned with making acute-angled edges, and while there

EXPERIMENTAL WORK IN RELATION TO THE STONE INDUSTRIES

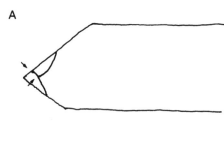

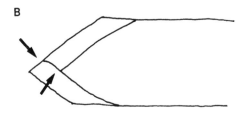

Fig. 10.7 Cross sections of quartzite slab bifaces (A and B) as compared to a quartzite biface made on a large flake (C). The arrows indicate the types of force required to detach re-sharpening flakes. The flake biface has sharper edge angles and is easier to retouch

are many ways of doing this with fine-grained materials (e.g. turning the edge and soft-hammer flaking) it is not easily achieved in the materials from Olduvai. While platform preparation and finger pressure on the dorsal areas of intended secondary flakes will help, one will not be able to flake such an acute edge as is possible with a flake blank tool, unless perhaps a great deal of time and preparation is spent on manufacture. An extension of this principle relates to the re-sharpening of a biface. Fig. 10.7 shows that in order to maintain a serviceable edge in the core-blank tool larger blows are needed, so that the flakes will carry across the surface rather than stop short and result in an unusable edge or the loss of a platform altogether. The flake-blank tool can be re-sharpened with little effort and little loss of raw material. This will vary from material to material although the general principle holds true for all.

Fig. 10.11 shows a series of graphs comparing experimentally made tools in different materials through several re-sharpening stages. Through time the flake-blank bifaces change in edge angle and come to approximate the core-blank tools in terms of edge angles and character. This is because of the gradual thickening of the tool in relation to its width as it is flaked to a smaller size.

The other important difference is that the manufacture times of bifaces from flakes is roughly half or two-thirds that of core/slab bifaces, because there is an initial preparatory stage for the core that is skipped or greatly reduced in flake-biface manufacture (Jones 1979). The time involved in removing the large flake from the outcrop or boulder in the first place is minimal. The large flake technique also makes available larger gross amounts of usable stone.

As has been discussed, there are certain measurable differences between slab and flake bifaces. What is not known, however, is the degree to which these differences were deliberately selected by early tool makers, or to which they represent an unconscious selection of blank type at the rock outcrop. Certainly, if long edges were needed the larger tools must be made. If acute-angled secondary and primary flake edges are needed then a flake is the best blank to choose. However, if the suitable slab blanks that I find on the Naibor Soit inselberg today are an unusual concentration, then it is possible that the striking of large flakes as tool blanks was on the whole the only practical way of making large tools in this material. The retention of primary flake edges was clearly deliberate on the part of the tool maker and must indicate their utility.

Biface shape

The shapes of the Olduvai bifaces have been studied in some detail by Dr Roe (this volume). My interest is not so much in the shapes of

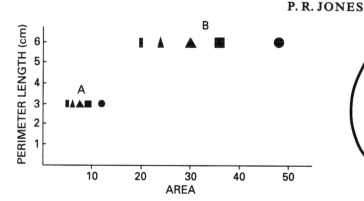

Fig. 10.8 The fourfold increase of shape and area as perimeter length is doubled (A to B) and also the greater area of a circle than that of a slim triangle of the same perimeter length. The symbols are not drawn to scale

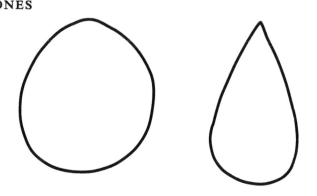

Fig. 10.9 The long, generally straight length of edge available on a triangular tool. This is in contrast to the totally curved edge of a disc shape which allows only a short length of edge to be used at any one time

individual bifaces as in the general shape that characterised handaxes the world over. There has been much discussion of the possible reasons for the widespread occurrence of this basic tear-drop shape, from its aerodynamic qualities (O'Brien 1981) to the suggestion that it is non-utilitarian and symbolic. My own study and discussion is based on the assumption that bifaces are functional and that they represent a collection of qualities that were of basic importance to tool users.

Fig. 10.8 describes the basic relationships of perimeter length to internal area for five basic shapes: a narrow oblong, two triangles, a square and a circle. The shapes are not drawn to scale and their positions have been calculated to show the differences in internal area for the same perimeter lengths (scheme A) and also for double their perimeter lengths (scheme B). The internal area for all shapes has to be quadrupled in order for the perimeter length to be doubled. Thus, if we consider the five shapes as possible biface plan shapes with flaked perimeters, we can see that the longer thinner shapes will produce more edge length per unit area. This will become even more important if we consider the corresponding increase in *volume* with size of solid shapes. In the case of the sphere and cube, the volume increases eightfold when plan perimeter is doubled. In the case of bifaces, which are basically thick flat-sided shapes, we can assume a rough mass area increase of six times as the perimeter length is doubled. This will naturally vary from stone to stone, and with the section of the tool, but will suffice as a rough measure.

Therefore, on paper, the most edge-efficient way to flake stone in order to get long edges is in long narrow shapes, which allow (in the case of Fig. 10.8) the same edge length but less than half the material of a sphere. A major technological problem exists in the flaking of these shapes in stone. First, there is the structure of the stone which will only allow certain minimum length/breadth/thickness ratios to be flakable and strong. Second, there is the phenomenon of end-shock which occurs when shock waves within a long narrow piece of stone coincide at some point and snap it in two. These shapes are not feasibly flaked in stone, though this edge-economic shape was used extensively at later periods when a new technique, that of blade production, was discovered.

Pointed triangular shapes are easy to flake, have very low mass per perimeter lengths, and have the advantage of long continuous stretches of edge. This latter point must also be an important consideration if the edges are to be functional. A circle has a continuous edge, but if it were to be used on any surface for cutting, only a small portion of it could actually be used at any one time. A triangular shape has long edges that can be used as a single cutting edge, see Fig. 10.9.

Many of the phonolite bifaces from HEB level 3 and the nephelinite and basalt tools from PDK Trench IV push the manufacture of long slim shapes close to their technological limits, and result in long edges per unit mass. A view of all the Acheulean assemblages, particularly from Bed IV, shows that in fact a basic aim seems to have been to get the longest edges possible for the tools of weight ranges of about 300 to 700 g. At HEB Level 3, the edge lengths are similar to those of other sites, but the weights are smaller.

A further important feature of the mass/edge length relationship occurs when bifaces get larger. Here shape becomes critical in terms of bulk if a larger tool with a longer edge is to remain manageable. The importance of this can be illustrated by trying to predict the weight differences of a tool of given edge length for handaxes of both tear-drop and circular shape.

One of the large quartzite bifaces from FLK Masek has a length/breadth ratio of about 1.2.5, it weighs 2.315 kg and has an edge length of 66 cm. The plan shape of the biface has an area of 238 sq cm, and a weight to area ratio can be calculated at 1 sq cm to 9.73 g. If a biface in the same material had been flaked to a circular shape in the same way, with exactly the same edge length, the total estimated weight (calculated by multiplying the new area by 9.73 g) would be 3.366 kg, representing a difference of over 1 kg of stone. One would therefore expect that the larger the biface, the narrower it needs to be made, and though the large biface sample from FLK Masek is small, this does appear to be the case.

To sum up the main points, I consider that if long units of edge are to be made in stone, the classic tear-drop shape of the handaxe is the most efficient tool design. Clearly, edge length alone is not enough, and the weight of the tool must also be important to its function. The average Acheulean handaxe is an optimum solution to functional and structural problems: namely, that long edges and weight are needed, but that tool mass increases at roughly six times the rate of edge increase. A point will be reached where too much weight is added for a small unit of edge increase. The Acheulean bifaces of Bed IV seem to make a very good compromise between all of these factors, and the majority make optimum use of them. The FLK Masek sample of bifaces, while triple the weight of most others, have only 50 per cent more edge length. In one sense though, if one needed a 2 kg artefact, no other way of flaking it could have resulted in more edge length.

The shape of bifaces, therefore, can be seen to be a by-product of two functional considerations, those of tool weight and usable edge length, and it is these which control all other aspects of the end-product itself.

The Developed Oldowan and the Acheulean

Discussions to date in the archaeological literature have compared and contrasted the biface samples from these two types of assemblages and generally concluded that the basic difference between them is that the Acheulean tool makers possessed the ability to strike off large flakes of stone and the Developed Oldowan tool makers did not (Leakey 1971, Clark 1970). Another view (Stiles 1977) suggests that the observed differences are due to raw material. These previous studies have been based on the general assumption that the two samples are functionally equivalent because they are of similar basic morphology. My own study describes several differences between the samples and shows that in technological approach they are quite different, thus supporting the views referred to earlier with technological data. However, a fundamental problem exists in that the only similarity between the two types of sample is basic plan shape. Other differences, however, are so consistent that there is no real reason to assume a functional equivalence between the two. The following discussion describes some of the basic differences between these two samples and discusses the various possible reasons for it. This discussion deals primarily with Bed IV since both assemblages occur there, whereas they do not in the Masek and post-Masek beds.

At Acheulean sites between 70 and 90 per cent of the bifaces are clearly made on large flakes, as opposed to only 10 to 20 per cent at Developed

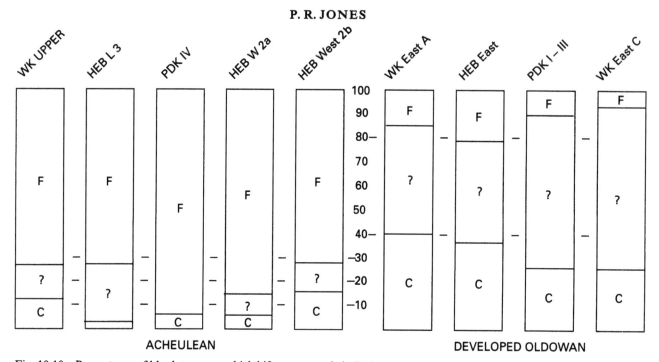

Fig. 10.10 Percentages of blank types on which bifaces are made in Bed IV: F, flake blank; C, core blank; ?, blank unknown

Oldowan sites (Fig. 10.10). The bifaces that are made as core tools, on selected slabs or cobbles, make up only 5 to 15 per cent of Acheulean biface assemblages, but account for 30 to 40 per cent of Developed Oldowan assemblages. The remaining group of bifaces whose blank type cannot be determined owing to the extensive flaking over the tool surface make up 0 to 20 per cent of Acheulean collections, and 40 to 50 per cent of Developed Oldowan samples. These figures are based on bifaces which clearly show the type of blank from which they were made. On the basis of morphology, however, it is clear that most of the 'unknown' category at Acheulean sites were probably made on flakes, while perhaps half of the equivalent Developed Oldowan group were made on cores. Interestingly, however, there is a small but definite overlap between the two types of collection in that 5 to 10 per cent of the Developed Oldowan samples consist of bifaces that are identical to the majority at many Acheulean sites, and less than 5 per cent of several Acheulean collections consist of small bifaces which are morphologically and technologically similar to the majority at Developed Oldowan sites.

A second and visually more important difference between the bifaces of these two assemblages is the overall size of the bifaces found. The Acheulean bifaces from Bed IV are on the whole in the weight range of 300 to 700 g, while at Developed Oldowan sites the range is from 50 to 300 g. The most notable exclusions from these ranges are in the case described above where the few larger bifaces at Developed Oldowan sites are totally Acheulean in appearance, and many smaller bifaces (particularly from WK Upper Channel) are of a Developed Oldowan type.

Related to the types of blanks chosen, and to the overall size of the tools, are the qualities and lengths of the tool edges. Most of the Developed Oldowan bifaces, being made on cores or extensively flaked, naturally have edges made up entirely of secondary retouch, whilst most of the Acheulean bifaces preserve lengths of primary flake edge. This is clearly seen when looking at the collections, and accordingly most Acheulean bifaces preserve acute edge angles and the Developed Oldowan sample preserve thick obtuse edge angles, owing to the blanks from which they were made. Edge lengths also vary with the size of the tool and calculations of grams weight per centimetre edge show that most of the Developed Oldowan biface sample have between 5 and 10 g per cm edge, while the Acheulean sample preserve between 10 and 20 g per cm edge. In terms of total edge length, the Acheulean samples tend to have edges of 20 to 50 cm, while the Developed Oldowan samples preserve edges between 10 and 30 cm.

Other differences include the very variable morphology of the Developed Oldowan biface sample, as compared to the general consistency of the Acheulean samples. The sample sizes of the Developed Oldowan occurrences are generally lower than most Acheulean samples, and while the Acheulean collections from any one site will tend to be dominated by one or maybe two materials, the Developed Oldowan collections will preserve roughly equal numbers of each material. Quartzite, however, has a notably lower occurrence at Developed Oldowan sites than at Acheulean sites, where it tends to be the dominant raw material for bifaces.

Many of the differences described above relate directly to the types of blank chosen for tool manufacture, and previous sections have discussed the mechanics of this in detail. One very significant similarity between the two groups is that both occur in the full range of raw materials that were available; most other artefact groups at Olduvai tend to be found in only one or two types of stone.

When considering the possible reasons for these two different types of collection, several decisions have to be made. The two groups are different in many respects but similar in some important ways (shape and use of raw material). There are several main interpretations that could be made, some of which have been suggested by other authors before:

(1) these artefact types were made for the same functions but by groups that possessed different technologies (Leakey 1971);
(2) they were made by the same group but for different functions (Jones 1981);
(3) they were made by the same group for the same functions, but from different types of tool blank;
(4) they are the same artefacts, but seen at different stages of their life. In other words, the Developed Oldowan bifaces are re-flaked Acheulean bifaces.

These four points simply state the main possibilities for these two artefact groups, but undoubtedly others could be included also.

Were these two artefact groups made for the same function? If so, why are they so different technologically? If it is assumed that these two collections represent different functions then the discussion can stop, and research should begin to elucidate what these different functions might be. I suggest that the functions of the two groups may indeed be very similar or overlapping, because the features that both groups possess are a long cutting edge manufacture in the full range of Olduvai raw materials.

If these two groups were made for the same set of functions, there are several options to explain their observed differences. The two samples could have been made by two culturally different groups, one of which did not have the ability to strike large flakes and had to make do with the selection of suitable slabs of stone. This assumes that the Developed Oldowan tool makers who lived in close proximity to the Acheulean tool makers from Middle Bed II through to the end of Bed IV never learned to strike large flakes. For this hypothesis, one has either two different types of hominid making different types of tool over this period, or two culturally distinct groups of the same type of hominid.

A third possibility is that these groups of artefacts were made by people who had restricted and different access to the raw material sources. This appears to be possible for the basalt, trachyandesite and nephelinite, but not for the phonolite or quartzite. The former group of raw materials occurs in river beds, and the further downstream one is located the smaller will be the available stones. Upstream one could knock off large flakes for biface manufacture, but downstream the tools would have to be made on small

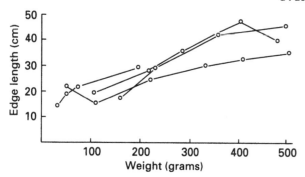

Fig. 10.11 The changing ratio of weight to edge length of handaxes as they are re-sharpened several times. The re-sharpening consisted of retouch aimed at renewing the tool edge as if it had been blunted through use. This graph provides an interesting comparison with the Acheulean and Developed Oldowan edge/weight ratios illustrated earlier

cobbles. This suggestion could be supported by the work of R. L. Hay who noted significant differences in the palaeogeographic location of Developed Oldowan and Acheulean sites in Bed II. The Developed Oldowan sites were seen to be within 1 km of the palaeo-lake shore, while the Acheulean sites occurred more than that distance from it. A problem exists, however, with the sources of both phonolite and quartzite, which are both very localised. Any tool maker would have had equal access to all of the potential blank types that these sources have to offer, and need not be restricted to the selection of small pieces. There is no evidence that the quartzite or phonolite Developed Oldowan biface blanks were small.

A fourth possibility exists, and that is that the bulk of the Developed Oldowan bifaces started out as typical Acheulean handaxes, but through use and the need to renew edges, or a general need to produce small flakes, they were flaked to their present shapes and discarded. This applies well to the phonolite and quartzite samples, but not to the basalt and trachyandesite collections. There is no evidence that the blanks for the Developed Oldowan quartzite and phonolite bifaces started out small; the bifaces in these two materials started out at the same size. This is further borne out by re-sharpening experiments on typical Acheulean bifaces. After three or four phases of re-sharpening, I was left with what could only be classified as a typical Developed Oldowan handaxe. Perhaps the Developed Oldowan sample is made up of more than one artefact type, or tool manufacture strategy. Certainly, if phonolite or quartzite bifaces are re-sharpened several times the edge angles become more obtuse, owing to the thickening of the section relative to width, and the results can look identical to the Developed Oldowan sample. Fig. 10.11 shows the sequence of changing weight/edge length for bifaces when re-sharpened several times.

This possibility can account for the occurrence of both samples in the same range of raw materials. While the Acheulean bifaces are made in the first place under similar conditions, that is from large flakes (and hence accounting for their consistency in morphology), the Developed Oldowan sample would have been re-flaked under many diverse situations which could account for the variable morphology of the sample.

This interpretation also has implications for the Acheulean sample of bifaces. It suggest that they are very much at the beginning of their functional lives. Certainly when looking at the manufacture times of these bifaces, very little time has been invested in them, and only a little retouch was needed to shape them. This latter is a manufacture process rather than the result of use or tool maintenance. Unless one suggests a function for these bifaces that leaves little visible damage to the tool, but still renders it useless, one is left with a collection of tools which are largely unused. If the functions of bifaces do ultimately result in edge damage, and if the edges were renewed by further flaking, one would then expect to find the worked-out remnants of these tools. Indeed, the phonolite, quartzite and many of the basalt Developed Oldowan bifaces do appear to be just this, that is, flaked out Acheulean bifaces. The big difference between the two assemblages is the differing treatment of the basalt, nephelinite and trachyandesite bifaces. In the Developed Oldowan assemblages, most of these are made as core tools, whereas in the Acheulean assemblages, most of them are made as large flake tools. It could perhaps be that this differing treatment is related to the more widespread riverine occurrence of these materials, and that many of these core tools were made opportunistically in areas where the larger boulders could not be found. The worked-out nature of all the other Developed Oldowan bifaces would support this by suggesting a scarcity of material in the areas where they were re-flaked.

This model suggests that the two assemblage types were made by the same groups of people, and it also accounts for the small but clear overlap of biface types between the two. Unfortunately, the Bed IV palaeogeographic evidence is quite different from that of Bed II, and all the Developed Oldowan collections occur within a few hundred metres of the major Acheulean sites. Indeed, there are some intriguing similarities between some of these neighbouring sites. The WK Upper Channel site is only 0·5 km away from the WK East A which is the largest Developed Oldowan collection, and in the same channel system. Both of these collections preserve large numbers of *outils écaillés*, punches and pitted anvils, artefacts which are rare at all other sites. The Developed Oldowan collection from HEB East is only 80 m from the HEB sequence of Acheulean assemblages, though in a lower channel. Like HEB Level 3 it contains a large number of phonolite bifaces, and like several of the HEB Acheulean collections it has trachyte bifaces of identical surface texture and colour.

Polyhedrons, spheroids and subspheroids

These three artefact types are considered together in this section because of the broad morphological similarities and the relationship suggested by their definitions. They have been defined by M. D. Leakey (1971, pp. 5–6) as follows:

Polyhedrons: These are angular tools with three or more working edges, usually intersecting. The edges project considerably when fresh, but, when extensively used, sometimes become so reduced that the specimens resemble subspheroids...
Spheroids: These include some stone balls, smoothly rounded over the whole exterior. Faceted specimens in

Table 10.1 *The archaeological distribution of polyhedrons, spheroids and subspheroids in the Olduvai sequence*

	Polyhedrons	Spheroids/ subspheroids
Bed I, Oldowan	54	32
Lower Bed II, Dev. Old. A	33	160
Middle and Upper Bed II, Dev. Old. B	47	833
Upper Bed IV, Dev. Old. C	—	33
Middle and Upper Bed II, Ach.	5	9
Bed III, Ach.		
Lower Bed IV, Ach.	20	63
Upper Bed IV, Ach.	2	11
Masek, Ach.	—	32

which the projecting ridges remain or have been only partly removed are more numerous.
Subspheroids: These are similar to the spheroids but less symmetrical and more angular.

The archaeological distribution of these tool types in the Olduvai sequence is shown in Table 10.1.

The Bed I collection is dominated by lava as the main raw material, with polyhedrons outnumbering subspheroids, and with only four identified spheroids (three from FLK North Levels and 1 and 2, and one from Level 5). The main occurrence is at DK where polyhedrons make up 20·8 per cent of the tool assemblage and where they are predominantly made from weathered nodules of lava.

Lower and Lower Middle II see two major changes in this pattern. Subspheroids become the dominant artefact of the three, and quartzite is the most frequently used material for the spheroids and subspheroids, although polyhedrons are still usually made from lava. For instance, there are only two lava spheroids as opposed to twenty-five quartzite ones. The two sites which contain the most spheroids are HWK East Level 3 and FLK N Sandy Conglomerate. Subspheroids comprise 22·4 and 22·7 per cent of the total tool assemblages respectively, whereas polyhedrons make up only 3·9 and 3·6 per cent. Polyhedrons, however, are far more variable in their size ranges

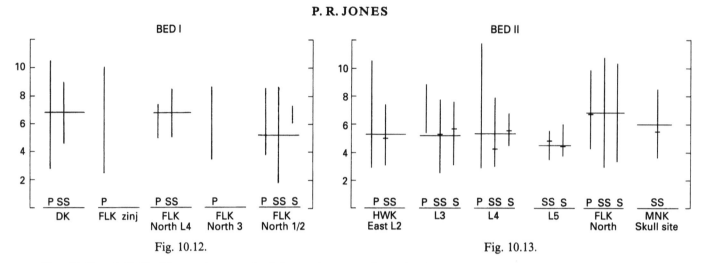

Figs. 10.12 and 10.13 The maximum and minimum size ranges from sites in Bed I, Lower and Lower Middle Bed II for polyhedrons (P), subspheroids (SS) and spheroids (S). The short cross bars indicate average measurements while the long cross bars indicate the median of the subspheroid range. Sample sizes of less than three are not shown. Data from Leakey 1971

than either spheroids or subspheroids, the significance of which will be shown below.

The Upper and Middle II sites do not change this picture dramatically, other than to increase the sample size substantially. MNK produced 143 subspheroids, of which 86 per cent were quartzite, and sixteen spheroids, all of which were quartzite. SHK yielded 258 subspheroids, 94·5 per cent of quartzite, and sixty spheroids, 98 per cent of quartzite. BK produced 386 subspheroids, 89·3 per cent of quartzite and sixty spheroids, 71·6 per cent of quartzite. The combined total for polyhedrons from these sites is thirty-eight, more than 75 per cent of which are made of lava and many of which can be seen to have been made on cobbles. Figs 10.12, 10.13 and 10.14 illustrate the size ranges of the Beds I and II samples.

A basic conclusion that can be drawn from these data is that polyhedrons, which commonly occur in lava, are most probably not technologically or functionally related to the spheroid or subspheroid groups, as is suggested in the definitions of these artefact types. Both spheroids and subspheroids usually occur in quartzite, and it seems probable that these two groups are related, and that the former represents a heavily battered extreme of the latter. The fact that the size range of spheroids is almost always smaller than that of subspheroids, that is, minimum measurements

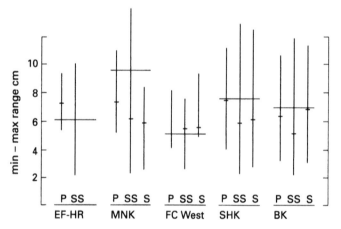

Fig. 10.14 The maximum and minimum size ranges from sites in Upper Middle and Upper Bed II for polyhedrons (P), subspheroids (SS) and spheroids (S). Cross bars as in Figs. 10.12 and 10.13

are larger and maximum measurements are smaller, suggests that while battering is applied to a large size range, only a small select group within that parent range is intensively battered. Thus it does not appear that the battering process can be considered a process of manufacture since one would expect the spheroids to be consistently smaller than the subspheroids. This does, however, suggest that the battering is part of an activity, and that while a large range could be and was used briefly, only a small mid-range was used

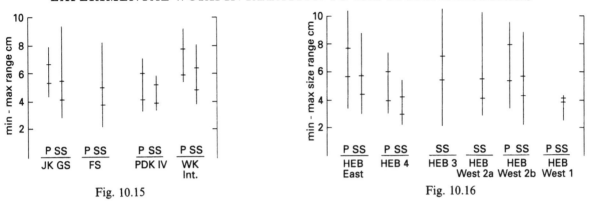

Fig. 10.15

Fig. 10.16

Figs. 10.15 and 10.16 The maximum and minimum size ranges of polyhedrons (P), and the subspheroid group (SS) from sites in Bed III and the base of Bed IV, and from Lower Bed IV. The upper and lower cross bars indicate average maximum and minimum measurements respectively

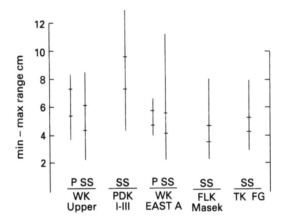

Fig. 10.17 The maximum and minimum size ranges for polyhedrons (P) and the subspheroid group (SS) for sites from Upper Bed IV, and for Masek and post Masek sites. The upper and lower cross bars indicate average maximum and minimum measurements respectively

intensively. This assumes, of course, that the Bed II sample of spheroids and subspheroids all preserve battering as is suggested by the definitions of these artefact types. This may not be the case, as the following sections will show.

The sample of polyhedrons, spheroids and subspheroids from Beds III, IV and the Masek Beds is much smaller than that of Beds I and II, with a total of only 242 specimens (Figs 10.15–10.17; see Table 10.2 for the breakdown by site). When dealing with the Beds III–IV and Masek specimens, M. D. Leakey has grouped the spheroid and subspheroid categories together, and in this section these specimens will be known as the subspheroid group. The polyhedron and subspheroid groups are found at both Developed Oldowan C and Acheulean sites. Of the four largest collections of subspheroids, two are Acheulean and two are Developed Oldowan. The subspheroid group makes up a far smaller proportion of the worked artefacts than it does in Bed II. The average percentage of the total number of artefacts for this group is 11 per cent for the main Acheulean assemblages and 6·6 per cent for the main Developed Oldowan collections. The polyhedron sample is too small (n = 48) to show any pattern. The largest collection is from HEB East, a Developed Oldowan occurrence. Half of the polyhedron sample is made on lava (basalt, trachyandesite and nephelinite) and the

Table 10.2 *Polyhedrons, spheroids and subspheroids in Beds III and IV, the Masek and post Masek sites*

	Polyhedrons	Spheroids/ subspheroids
TK Fish Gully	—	4
HK	—	1
FLK Masek	—	31
HEB West Level 1	2	3
PDK Trenches I–III	—	5
WK East C	—	2
WK East A	3	21
WK Upper Channel	7	25
WK Intermediate Channel	2	4
HEB West Level 2b	4	9
HEB West Level 2a	2	13
HEB Level 3	—	10
HEB Level 4	3	11
HEB East	18	23
PDK Trench IV	4	3
JK ferruginous sand	—	14
JK grey sand	3	20

other half in quartzite, while 92.4 per cent of the subspheroid group is made on quartzite.

Observations

During my analysis of the Beds III, IV and Masek samples of polyhedrons and subspheroids, I noted very little technological unity, and only a slight morphological unity within these classes. I found that less than half the subspheroid group showed any battering, and that when it was present it tended to be localised in one or two areas of the stone that projected slightly. Of the unbattered part of the collections, some appeared to be simple rolled and abraded pieces of stone; others consisted of heavily flaked and shaped stones; and others seemed to consist of fresh, angular chunks of quartzite. The polyhedron group consisted largely of flaked cobbles and angular chunks of lava. The most unusual occurrence was at FLK Masek, where at the small

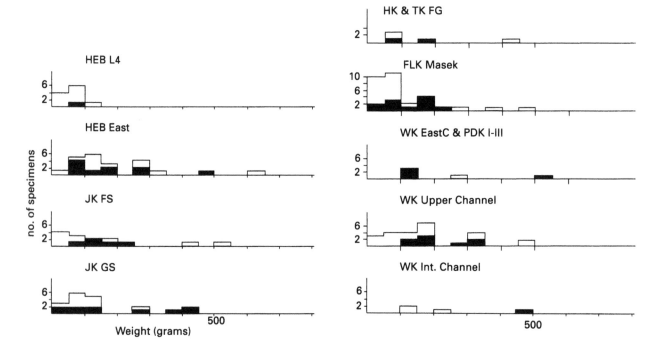

Fig. 10.18 Weight frequency charts for subspheroids from Bed III, IV, Masek and post Masek sites. The black areas indicate numbers of specimens with battering

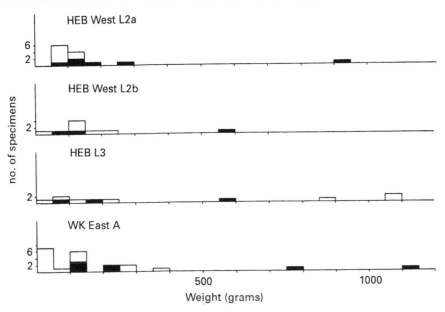

Fig. 10.19 Weight frequency charts for subspheroids from Bed IV. Black areas indicate numbers of specimens with battering

end of the size range of the subspheroid group, the sample was technologically and morphologically consistent. These consisted of many small pieces of quartzite that clearly had been flaked to their present size, and a small number of these showed some abrasion. A few specimens from other sites were similar to these. Figs. 10.18 and 10.19 show the weight ranges of the subspheroid group, with the battered specimens coloured in black. The weights are seen to be quite variable, but at most sites a large proportion weigh between 100 and 200 g. The main exceptions to this are FLK Masek and HEB Level 4 where the majority of specimens are smaller.

One preliminary observation was that the battering noted on the subspheroid group was identical in type and location to that commonly found on hammerstones that have been used for tool manufacture. It was also seen that three agencies contributed to the nodular nature of the sample as a whole: (1) localised battering down on projecting portions, (2) flaking, and (3) pieces naturally shaped. The battering was seen to occur on both the natural and flaked chunks of quartzite.

The experimental work was aimed at determining the types of action that could best reproduce the battering seen on a portion of the specimens, and to determine the starting piece or blank of which the basic tool consisted.

Experiments

In order to clarify the issue of why the subspheroid group possessed a high degree of subsphericality with only small amounts of battering to shape them, I made a study of the accumulated debitage on my flaking floor. As a means of defining the 'chunkiness' or subsphericality of the archaeological material and my own experimental material, I devised an index which is arrived at by dividing the maximum dimension of a piece by its smallest dimension. Thus, a perfect sphere has an index of 1, a hen's egg about 1·5, and a matchbox about 4·5. This index does not express roundness, which is a condition of the edges or margins, but it expresses the relationship between the largest and smallest dimension of a piece, measured roughly across its centre such that the smaller the index, the 'chunkier' is the piece.

My study of the debitage on my flaking floor showed that quartzite debitage contains naturally a chunky/globular element that is virtually miss-

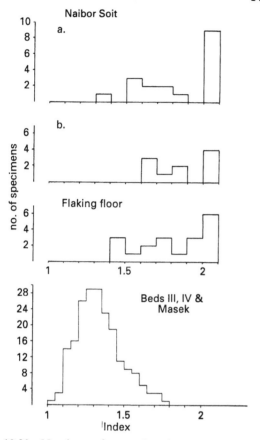

Fig. 10.20 Numbers of quartzite pieces with spherical index of 2 and less from: (1) Naibor Soit (a) bottom of the slope and (b) upper slope. In each case the sample was collected from within 1 sq m; (2) my modern flaking floor; and (3) the archaeological sample of spheroids and subspheroids

ing from the debitage of the other materials. To document this I made a collection of the 'chunkiest' pieces in each of the four main materials represented: quartzite, phonolite, basalt and nephelinite, and measured their indices. The collection was limited to pieces of between 2·5 cm and 12 cm in minimum and maximum dimensions. Thirteen pieces of quartzite were found with indices of less than 2·0; six pieces of basalt were found with indices of less than 2·5 but none of less than 2; the five best-scoring pieces of nephelinite averaged 2·17; for phonolite the average was 2·95. Fig. 10.20 shows the weight distribution of the twenty best-scoring subspherical pieces of quartzite from the flaking floor, in comparison with the archaeological samples. Thus, there is a 'chunky' subspherical element to quartzite debitage that is not found in other materials. This would naturally stand out during the analysis of an assemblage, and some of the subspheroid sample must consist of this 'chunky' debitage.

Further experiments were carried out to simulate the battering that was found on many of the archaeological subspheroid group. Since the minimal battering on the archaeological specimens showed the original blank to be in many cases a 'chunky' piece of debitage of the type noted above, I used these for my experiments.

It was soon seen that chunks of quartzite, when used as hammerstones, quickly sustained the characteristic damage resulting from this activity. Even a few blows with a fresh angular chunk of quartzite when used as a hammerstone for tool manufacture produced a small but readily noticeable area of crushed damage. Sixty-five seconds of secondary flaking resulted in a damaged area measuring 1 cm by 0·5 cm, while similar flaking with an angular piece of basalt resulted in a flaking rather than a crushing of the edge. Quartzite was found to be easily and more noticeably damaged for a given period of battering than any other material.

More intensive use of quartzite hammerstones for tool manufacture resulted in the manufacture of well-rounded spheroids. One such experiment involved an angular chunk of quartzite weighing 334·5 gr with an approximate initial index of 1·6. This stone was used for ordinary secondary flaking for a total accumulated time of about one hour, producing four handaxes as well as flaking other blocks of stone. At the end of this time it weighed 296·5 gr and had an index of 1·2. At this point, the hammerstone resembled in morphology and character of damage some of the most intensively battered spheroids of Bed IV.

During this experimental use of quartzite as hammerstones, several observations were made on the usefulness of this material for tool manufacture. The fact that the quartzite crushes easily means that this material has a soft-hammer quality. The crushed area of damage also resulted in a slightly larger area of hammerstone striking

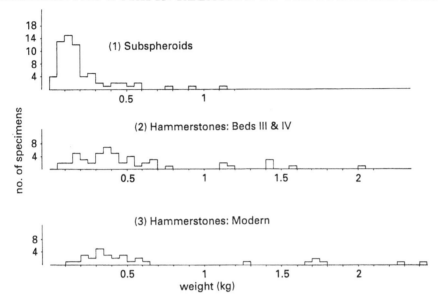

Fig. 10.21 Weight frequencies for the Beds III, IV and Masek samples of (1) subspheroids and (2) hammerstones compared with (3) hammerstones from PRJ flaking floor

the core, and this resulted also in soft-hammer features on the flakes. Quartzite was found to be particularly useful for the flaking of phonolite where the softer and larger area of the hammerstone impact allowed a type of delicate and thin flaking that was not easily obtained with other types of hammerstone. Over a long period of use, the quartzite hammerstones tended towards a spherical shape because the slight projections of a hammerstone, being the areas that are used for striking, would be continually worn down.

Naturally, the use of quartzite chunks as hammerstones for tool manufacture is not the only activity for which hammerstones may have been used. What does appear to be clear, however, is that the basic action consists of battering with the quartzite piece in the hammer position. This results in the localised and slightly convex damage found on the archaeological material. Experiments involving the quartzite piece in the anvil position produced similar damage, but generally not on the projecting parts of the stone only, and tending to be much flatter than that seen in the archaeological sample and produced on hammerstones.

The archaeological sample only rarely shows all-round intensive battering of a stone; most stones show only one or two small areas of damage which, if caused by use for tool manufacture, would be the result of only five to fifteen minutes of flaking. Fig. 10.21 shows the collected weight distribution of all the Beds III, IV and Masek subspheroids that showed battering. The majority of these are between 50 and 200 gr in weight, about 200 gr lighter than the mode weight for the Beds III, IV and Masek specimens actually identified as hammerstones. Hammerstones are defined by M. D. Leakey as follows: 'The hammerstones consist of water-worn cobblestones (generally lava) with pitting, bruising and slight shattering at the extremities or on other projecting parts' (1971, p. 7).

This definition of hammerstones will tend to exclude any angular quartzite specimens, even though they show the same general type and placement of damage. Fig. 10.21 shows the weight distribution of the collected hammerstones from my experimental flaking floor. These modern hammerstones, like the archaeological sample, are mostly river cobbles of basalt, trachyandesite and nephelinite. These two weight distributions are remarkably similar, and what is most interesting is the few specimens which weigh beyond 1 and 2·5 kg. Hammerstones in my sample in this weight range were used for striking large flakes for biface manufacture, while the smaller ones were used for secondary flaking and retouch. The equivalent high weight-range hammerstones from the archaeological sample come from Acheulean sites: WK Upper Channel and HEB West Level 2b. The weight distribution of my smaller hammerstones is very much the same as that of the archaeological sample with the mode occurring between 300 and 400 gr. Both of these distributions, however, are quite different from that of the battered spheroids. Thus, while a portion of the spheroids could be functionally equivalent to battered lava cobbles used for tool manufacture, the majority of them are not. My own sample of hammerstones was mostly used for biface and chopper manufacture, and general large secondary flaking. I was only rarely concerned with light-duty retouch of scrapers or the making of small flake tools. The lightweight portion of the battered spheroids could have been used for this type of light-duty flaking.

Discussions and conclusions

The sample presently defined as the subspheroid group from Beds III, IV and the Masek Beds can be subdivided into three mutually exclusive categories: (1) simple chunks of quartzite debitage, (2) cores and broken tools, and (3) pieces that seem to have been deliberately flaked to a subspherical shape. Superimposed on this to varying degrees is battering, of a type that suggests it is a result of utilisation rather than part of the process of manufacture of a tool. The two major suggestions published to date with regard to the functions of spheroids and subspheroids have been (1) as bolas stones for hunting (Leakey 1931, p. 39) and (2) as pounders for vegetable processing (Clark 1970).

The first suggestion implies that the battering is part of the manufacture process, while the latter makes it the result of the basic activity. Today, South American bolas stones (which are commonly made on quartzite) are found to vary little from a functionally optimum weight of 500 gr. Thus, at best this interpretation could apply to only a small portion of the archaeological sample or perhaps to spheroids from later periods. The use of these stones as pounders I also find to be unlikely since, as simple pounders, there would be no need to choose quartzite preferentially over other materials; one would also expect heavier-duty stones to be used as pounders. And one might find the resulting quartz dust to be a health hazard. This latter interpretation, however, might be more applicable to the larger samples from Bed II.

My own preference for the function of the battered portion of the subspheroid group is that of hammerstones for tool manufacture. Hammerstones are very important artefacts, and as important an element of tool manufacture as the tool blank itself. Like most good lithic materials, hammerstones do not occur everywhere and most of my flint-knapping colleagues curate their hammerstones. During my experiments, I was rarely at a loss for good hammerstones since I was able to bring as many as I needed to my flaking area by vehicle. If, however, one were in the position of needing a hammerstone at many of the Palaeolithic sites in Beds III and IV, one important group of suitable materials that would be immediately available would be chunks of quartzite debitage. While the basalt cobbles were also used, the quartzite shows light-duty change far more easily and would thus stand out while similarly utilised cobbles would show no damage.

Prolonged use of quartzite proved it to be most suitable for hammerstones in tool manufacture and, unlike many of the lava cobbles, these hammerstones rarely fractured while being used. It seems possible that, useful as hammerstones are, chunks of quartzite could have been deliberately shaped with a view to use for this purpose. Some of the flaked cores commented on earlier, which appear to have been flaked deliberately to a roughly subspherical shape, could represent the trimming of blocks of quartzite to make them more useful as hammerstones.

There does remain a small component of this whole subspheroid group that consists of very small flaked pieces. These are particularly evident in FLK Masek where they make up almost 50 per cent of the subspheroid sample. These specimens appear to be very small for cores and too small to be trimmed pieces for future use as hammerstones. Some of them are not unlike other small pieces found in the discoid and sometimes biface groups from other sites. On the whole, it appears that quartzite was often flaked down to a small size, and this collection of specimens may represent some specific activity of their own.

Outils écaillés, punches and pitted anvils

These three artefact groups are considered together. The reason for this is that the initial experimental work, which was aimed at replicating and hence explaining the outils écaillés, indicated a strong technological association between them and the other two tool categories. Outils écaillés have been defined by M. D. Leakey (1971, p. 7) as follows: 'Both single and double ended specimens occur. They exhibit the scaled utilisation characteristic of these tools. The edges are blunted and one face is usually slightly concave, whilst the opposite side is straight or slightly convex.' Although this describes the tools, it does not attempt to explain their function or say how they reached the condition in which they are found. In fact, there is no satisfactory explanation of them in the literature although they occur at many stages of prehistory. This is why I made them the object of a special experimental study.

Pitted anvils can be defined as manuports, generally flattened oval lava cobbles ranging in size from a few centimetres across to over 20 cm, which have pits or depressions, usually oval in shape, 2 to 3 cm long and a few millimetres deep. Anvils are found with many different arrangements of such pits, from a single pit on one face to pairs of pits on both faces. These are discussed in more detail below. There are rarely more than two pits on one face. Some of these specimens are illustrated in Fig. 10.31.

Punches are short cylindrical pieces of quartzite, usually between 2 and 4 cm long, with roughly pointed and sometimes crushed ends.

The archaeological distributions of these tool

Table 10.3 *The archaeological distribution of* outils écaillés, *punches and pitted anvils in the Olduvai sequence*

	Outils écaillés	Punches	Pitted anvils
Bed I, Oldowan	—	—	—
Lower Bed II, Dev. Old. A	—	—	—
Middle and Upper Bed II, Dev. Old. B	62	—	—
Middle and Upper Bed II, Ach.	2	—	—
Bed III	8	2	13
Lower Bed IV, Ach.	15	1	22
Upper Bed IV, Dev. Old. C	32	147	66
Upper Bed IV, Ach.	9	2	41
Masek, Ach.	3	5	5

Table 10.4 Outils écaillés, *punches and pitted anvils in Bed III, Lower and Upper Bed IV*

	Outils écaillés	Punches	Pitted anvils
HEB West Level 1	2	—	—
PDK Trenches I–III	3	20	14
WK East C	13	26	15
WK East A	62	68	68
WK Upper Channel	33	14	106
WK Intermediate Channel	3	1	—
HEB West Level 1	2	—	—
HEB West Level 2b	5	—	4
HEB West Level 2a	2	1	3
HEB Level 3	1	—	3
HEB Level 4	—		
HEB East	4	—	3
PDK Trench IV	—	1	2
JK ferruginous sand	1	2	11
JK grey sand	7	—	2

types in the Olduvai sequence is shown in Table 10.3 (from Leakey 1975a):

Only three Bed II sites contain *outils écaillés*, SKH and BK (both Developed Oldowan B) and the Acheulean site of TK, Lower Occupation Floor. The largest occurrence is at BK and even there they make up less than 5 per cent of the tool component (n = 5). None has been found at EF-HR, the principal Bed II Acheulean occurrence,

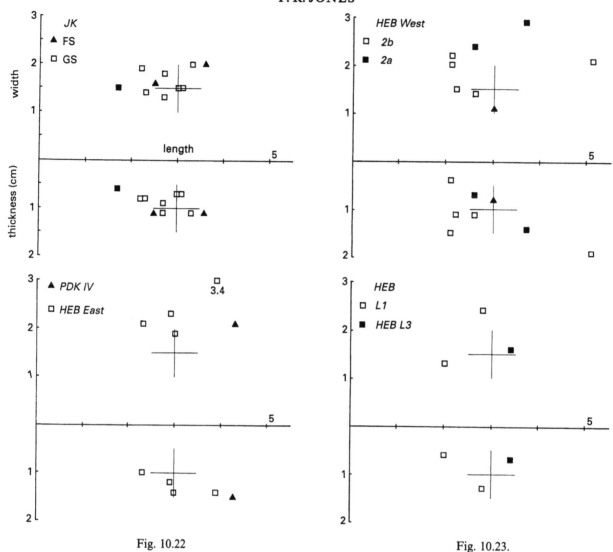

Fig. 10.22   Fig. 10.23.

Figs. 10.22 and 10.23 Scatter diagrams showing the width/length and thickness/length ratios for the *outils écaillés* (squares) and punches (triangles) from JK, PDK Trench IV, HEB East, HEB West Levels 1, 2b and 2a and HEB Level 3

and no punches or pitted anvils have yet been recognised in any Bed I and Bed II assemblage.

Punches and pitted anvils are present in Bed III and Lower Bed IV in small numbers, but they only become an important component within their parent assemblages in Upper Bed IV (see Table 10.4). WK East A and WK Upper Channel are the main sites, with WK East C and PDK Trenches I–III also preserving a substantial number. Figs. 10.22 to 10.27 present length/width and length/thickness scattergrams for all the Beds III and IV specimens.

There does not appear to be a distinctive pattern in the occurrence of these artefact types, and they occur in similar quantities in both Acheulean and Developed Oldowan sites. There is a substantial increase in number in Upper Bed IV and the majority of all of the specimens come from the small area of the WK and PDK sites (see Figs. 10.24–10.26).

With reference to Figs. 10.22 to 10.27, the basic distribution and size ranges appear very similar at each site. The crosses on the charts are to allow an easier comparison of these distributions. WK

EXPERIMENTAL WORK IN RELATION TO THE STONE INDUSTRIES

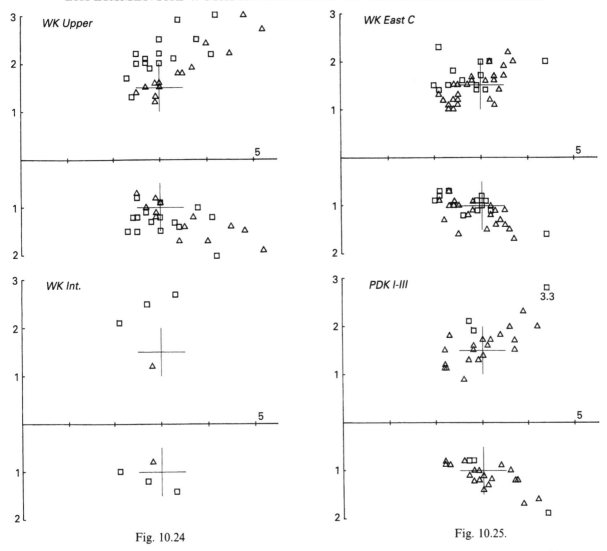

Fig. 10.24   Fig. 10.25.

Figs. 10.24 and 10.25   Scatter diagrams to show the width/length and thickness/length ratios of *outils écaillés* (squares) and punches (triangles) from WK Upper and Intermediate Channels, WK East C and PDK Trenches I–III

Upper Channel, WK East A, WK East C and PDK Trenches I–III, the sites with the largest samples, all have virtually identical patterns of distribution. The punch samples tend to show a narrower range of length/width and length/thickness ratios than do the *outils écaillés*; given the nature of punches, this is perhaps predictable.

Experiments

A series of experiments was carried out to try to determine the possible causes of the damage, apparently the result of use, which is found on the *outils écaillés*. The hammer and anvil stones used for the experiments were of basalt or trachyandesite, and quartzite from Naibor Soit was used for *outils écaillés* replication since all the original specimens are found in this material. The preliminary experiments, which are described first, were short and simple, carried out only to provide broad guidelines for further work. The main aim of these experiments was to duplicate the damage characteristic of the *outils écaillés*.

I collected quartzite flakes of various sizes from

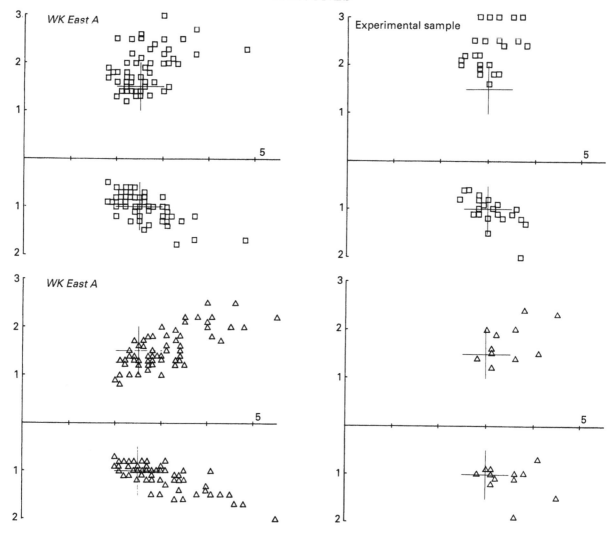

Fig. 10.26 Width/length and thickness/length ratios for *outils écaillés* (squares) and punches (triangles) from WK East A

Fig. 10.27 Comparable ratios for *outils écaillés* and punches from experimental flaking

the debitage on my flaking floor, held them vertically on the anvil stone and battered them with a hammerstone. The quartzite was crushed and shattered with many small flakes and slivers being detached from the upper edge of the flake. The bottom edge was crushed, and if I continued battering after these effects were achieved, I often broke the flake lengthwise. Sometimes a flat surface was produced from which no further flakes were detached, and in these cases the only noticeable product was quartz dust. Both anvil and hammerstone showed localised damage, but only after the battering of many flakes were pits produced in either the hammer or the anvil stone.

Experiments were carried out with the battering of larger blocks of quartzite in an effort to produce large and potentially usable flakes by this method as none seems to be produced when quartzite flakes were used. Substantial flakes were rarely removed and more often a flat surface developed at either end of the block. When great amounts of force were applied by means of very large lava stones the quartzite flakes or blocks usually split from the upper point of impact to their contact with the anvil. In no cases were flakes removed that could not more easily have been produced by normal direct percussion.

I also used bone and wood as hammers and anvils, and some splitting experiments were carried out using stone flakes as wedges. When flakes became embedded in either of these materials, they tended to snap transversely – probably as a result of some type of shock rebound effect. When flakes were battered with either wood or bone very little damage was incurred at the upper end of the quartzite flake. The lower end of the flake did show some crushing and damage could be seen on the anvil stone. The bone or wooden hammer sustained considerable damage, the distribution of which was related to the nature of the upper end of the flake.

After completion of these preliminary experiments, I returned to the first exercise in which flakes were battered between two stones (hammer and anvil), since this situation had most clearly and readily duplicated the 'characteristic scaled utilisation' that was a feature of the archaeological specimens. Flake groups of different sizes and shapes were battered in this way and the resulting pieces collected.

While the initial experiments had been aimed only at replicating certain types of characteristic damage found on small archaeological pieces of quartzite, the experiments in battering quartzite between two stones had also produced two interesting classes of by-products. The first of these consisted of small cylindrical pieces of quartzite with pointed ends, which when found in the archaeological assemblages had so far been called punches; and the second was a selection of pitted stones of various sizes that were very similar to the archaeological category of 'pitted anvils'. An examination of the archaeological associations of these types shows that the pitted anvils are not found at sites without either *outils écaillés* or punches, and usually both are present. However, the opposite is not always true. The WK Intermediate Channel and SHK sites have only a few *outils écaillés* (and one punch in the former) and no pitted anvils, though these artefact types had not been recognised at the time SHK was excavated. A recent surface survey near the SHK site did reveal at least one pitted anvil. If, however, the absence of pitted anvils at these two sites were real, I should not regard it as significant, since my subsequent experimentation has shown that it is only after a substantial production of *outils écaillés* and punches that readily identifiable pitted anvils or hammerstones result.

Formation of concavo-convex edges

As the quartzite flakes were being battered in my experiments, I noted that concavo-convex edges characteristic of *outils écaillés* and mentioned in Dr Leakey's definition would often form on the upper and lower edges of the flake, where it made contact with the hammer and anvil stones. A flake is rarely symmetrical in long or short cross section and the edge angle around the whole perimeter is highly variable. Thus when the flake is battered, small flakes and fragments will be more easily removed from one face than the other (as in Fig. 10.28 A–B). Continued removal of these fragments from one face by battering will have the effect of moving the edge at that particular spot towards the other face. Eventually

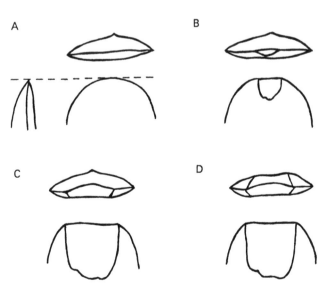

Fig. 10.28 How the concavo-convex edge is formed through battering the end of a flake

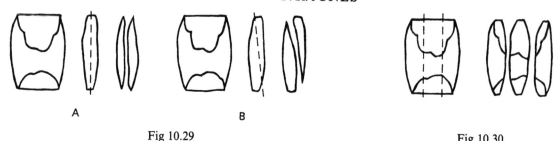

Figs. 10.29 and 10.30 The three main ways in which flakes split while being battered. Figure 29 A and B result in *outils écaillés*, while Fig. 10.30 results in punches

a point of equilibrium is reached where both faces now meet the edge at the same angle and further battering, therefore, removes fragments from both of them. The edge will almost always be concavo-convex at this point, the amount of curvature depending on the morphology of the original flake (see Fig. 10.28 C–D).

Flake breakage during battering

With continued battering flakes may break or split in several different ways:

1. Flakes can split longitudinally at right angles to their thickness, producing two pieces of about the same width and half the original thickness of the flake (see Fig. 10.29 A–B). A common variation of this is when the split does not entirely follow the long axis of the flake but turns out before reaching the bottom end (Fig. 10.30). When flakes split during the battering, the upper and lower edges of the flake will fragment. This results in small flake scars on the fresh split surface at either end.

2. Flakes can also split longitudinally in a plane approximately at right angles to the dorsal surface, producing two or sometimes three pieces of equal thickness but only half or a third of the original width (Fig. 10.30). Again the edges generally fragment at either end of the split and often the crushing on the original edge is totally removed. The ends of these pieces tend to be quite pointed and do not show the usual signs of extensive battering though they do have small scars near either end. They do not resemble an ordinary direct percussion flaking debitage. I have noticed that the thinner the flake in relation to its width the more common this type of fracture is.

3. Flakes can also split in various combinations of the above. Breakage is almost always longitudinal from point of impact to anvil contact: transverse breakage is very rare and when it does occur can usually be attributed to some original flaw in the flake.

Flake breakage of Type 1 can result in both double and single-ended *outils écaillés*, while breakage Type 2 results in the cylindrical pieces called punches. Combinations and variations on these breakage types result in other forms whose damage patterns obviously have bipolar origins and these are varied enough to provide an explanation for all 'atypical specimens' noted by M. D. Leakey from WK East A.

Formation of pits on anvils and hammers

During these experiments the anvils and hammers sustained considerable damage. I found that the hammerstones sustained more damage than did the anvils, and pits were produced at a rate of more than five on a hammerstone to one on an anvil. This is for two main reasons: (1) hammerstones are always directly striking the flakes, while the anvils only receive a force which has been transmitted through the flake; (2) the same area of the hammerstone tends to hit the flake while the flake often moves about on the anvil.

For these reasons also, larger pits tend to be found on anvils while the hammerstones have smaller but deeper pits. Another factor is that

EXPERIMENTAL WORK IN RELATION TO THE STONE INDUSTRIES

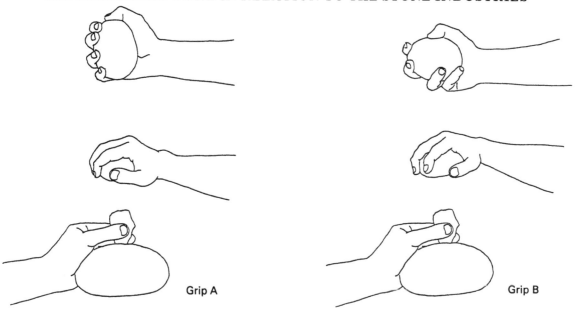

Fig. 10.31 The two main ways in which an oval cobble can be held for bipolar flake battering

anvils can be used for a longer period than hammerstones, which are not easy to use when a face already has two pits on it. The general area of damage tended to be oval because of the lenticular cross section of most flakes. However, there was a secondary deeper area of damage in the centre of the depression of a more circular shape.

Particularly striking during the experimental work was the occurrence of twin oval pits identical in size, morphology and position on the hammerstones of many of the archaeological specimens. These were produced for the first time quite by accident during the experiments. Subsequent study indicates some simple reasons for the occurrence of these twin pits.

Twin pit formation

Twin pits are commonly produced on elongate, flattish cobbles that have been used as hammerstones. I found that cobbles of this type are most easily held in one of two ways if they are to be used efficiently as hammerstones: Grip A and Grip B (see Fig. 10.31).

For Grip A the width of the stone is gripped between the heel of the hand and all fingers, whilst with Grip B the length of the stone is gripped between the heel of the hand and the first two fingers, with the thumb and third finger supporting either side. These are grips that I naturally adopted when starting the experiments and I never found any more effective alternative.

When the hammerstone, held by either of these methods, is used to batter flakes for some time, the resulting damage to the hammerstone consists of an oval pit which is set at an angle to both the long and wide axis of the hammerstone (see Fig. 10.32a). Continued battering deepens the pit and when it becomes too difficult to strike the flake properly a new surface is needed. The hammerstone can be turned over to use the other side, or the grip can be changed and a new area of the same surface used.

The relative positions and angles of the twin pits formed on the hammerstones depend on how the grip is changed. If the hammerstone is rotated 180° from Grip A (so that the same grip is used) the resulting pits will be parallel, and at an angle of about 45° to both the long and wide axis of the stone. The pits will also be parallel if the stone is rotated 180° from Grip B (see Fig. 10.32b). If, however, the hammerstone is merely rotated about 90° from either of these basic grips, in other words if there is a change from one type of grip to

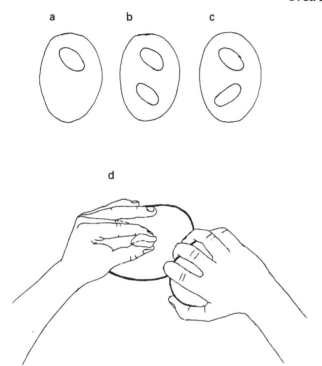

Fig. 10.32 The angled pits produced on cobbles used for bipolar battering. When using a hammerstone for this purpose one tends to use one end to strike the flake, rather than the centre. Intensive use of a hammerstone and a change of grip result in angled parallel or convergent pits identical to those found in Bed IV

the other, the pits that result will be converging (at 90° to each other). This may sound complex but it works, and is a function of the oval shape of the stone, the way force is best delivered by one end of it rather than the middle, and the relative angles of the hand holding the flake and the hand holding the hammerstone (Fig. 10.32 c–d).

Summary and discussion

The various experiments described show how the *outils écaillés*, punches and pitted anvils can all be produced by battering a series of small quartzite flakes between anvil and hammerstone (Table 10.5).

The archaeological sample of punches and *outils écaillés* as a whole seems to be very consistent in size and shape. The modern sample is different only in that the minimum dimensions are somewhat larger (on average less than 0.5 cm) than the archaeological sample. The size limiting factor in the experimental case was the width of fore-finger and thumb, which were themselves battered when the flake got to a certain size. The experiments have indicated that *outils écaillés* themselves or other small flakes are virtually useless for splitting wood or bone, and the damage incurred in the attempt was quite different from that found in the archaeological sample.

I do not consider the *outils écaillés* or punches to be worked-out cores from which a series of useful flakes have been removed. While the skilful bipolar battering of larger blocks of quartzite can produce flakes, the resulting cores do not resemble *outils écaillés* or punches. Furthermore, as the small flakes were battered and turned into the *outils écaillés* or punches no flakes or splinters were removed that could not be produced by direct percussion. While the bipolar battering of small pieces might be resorted to in areas where the only available raw materials occurred in sizes too small to flake normally, this does not apply to the Olduvai sites where these specimens occurred. The small size of the pits found on hammerstones (rarely more than 3 cm across) also attests to the fact that small pieces were selected for battering.

Thus the picture emerges of the selection of small quartzite flakes which were battered for a short while each, and discarded when they either broke or became too small to hold easily. I suggest that this battering action must have been part of the relevant activity and not a tool manufacture process. This means that the three tool types, *outils écaillés*, punches and pitted anvils, are all most probably the accidental by-products of a single, brief activity.

Having defined the basic action involved, attempts were made to determine the possible reason for this activity. This is quite difficult and might perhaps be solved only by the observation of some ethnographic analogy or the association of these types with some other feature which might not yet have been found or recognised. Some preliminary experiments of my own show that dried rawhide or meat can be difficult to cut by conventional means, and can be cut (somewhat laboriously) by battering quartzite flakes through

EXPERIMENTAL WORK IN RELATION TO THE STONE INDUSTRIES

Table 10.5 *Results of experimental battering of flakes using bipolar techniques*

	PIT 1	PIT 2	PIT 3	PIT 4	PIT 5	Mean of all Pits
Generalised area of pit in cm	3 × 2 × 0.5	3.7 × 2 ×	3.5 × 2.2 × 0.6	2.7 × 1.4 × 0.3	3 × 2.1 × 0.7	
Area of intense damage in cm	2.5 × 1.5	2 × 0.6	2 × 0.7	2 × 1	2.1 × 1	
Number of flakes used	3	5	4	2	14	5.6
Average flake size in cm	4.8 × 4.5 × 1.4	5.7 × 4.1 × 1.5	5.6 × 3.8 × 1.4	5 × 3.9 × 1.4	4.8 × 3.6 × 1.3	
Time taken to produce pit in seconds	67	91	80	49	138	85
By-products per pit:						
punches	3	4	3	—	3	2.6
outils écaillés	3	3	3	2	11	4.4
undamaged flake	2	4	5	1	5	3.4

them. It must be significant that all the Olduvai specimens so far identified are in quartzite. Experimental battering of both basalt and phonolite flakes produced similar results but with a lower occurrence of punches. These materials, however, do not appear to have been used in this way at any of the Olduvai sites. This may be an important point since some function may be found for which quartzite is seen to be superior to the other materials. What that function is remains to be seen.

I am not suggesting that all artefacts described as *outils écaillés* or punches in other areas or other time ranges are produced in the same way. Undoubtedly, some amongst such tools must represent different activities. Experimentation in the relevant raw materials should be carried out. Hopefully this section on the *outils écaillés*, punches and pitted anvils from Olduvai will at least stimulate discussion and experiments in other areas. If it is found that all these specimens are produced in the same way and that they do indeed represent the same activity, then it must have been a frequent and important one given their common occurrence through most of the Middle and Late Stone Age in Africa.

Experimental use of stone tools

During the period of my work at Olduvai I was able to carry out many experiments of different types involving the use of stone tools. These were sometimes controlled and carefully recorded, but were mostly opportunistic in nature. This was particularly the case with butchery when, for example, a goat suddenly became available or an animal had died in the vicinity of the camp. The bulk of this experimentation was aimed at learning the relative efficiencies of the different Olduvai raw materials for a variety of tasks, and to try to learn the qualities that are desirable in tools for different types of task.

There are many problems connected with experimental work of this type, and one of the biggest of these is the difficulty of maintaining adequate experimental controls to allow the collection of quantitative and comparative data. While such data can be collected from an experiment, a problem exists in accurately interpreting them and relating them to other experiments. Many of my experimental results and conclusions are therefore expressed in qualitative terms which are based on the results of many experiments and my own observations.

A further problem exists in the type of experiment that is chosen and with the experience of the experimenter. One cannot try every conceivable activity that a stone tool might have been used for, yet casual experimentation may give very misleading results. Any experienced craftsman will know that having the right tool for a given job is not enough to ensure good results: the tool user must also be experienced in order to get the best results from a tool. There is also the danger, however, that an experimenter will become adept at using a stone tool for some obscure and unlikely function through the practice and experience gained during their own intensive experiments.

The following sections briefly describe a series of experiments and results. Many of these deal with aspects of elephant butchery. This is because the previous small-animal butchery has already been published elsewhere (Jones 1980) and because the butchery of elephants would possibly consist of the most difficult butchery tasks that could be expected of stone artefacts.

By way of combating some of the problems described above, I have decided to focus on edge types rather than artefact types, the aim being to establish the qualities of the different edge types in the different raw materials, and the qualities that are required for different tasks.

Experiments

Several experiments were carried out in order to compare the edges of flakes in the different raw materials for different types of task.

Elephant skin penetration

Two experiments were carried out where four unretouched flakes were used to penetrate elephant skin on fresh carcasses and to make incisions of about 15 cm in length. Both experiments were carried out high on the shoulder of mature elephants where the skin was approximately 1·5 cm thick. These two experiments were separated by a period of eight days, during which time I had gained much experience in the course of butchering five elephants.

Experiment 1	Time	Experiment 2	Time
Quartzite	1 min 10 sec	Quartzite	11 sec
Phonolite	1 min 40 sec	Phonolite	42 sec
Basalt	2 min 10 sec	Basalt	35 sec
Chert	1 min 10 sec	Chert	33 sec

These data are presented graphically in Fig. 10.33. The experiments were carried out using the different flakes in the same sequence, and they

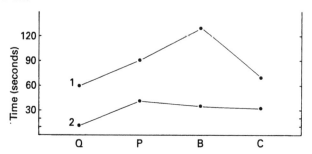

Fig. 10.33 The time required to penetrate elephant skin 1·5 cm thick with un-retouched flakes of quartzite (Q), phonolite (P), basalt (B) and chert (C). Sequence No. 1 was the first elephant I experimented on while No. 2 was carried out eight days later with similar flakes on my fifth elephant

were all of approximately the same size and shape. The results clearly show the effects of experience and practice but, more importantly, the relative order for the different raw materials remains more or less constant. Quartz is shown to be the most useful, followed closely by chert. The basalt and phonolite flakes come in third and fourth places.

For this activity the edge qualities that were important were the sharpness and strength. The quartzite flakes may have had the added benefits of serrated and saw-toothed edges. The cutting rates that were suggested by this data cannot, however, be maintained for long working periods, as the following section suggests.

Elephant skin cutting

Primary flake edges, particularly those of quartzite, were found to be the most useful for skin cutting; however, there are several other important associated qualities that aid this task. These are the weight of the tool, the length of the cutting edge, and the tension of the skin that is being cut. A difference was also noted in the types of material and edges that were found most useful for short-term work as opposed to those best suited to longer-term work.

Fig. 10.34 presents data for the rates of skin cutting and the length of skin cut for the primary flake edges of three different materials. Several interesting points stand out. Blunting is clearly seen in the reduction of the cutting rate through time of the basalt and phonolite edges. The

EXPERIMENTAL WORK IN RELATION TO THE STONE INDUSTRIES

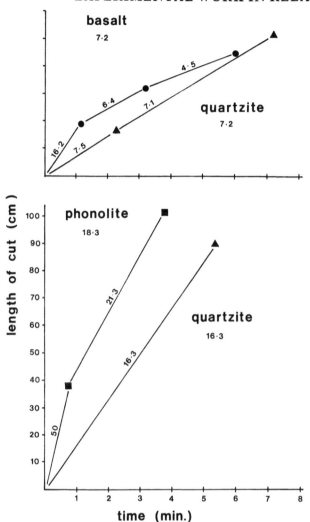

Fig. 10.34 Rates of skin cutting by different flake types. Numbers on the lines indicate cm cut per minute. Blunting on the basalt flake can be seen in decreasing cutting rates through time

comparable data for quartzite show only a very small reduction in cutting rate, as opposed to the 50 per cent reduction in both of the other materials after only one minute's work. Interestingly, in Fig. 10.34 the total average rate of cut is the same for both basalt and quartzite edges. However, had the experiment continued I would predict a continued reduction in rate for basalt, but only a small reduction in the rate of the quartzite edge. The high rate of cut seen for the phonolite tool is not really directly comparable to the rates of the other materials, since this was due to the high tension and perhaps even weakening of the skin of a 48-hour-old bloated carcass. The high rate of cut for the quartzite tool in the same figure is because the belly skin being cut was only a few millimetres thick.

By way of interpretation of these data, the basalt and phonolite primary flake edges were both thin, fairly straight and even, whereas the primary flake edges of the quartzite, while being thin and straight, were jagged. The quartzite, however, is the hardest material and will hence preserve its sharpness longer than the other two. The data show that the thin and even nature of the edges in the two softer materials are a considerable asset for small amounts of this type of work, while the jagged edges of the quartzite impair fast cutting of skin. Through time the other two materials blunt considerably, while quartzite continues cutting at the same rate.

Meat cutting

Large cutting tools were found to be the most useful for this activity with little difference noted between primary and retouched flake edges. I often found that several initial strokes could cut meat very well over a large area. Several strokes recorded for a quartzite tool, for example, opened a cut 35 cm long and 4 cm deep on the first stroke, and extended it a further 3 cm deeper on the second. This works out at 245 cm cut in about 15 seconds, or about 1,000 sq cm per minute. This required little effort, and was largely due to the combined effects of the weight of the tool and the length of the cutting edge.

The actual process of removing meat, as opposed to just cutting it, is a different matter. Over a long working time, when one is cutting tendons and working where the approach is more awkward, the overall numbers of centimetres cut per minute naturally drops. Table 10.6 presents data on the amount of meat cut, time taken and amounts removed for a series of different tools and different tasks. In all I found that the serrated quartzite cutting edges were the sharpest and longest lasting. Basalt edges blunted the most quickly during these tasks, but still had a long

Table 10.6 *Experimental butchery of large carcasses with stone tools showing the amount of meat removed and time involved. All of the cutting was done without re-sharpening any of the tools, and it was noted that steel knives used by other workers at the butchery site had to be re-sharpened every fifteen to twenty minutes*

	Meat removed	Time	Sq cm cut per minute
1	Phonolite cleaver used to remove meat from elephant femur:		
	16 kg	11.5 min	113
2	Large basalt flake used to remove meat from elephant femur:		
	19 kg	6.5 min	225
3	Phonolite handaxe used to remove hind leg of mature cape buffalo:		
	20 kg	6.0 min	226
4	Quartzite handaxe used to remove meat from elephant shoulder:		
	36 kg	40.0 min	74
5	Quartzite chopper used to remove elephant trunk:		
	55 kg	20.0 min	59

initial period of utility. No problem was encountered while holding these large tools in bare hands. The force on the edge against the palm was spread along the length of the edge, and I was never cut by the large tools when using them.

Tuber peeling and cutting

Experiments were also carried out to determine the most useful edge qualities for peeling and cutting up tubers. The tubers used were those of *Ipomoea longituba* Hall f. (Family: Convolvulaceae) which occur in the Olduvai area and are eaten raw by various tribes in Tanzania (Annie Vincent pers. comm.). It was quickly realised that for this task the primary flake edges of phonolite and basalt were the easiest to use and the quickest. This was the observation of two independent witnesses as well as myself, experimenting with a selection of flakes of different types. Since the tubers are quite soft no blunting was noted during these experiments, although judging from other experiments one would expect the phonolite and basalt edges to be the first to blunt. However, the time required and the amount of tubers necessary to illustrate this were not available. The irregular edges of the quartzite flakes were found to be a hindrance in these activities, presumably because the edge could not all cut along one precise plane as the other materials did and the resulting resistance required more effort.

Wood working

No primary flake edges were found to be able to whittle any of the dry local hard woods. When green branches were sharpened to a point (the idea being to use the tools for some specific task, in this case to make a digging stick, rather than just aimless experiments) it was found that only the finest-grained and hardest phonolite and basalt edges could actually whittle the wood in the manner that steel edges can. The principal action that was found to shape the wood relatively easily was by scraping or rasping it, and for this right-angled edges, either flaked or unflaked, were particularly useful. For working periods of up to fifteen minutes, little difference was noted in the qualities of the various material types except that the phonolite and basalt gave a much smoother finish because of their straighter edges.

The easiest way of working wood was to use a larger stone tool and a chopping action to shape the end, and then smooth it later with a smaller tool. For the chopping action, the basalt edges and thick primary flake edges were found particularly useful. The main advantage of this was clearly the weight of the tool. Bifaces in all of the materials were able to shape the end of a small branch to a point in between three and ten minutes. The edges of the retouched tools used blunted and chipped in the area that had been used most intensively, while the primary flake edges scraped off. It was noted that for the chopping action one tended to use only a few short lengths of edge on any one tool. This was due to the problem of holding the tool and applying force to it which left only a few working edges. For woodwork of this type, the common biface design is not the most useful as only a small portion of the long edge can be used at any one time. The impact necessary in this activity caused problems with the holding of these tools in bare hands. Generally, the thicker the edge (i.e. the closer it is to 90°), the longer it will last for this type of work. Basalt and quartzite were found to be the most useful for these types of impact actions while the phonolite tools sustained most damage.

The qualities needed in tools for this activity are steep-angled edges in either basalt or quartzite. The edges can be quite short and still be effective, but the rest of the tool must be such that it can sustain the forces that are applied to it.

Summary

All of the edge types described can be used for all the activity types discussed at least for a short time, or through necessity. The difficult part in the assessment of the experiments is to determine which artefacts are the most useful or the most time and material efficient for certain tasks. The conclusions that are drawn may be different between short-term and long-term use.

Primary flake edges are not useful for impact action activities because the edges are so quickly damaged. A rough hierarchy for the raw materials for this task would be basalt, followed by phonolite and then quartzite.

Primary flake edges are useful for skin cutting, which can be considered a tough-pressure activity; and for tuber peeling and cutting, a light-pressure activity. The quartzite edges are most useful for the former, while the even edges of basalt and phonolite are better for the latter.

One important principle that holds, virtually no matter what the task, is that the larger and the heavier duty the task, the larger the tool needed for easy use. This is not as important for short-term work, but becomes more so for longer-term work where, with stone artefacts, there are not only the weight and long edges that can be brought to bear, but the extra material and potential edges that a larger tool has. A 600 g handaxe has greater functional long-term potential than twelve 50 g flakes, even though the latter may have a total of 1 m of primary flake edge. This greater potential of larger tools must have been of great significance to early tool users who lived in areas where material sources were few and very localised. Lack of suitable raw material at certain critical times may have had serious effects on a group of people, and the possession or not of material may at times have been of life or death significance.

Discussion

The previous sections of this chapter are self-contained with regard to the artefact definitions, their occurrence at Olduvai, experimental manufacture and conclusions. In this final section I want to discuss in more general terms the implications of some of my experimental results. My comments relate mainly to the relationship between Developed Oldowan and Acheulean sites and to the importance of bifaces to these early hominid groups. In the following pages I will briefly summarise the relevant information which has been described in detail in earlier sections.

First, the bifaces from Acheulean assemblages are large tools that were consistently made to the form in which we find them. The minimal retouch on many of these large flake blanks indicates that little time was spent on their manufacture and also suggests that the tools were not intensively used or reworked from the time they were first made. These artefacts do deserve the term 'formal tool type', and they are certainly in a different category from the many other artefact types that were created through use and that actually represent consistent types of damage (e.g. subspheroids, spheroids, *outils écaillés*, punches and pitted anvils). The Developed Oldowan bifaces, on the other hand, have a very variable morphology within sites and have been so extensively flaked that for more than 75 per cent of the entire Bed IV sample the blank type cannot be determined. Hence we cannot easily assess the amount of time that was spent on manufacture, nor can we say that these tools were deliberately made to the shape in which we find them. There are, however, two very important similarities between these two samples of bifaces: first, both samples are made in roughly the same manner, i.e. using the same basic set of techniques, to the same basic plan shape. Second, both samples are made in the same range of raw materials. Naturally it is the same basic plan shape that puts both artefact groups together as 'bifaces', but it is the fact that both samples occur in the same range of raw materials, and not principally in one (like all other tool types from the Olduvai collections), that really suggests that their functions were similar.

One technological difference between the two samples is in the blank types commonly used for biface manufacture and the degree of flaking seen on them. Blank type can be definitely assigned to about 85 per cent of the Acheulean sample and to about 45 per cent of the Developed Oldowan sample. Of the known Acheulean blanks more than 75 per cent were large flakes and of the Developed Oldowan sample about 30 per cent were cores. Only about 15 per cent of the Acheulean biface sample has been flaked to such a degree as to obscure the blank type, while this category makes up about 55 per cent of the Developed Oldowan sample. The majority of the extensively flaked Developed Oldowan tools are made in phonolite or quartzite, while most of the core tools are made on lava cobbles. Of principal concern here is to determine whether the Developed Oldowan bifaces represent tools initially made to the size and shape in which we find them, i.e. the end product in themselves (as is suggested for the sample of Acheulean bifaces) or whether they represent something else. Naturally there are many possible explanations for the Developed Oldowan sample of bifaces, and no matter how many I list here other workers would be able to suggest more. The differences are not due to the raw materials in which these tools are made, and need not be greatly influenced by the presence or absence of the Large Flake Technique since 55 per cent of the Developed Oldowan biface blanks cannot be determined. My own conclusion is that the majority of the Developed Oldowan sample consists of re-sharpened and re-flaked Acheulean handaxes.

Figs. 10.2, 10.3 and 10.4 show the edge length/weight ratios for the entire Bed IV sample of Developed Oldowan bifaces and Acheulean bifaces. The two samples can be seen to overlap to a great extent since each assemblage type does contain some artefacts which are typical of the other. These distributions can be compared to the changing edge length/weight ratios of four bifaces through several re-sharpening stages. The experimentally made tools are seen to move from the 'core' area of the Acheulean scatter to the 'core' area of the Developed Oldowan scatter. While handaxe re-sharpening can account for the quartzite and phonolite Developed Oldowan bifaces it cannot account for all of the basalt tools since it can be clearly seen on many of these that the tool blank consisted of a water-worn cobble scarcely larger than the final tool itself. My interpretation of this is that the hominids had made the best possible use of this material type in areas where it was not possible to make larger bifaces. In other words the hominids were coming from a material-scarce area which had necessitated the re-sharpening of their handaxes and the manufacture of others from whatever materials had been available. It is interesting to note that quartzite, which accounts for well over 30 per cent of the Bed IV Acheulean bifaces, makes up a much smaller percentage of the Developed Oldowan collections. Quartzite is the preferred material for small light-duty artefacts and the majority of all debitage from most sites is in quartzite. This particular preference for quartzite light-duty tools might have resulted in the re-flaking of quartzite bifaces in advance of other materials since the flakes would be more useful.

A simple study of the common shape of bifaces shows how maximum edge lengths are obtained for these units of stone. Handaxes made to a more circular shape are heavier per unit edge, while longer slimmer handaxes are harder to make, are fragile, and have less re-sharpening potential. The edge length/weight graph in Fig. 10.16 shows how biface weight increases with edge length. The majority of the Olduvai bifaces are seen to occupy an area defined by edge lengths of at least 20 cm and total tool weight of generally less than 900 g. Within those boundaries there is a definite cluster of bifaces which range from 25 to 45 cm edge length, and from 400 to 650 g in weight. If edge lengths are doubled the tool weight increases about sixfold. This is seen in Figs 10.2–10.3 where the bifaces of 800 to 1,000 g in weight have on average cutting edges only a few centimetres longer than do the bifaces in the weight range 400 to 600 g. As has already been discussed, however, the stone itself is a functional part of the tool since it can be re-flaked. The bifaces from FLK Masek in the Masek Beds, almost half a million years younger than the Bed IV sample, have an average weight of over 1,000 g, and average edge lengths of over 50 cm. These bifaces are made to obtain the longest possible edges for stones of their bulk, but the emphasis of this technology does seem to be large size. This is a pattern also seen at Olorgesailie and Isimila, where very large tools were made at a late stage of the Acheulean. I would suggest that the emphasis here is stone bulk, rather than the long edges, and that this could mean greater mobility of hominids resulting in the need for more material transport.

The handaxe is thus seen to be: (1) the transported unit of stone from the source to living or use areas; (2) a tool with very long cutting edges capable of a variety of tasks, and which can be re-sharpened a number of times; (3) a unit of stone with maximum manageable weight from which flakes can be removed or which can be flaked into some other form. In short, it is a well-designed tool aimed at the problems of living in and travelling through material-scarce areas.

A group of hominids with a seasonal round exploiting a variety of different food sources would tend to be doing the same types of activities in the same general area year after year. The use of stone cutting edges was presumably of importance to their lives (though not necessarily applied to all of their activities). Casual collection of stone whenever it was available, curation and maintenance when it was scarce, and discard when it ceased to be of further value would result in an annual distribution of various types of artefacts within the hominid range. Hominids would have been casual with stone when near sources, or when they knew they were going to be near sources soon. Handaxes would have increased in value away from stone sources and decreased in value as other sources were approached.

I have presented a very functional and technological approach to the interpretation of handaxes. Much of this may be unacceptable to those researchers who are looking for other things in the forms of handaxes. I feel, however, that the fact that this type of artefact persisted for over a million years in sub-Saharan Africa argues that it must be a tool made up of purely functional ingredients, and not a symbolic form or purely a representation of an organised mind by *Homo erectus*. Naturally my research described here was directed specifically towards the collections of handaxes from Olduvai, and while many of the features described will apply to all other bifaces, the observed pattern of Developed Oldowan and Acheulean sites is quite different in Bed II from that in Bed IV, and it is hard to suggest any model that adequately accounts for both. One can suggest, having established the technological relationship of the two assemblage types, that the Developed Oldowan assemblages with their higher number of tool types and greater percentage of debitage represent important activity areas involving the use and maintenance of tools. By my model the sites that predominantly contain handaxes are discard areas representing few if any other activities. I do not envisage that Acheulean assemblages represent a simultaneous mass discard of artefacts, but that many may represent annual accumulations.

One major way that some of these ideas could be tested would be to look at the technology found in areas of the same age but where good raw materials in a wide variety of sizes and shapes occurred throughout the landscape. This situation should, if my model is correct, have eliminated the need for planning in the technology of that area, and all tools could have been made whenever the need arose. I would not expect to find tools such as handaxes in these areas. Further work on the Olduvai assemblages and the co-occurrence of the various tool types may help to clarify some of these problems. The Olduvai sequence as a whole is a particularly useful one because of the long time-scale represented, and the variety of raw materials used. The further fact that the major material sources are known and still available turned out to be of great importance to my own study.

11

SUMMARY AND OVERVIEW

DEREK ROE

Introduction

With this volume, Dr Mary Leakey completes the presentation of her long years of work at Olduvai Gorge, though much has still to be written by others in the way of specialist reports on some of the finds she and her colleagues have made. Her achievements at Olduvai have been remarkable by any reckoning, and the quest has been spread out over more than fifty years. As she has recorded elsewhere (1975b, 1984), she first went to Olduvai with Louis Leakey in 1935. There were various visits and expeditions in the 1940s and 1950s, but it was her discovery of the '*Zinjanthropus*' hominid fossil in 1959 that really changed research at the Gorge into the regularly funded, full-scale scientific operation, with wide international participation, that it became in the 1960s and early 1970s. Olduvai was Mary Leakey's real home from 1968 until 1984.

The discoveries and work published in this volume belong to the full period of her involvement with Olduvai, and indeed even longer, since some of the sites, notably JK and HK, were first examined during Louis Leakey's 1931–2 Olduvai expedition. But at the heart of it all lie the definitive excavations in Beds III and IV, and at certain sites of post Bed IV age, carried out in 1968–71, under Mary Leakey's own direction. In just the same way, her previous volume in the Olduvai series, describing Beds I and II (1971), gave information gathered over many years, but its core was provided by the great series of excavations of 1961–3. The new volume follows the pattern of its predecessor as closely as the nature of the finds allows.

That it has taken a full twenty years since the main excavations ended to assemble the results into this volume should surprise no one. The enormous quantities of lithic material recovered would have daunted many people to the extent of never publishing a full report at all: it will not escape notice that a few colleagues who took part in the main period of work have indeed failed to produce reports that were expected for this volume. Apart from the sheer quantity of material, new developments in Quaternary research techniques and new approaches in Palaeolithic archaeology since 1971 somewhat altered the writing up process as it proceeded, and added unexpected new dimensions to the task of description and interpretation. It should also be recalled that the geology of Olduvai from Bed III onwards was fully reported by Richard L. Hay in his important monograph *Geology of the Olduvai Gorge*, as early as 1976. That volume needs to be consulted alongside this present one, and only a working summary of the geology has accordingly been included here (in chapter 1).

If anyone were thinking that Dr Leakey herself has taken an unduly long time to get this volume to press, let them first consider her lion's share of the text and the amounts of information included in the accounts of the various sites and the classification of the tens of thousands of stone artefacts, which make up chapters 2–6 of this book, followed by the faunal lists in chapter 7, and documented by her magnificent illustrations. It would have been no mean achievement to produce all of that, virtually single-handed, in the available unencumbered research time of fifteen years or so. But then reflect that from 1976 to

1981 Dr Leakey was also running the remarkable and unexpected Laetoli research project, with its dramatic finds of early hominid remains and the unique composite hominid footprint trail of Pliocene age, accompanied by a mass of faunal, environmental and geological data – discoveries that demanded the highest priority and the assembling of international expertise on a huge scale. She has already with J. M. Harris compiled, edited and published the large Laetoli report (Leakey and Harris 1987), quite apart from supervising the presentation of the Laetoli discoveries as they occurred, in many other ways around the world.

With all that going on, and with an autobiography and a book on Tanzanian rock art also produced during the 1980s, Mary Leakey has still found time to get Olduvai Beds III and IV into print. Only partly to avoid the cliché, I will describe this as an achievement of supreme scholarship, rather than calling it a labour of love. Olduvai Gorge is a magical place, and almost everything about its archaeology is special, but I think perceptive readers will pick up here and there – in the first part of the introduction, for example – just a hint of Mary Leakey's personal disappointment with the quality of the sites in these higher levels at Olduvai. One could only be disappointed, so far as I can see, in strictly relative terms and because one had also been the excavator of the best of the Bed I and II sites at Olduvai and of Laetoli. In many less-favoured parts of the Old World, Palaeolithic archaeologists would be more than happy with the stratified succession of sites, the prolific quantities of artefacts, the occasional hominid remains, the fauna and so forth, that these higher levels have produced. But the Olduvai sequence, from Bed III upwards, is largely a record of Acheulean industries in stream channels, and those have never been amongst Mary Leakey's favourite kinds of archaeological occurrence. These upper beds have none of the primary-context, undisturbed, lake-margin 'floor' sites that Beds I and II yielded so abundantly; there is much less fauna in the higher levels, and the bones are less varied and also less well preserved. The sites, too, show less variety, and therefore provide less sense of a human population's practical exploitation of the different resources and features of a Pleistocene landscape. Hominid remains are scarcer, and the archaeological occurrences less clearly related to datable volcanic horizons. But all this is relative, and I will turn shortly to the positive features. The point here is simply to emphasise the rigorous scholarship of Dr Leakey in completing her report to the highest standards of thoroughness as a matter of course, regardless of any disenchantment she may herself have felt with the material.

My task – and privilege – in this closing chapter is to summarise and assess the book's contents, as someone who has taken part in the project but has not been too close to the work to consider it in its wider context. I first visited Olduvai in 1969, and last in 1983, with a scatter of other visits in between: in 1969, I saw several of the key excavations described in this volume in full swing, while by 1983 the process of finally closing down the camp had already begun. Now I find myself setting out to write an extended abstract of the whole text, together with what amounts to a review of the book, such as it would hardly have been appropriate for the principal author to provide. I am grateful to Mary Leakey for the opportunity to do both these things, and for all her kindness over the past twenty-one years.

Preliminaries: definitions, geology, environment, dating

The introductory section sets the scene and establishes the terminology for this publication: it is important to note that the relationships of the sites and the stratigraphic units are now significantly better understood than they were when the main digging was in progress. The accounts given here add substantially to, and supersede, even the interim reports of M. D. Leakey (1975a, 1976). Of some of the sites, little had been heard since they were mentioned by L. S. B. Leakey in 1951, and in these cases perception of the stratigraphy has often changed totally, most notably for the handaxe assemblages from HK and TK Fish Gully, formerly attributed to Bed IV but now

SUMMARY AND OVERVIEW

shown to lie in disturbed contexts in sediments actually formed later than the deposition of the Masek Beds. The major units studied in the present volume are Bed III, Bed IV and the Masek Beds, plus the post Masek disturbed occurrences just mentioned. The full sequence at Olduvai is completed by the Ndutu Beds and Naisiusiu Beds, both of which contain a little archaeological material which has been briefly reported elsewhere and is not considered here. Effectively, then, the archaeological occurrences described in this volume are all either Acheulean, or else attributed to a late stage of the Developed Oldowan. This continuation of the Developed Oldowan from Bed II into Bed IV as 'Developed Oldowan C' had previously been reported by Dr Leakey (1975a), and was one of the unexpected results of the Bed IV excavations of 1969–71. Considerable discussion will be found in various sections of the present volume concerning the nature and significance of 'Acheulean' and 'Developed Oldowan' as broadly contemporary industrial facies in Beds III and IV. Only one substantial site is known in the Masek Beds, at FLK, and that is Acheulean, so as things stand Bed IV marks the end of the long Oldowan tradition, if that is what it is.

Within all the major stratigraphic units, sub-divisions can be recognised, and these are explained in chapter 1 (for greater detail, see Hay 1976). Those in Bed IV are the most important, so far as establishing the archaeological sequence is concerned, and Bed IV can be divided as follows:

Base of Bed IV: the III–IV interface, and deposits up to and including Tuff IVa
Lower Bed IV: deposits between Tuff IVa and Tuff IVb
Upper Bed IV: Tuff IVb and deposits up to the base of Masek.

The cliffs at JK provide a type section for Beds III and IV, but it should be noted that, over much of the gorge, Beds III and IV are actually indistinguishable, and here the designation 'Beds III–IV undivided' is used.

The localities of special archaeological interest are fewer in number than was the case in Beds I and II. So far as Bed III is concerned, only JK has produced *in situ* (though not primary-context) material. The excavations by Dr Maxine Kleindienst, and later by Mary Leakey herself, revealed four artefact-bearing horizons, plus the remarkable JK pits. For Bed IV, the area known as WK (including WK East and WK Hippo Cliff) has produced several important sites at various different levels; so has the HEB area, while PDK produced two occurrences within Bed IV, one early and one late. For the stratigraphic units younger than Bed IV, there are just three single excavated sites: FLK (Masek), and the disturbed levels formed in post Masek times at HK and TK Fish Gully.

The age and environmental significance of the Olduvai sequence in Beds III, IV and Masek are also discussed in chapter 1. Magnetic polarity studies show that the boundary between the Matuyama reversed and Brunhes normal epochs occurs in Bed IV, apparently no earlier than Tuff IVb and perhaps even above it. On this basis, probable ages would be: Bed III, c 1.15–0.8 m.y.; Bed IV, c 0.8–0.6 m.y.; Masek, c 0.6–0.4 m.y. Important tectonic events changed the palaeogeography of the Olduvai basin at the end of Bed II times, so that Beds III and IV represent an alluvial plain environment with grassland, some scrub and rather sparse trees, except where strips of forest may have marked major water courses, including the main drainageway, whose precise location was affected from time to time by further faulting episodes. The climate was semi-arid. The Masek Beds were formed in not dissimilar circumstances, perhaps rather more arid, with the additional factor of prolonged activity by the Kerimasi volcano. It was only after the deposition of the Masek Beds that erosion, following more faulting, first created a gorge at Olduvai.

Given the environmental setting, it is hardly surprising that most of the concentrations of artefacts occur in contemporary stream channels. In the shade by the edge of a fresh-water stream is exactly where human groups might be expected to establish themselves, with a variety of resources ready to hand; but inevitably, the archaeological result is disturbed assemblages of dubious integrity, the loss of most of the more fragile evidence that would have accompanied the stone tools, and the absence of such patterning as might characterise one or another kind of working or living site. Much of the post-excavational work accordingly becomes an exercise in the classification and study of stone tools and the waste products of their manufacture. Stone artefacts thus dominate this book in a way that not even the most ardent student of such evidence would have chosen. In her introductory section, Dr Leakey sets out briefly her classificatory system: she has deliberately retained the scheme used for the volume on Beds I and II, with only a few modifications (e.g. within the polyhedron, spheroid, subspheroid group), in order to facilitate comparisons with the earlier assemblages. A few additions are necessary, notably the pitted anvils and hammerstones and the punches, which were virtually absent in levels older than Bed III. On the other hand, the number of classes of choppers is actually reduced, since only two kinds occur in Beds III and IV.

Chapters 2–6 are a painstaking presentation of all the sites as excavated, and the assemblages from them, from oldest to youngest. I shall only pick out a few salient points here, and no one should suppose that reading a few summarising paragraphs will be an adequate substitute for studying the text of these chapters in full.

Bed III

In Bed III there is enough material, albeit from secondary contexts, to show that both Oldowan and Acheulean tools are present; fragmentary hominid remains (the femur and tibia of O.H.34) were found with Acheulean at JK West. None of the stone tool assemblages was very informative, but JK also produced an extraordinary complex of some twenty-seven pits, associated with grooves and furrows which appear to be channels and runnels for water. All these features are interpreted as artificial, and some probable digging-stick marks and human finger-marks on the sides of the pits are described. Animal footprints are also present. The pits themselves are all described in chapter 2 (see also the photographic mosaic by R. I. M. Campbell, and Dr C. Nyamweru's area survey); in the fillings of some of them, artefacts and bone fragments occurred.

Various explanations for this unprecedented site are considered, and it is tentatively concluded that the features have resulted from a simple form of salt working, the salt being produced by the evaporation of highly alkaline water channelled into the pits by means of the grooves and furrows. No Lower Palaeolithic parallels anywhere can be quoted for the JK pits, and the nearest parallels of any kind are from modern, simple, trona working by the Meru people of the Magado crater in northern Kenya, and from the central Sahara at Teguida-n-Tisent. The Magado pits were studied for this volume as carefully as their owners would allow, and the results are reported. It will be interesting to see whether this account of the JK pits will lead in due course to the recognition of comparable Lower Palaeolithic occurrences and to a fuller understanding of the features as preserved. Meanwhile, if the tentative explanation is indeed close to the mark – and it is not insisted on by Dr Leakey – here is some surprising evidence to add to the range of human activities and resource exploitation about one million years ago.

Bed IV

The archaeology of Bed IV consists almost entirely of stone tool assemblages recovered from stream channels. At most of the sites, faunal remains are relatively scarce and in poor condition, though there are exceptions to this, such as WK Hippo Cliff, at the Bed III–IV interface, where substantial parts of a skeleton of *Hippopotamus gorgops* are well preserved, found with a small number of stone tools which may or may not be directly associated, including five handaxes, five choppers, two spheroids lying together and two scrapers, with only thirty-two items of debitage. PDK Trench IV, the oldest of

SUMMARY AND OVERVIEW

the Bed IV sites, yielded mainly fresh artefacts, spread over the surface of a hard conglomerate rather than obviously lying in a stream channel, for once; the density of the artefacts was low and they may represent the outer scatter of a site, or else not very intense occupation of the area.

The area of HEB (with HEB West and HEB East) produced several levels with artefacts in Lower Bed IV, and a correlation table for the occurrences is provided (see p. 54). Most are Acheulean, but they show considerable differences in both the character of the finished tools and the selection of raw materials. Level 3 was a striking industry, in which special use was made of the green phonolite from Engelosin, some 9 km away, for the manufacture of highly finished handaxes and cleavers. The bifaces in Levels 2a and 2b, which are of basalt, trachyandesite and quartzite, are very different to look at, almost like roughouts; here, for whatever reason, the phonolite was virtually ignored for purposes of biface manufacture. These observations were amongst those which prompted Dr Leakey to have various special studies made of biface morphology, and of the effect of different raw materials on implement technology: the results of this work will be found in chapters 8–10.

The WK area also produced several artefact-bearing levels, all channels; they are more spread out in time than the HEB sites, and represent between them all three subdivisions of Bed IV. The principal Acheulean site is the WK Upper Channel, in Upper Bed IV, which yielded fragmentary *Homo erectus* remains along with artefacts and fauna densely packed into a stream channel and hollows in the eroded surface associated with it. A total of 10,904 artefacts were recovered: 429 tools, 428 utilised pieces and 10,047 items of debitage. The tools included seventy-eight complete handaxes and fifty-nine cleavers, with lava and quartzite as the preferred raw materials. Stream channel or not, the bifaces look like a single series, made by consistent technological processes involving systematic production of the flake blanks on which they were fashioned: a lava core that had yielded such flake blanks was amongst the finds, and there was a large series of the pitted anvils and hammerstones that are interpreted as the tools used in bipolar flaking of certain of the Olduvai rocks, especially quartzite. The WK Upper Channel is in many ways highly typical of these Olduvai Beds III–IV sites as regards their nature, though it is more prolific than most: no shortage of finds, but a blurring of the information because of the nature of the context.

The same area, at WK East A and C, yielded 'Developed Oldowan' assemblages, in channel deposits high in Bed IV (later than Tuff IVb). PDK Trenches I–III, only a short distance away, produced closely similar archaeological material, and it seems likely that all these occurrences may belong to a single channel or to a series of related contemporary channels. The specialist studies of the Olduvai Bed IV bifaces subsequently suggested that the site of HEB East belongs with them as a 'Developed Oldowan' occurrence, rather than an Acheulean one as originally supposed. Dr Leakey's detailed classifications of the artefacts from these sites make clear the ways in which the assemblages of this set differ from the broadly contemporary (or in many cases slightly older) Acheulean ones. There are large numbers of battered core fragments, and many hollow or notched scrapers reminiscent of those in the Bed II Developed Oldowan tool-kits. The handaxes are small, crude, pointed implements, mostly made from cobbles or chunks of rock and hardly ever on large flake blanks. Cleavers are absent. On the basis of her observations of the artefact typology and technology at these sites, Dr Leakey designated them 'Developed Oldowan C' and regarded them as a continuation of the Bed II Developed Oldowan tradition, so well seen at such sites as BK. She suggested that they were perhaps made by a different human population from *Homo erectus*, as found at the WK Upper Channel, who may have made Acheulean industries. Her hope that one of the Developed Oldowan C sites might produce hominid remains to test this hypothesis was, sadly, never realised. But whatever explanation for the differences between the 'Developed Oldowan C' and 'Acheulean' industries of Bed IV may be preferred by any particular observer, those differences do appear to be quantifiably real, whether one is considering assemblage composition, or the morphology and technology within particular tool classes, or the selection of raw material. The question of possible explanations is discussed by other contributors elsewhere in the volume, particularly by Peter Jones in chapter 10.

Masek and younger sites

The Masek Beds have produced only one archaeological site worth excavating, that at FLK, but the industry proved to be a very remarkable one, specialising in use of the white quartzite of nearby Naibor Soit. Amongst the formal tools, the large handaxes are a considerable *tour de force*, perhaps by a single craftsman, in a material that is not easy to work and requires different knapping techniques from those appropriate for phonolite or basalt. As regards the two post Masek occurrences, HK and TK Fish Gully, the actual age of the industries remains obscure, as opposed to the age of the secondary contexts in which they were found. The 1931 finds at HK included most of a hippopotamus skeleton, but its relationship to the artefacts could not be determined in the more recent work there. The TK Fish Gully site, first noted in 1932, was still regarded as an *in situ* occurrence after Waechter's excavation there in 1962, but more digging in 1970 led to the recognition that, as with HK, the artefacts are derived in a late context. Both these industries demonstrate efficient and controlled use of the Naibor Soit quartzite, which dominates the raw material in each case. It is likely that they originally belonged in Bed IV or possibly Masek deposits, in the area subsequently removed by the formation of the gorge itself.

Fauna

In chapter 7, Dr Leakey summarises the fauna from Beds III, IV and Masek and provides lists, on the basis of work carried out and already published elsewhere by Dr and Mrs Gentry, J. M. Harris, T. D. White and J.-J. Jaeger. Reference should also be made to L. S. B. Leakey's publication of the Olduvai fauna in the first volume of the Olduvai monograph series (1965), though some of the nomenclature is now obsolete. Regrettably, the faunal assemblages from the sites in Bed III and later, unlike those in the earlier levels, are not primary-context collections referable to the food-gathering activities of humans and carnivores at individual locations. Much of the bone material referred to here consists of small fragments and, though some of the faunal assemblages may well represent human food debris or accumulations by animal predators, the stream channel contexts mean that the remains are incomplete, disturbed and very probably mixed with items added by purely fluviatile processes.

Bovidae are almost always most numerous at any site; hippopotamus and catfish are widely present. The latter may well have been a regular human food source, but the stream channel situations leave this open to doubt. Some useful environmental information can be gleaned on occasion: for example, there are good rodent remains from HEB West (Lower Bed IV), some reflecting dense vegetation with permanent water and others dry acacia savannah. This is what one would expect, where water courses lined by trees and bushes cross the dry grasslands. The overlying Masek Beds fauna, especially the bovids, suggest that at this stage there were less open conditions than when Bed IV was forming, with more cover from trees and bushes.

Studies of the stone tool assemblages

Chapter 7 completes Dr Leakey's presentation of the primary data gathered from the excavated sites in Beds III, IV and Masek. The hominid remains are not included: preliminary reports appeared when they were first discovered, and a general study of the Olduvai hominid remains is in course of preparation. The text now turns to a series of specialist reports on aspects of the stone artefacts, commissioned by Dr Leakey at various stages of the whole operation. As was the case with chapters 1–7, these chapters need to be read carefully in full, and those who want to know what they say should not depend solely on this summary.

In chapter 8, Roe (if I may refer to myself dispassionately in the third person), presents a metrical analysis of the Olduvai handaxes and cleavers in terms of their sizes, plan forms and profiles, based on methods he devised for British flint handaxes in the 1960s. If that all seems a bit out of date, that was what he was asked to do in 1969, and he had basically completed the task by 1974, in tandem with a similar long report on the Kalambo Falls (Zambia) large cutting tool assemblages, which has also been awaiting publication for over twenty years. His results were interesting enough to stimulate additional and more elaborate treatment of the Olduvai industries (see Callow, chapter 9). As he had done for selected British Acheulean assemblages, Roe produced tables and diagrams which include visual presentations of the range of biface shapes for each of the Olduvai sites studied: from these, the presence or absence of strong shape preferences at any site can readily be studied. However old-fashioned, these diagrams work perfectly well within their acknowledged limits, and the metrical analyses do also make it possible formally to compare site with site and, having done so, to say whether there are obvious similarities or differences when whole sets of handaxes or cleavers are compared, rather than individual implements. It is worth noting at this point that Roe and, subsequently, Callow were asked to include a few of the Bed II assemblages which had produced bifaces in sufficient numbers (EF-HR, BK, MLK, SHK and the Upper and Lower floors of TK). The reasons for this were, first, general interest in the 1970s in any consistent trends that might be revealed in biface morphology over long periods of time and, second, the need to study in objective metrical terms how the supposed Developed Oldowan C from Bed IV compared with Developed Oldowan B from Bed II.

Metrical analyses and morphological statements are one thing: interpretation of their significance is another, and the interests of archaeologists continually change as time passes and research makes new things possible. The early 1990s are a different research world from the early 1970s. So much the better: provided the objective data are there, the interpretive use that is made of them can change as often as new ideas emerge. Sadly, one encounters in some quarters a tendency to bring in more fashionable interpretations without any reference back to the primary data, and it seems to me that there are dangers in that. Perhaps, then, all those measurements Roe took between 1969 and 1972 – by hand, because the devices that would now do it for him did not exist – retain some usefulness.

Since, in the event, this volume was not published in 1974, some benefits have already accrued. First, Callow was able to upgrade significantly the statistical treatment of the figures obtained by Roe, and to produce the statistical tables and even the shape diagrams mechanically. Second, the results obtained by Roe and Callow posed certain archaeological questions and interpretive problems, which were of the greatest interest to Peter Jones, author of the important chapter 10 of this volume. Arriving at Olduvai in 1976, and being already one of the best experimental knappers of flint anywhere in the world, Jones was the ideal person to study biface morphology and technology in the Olduvai assemblages from a completely different angle: that of purely practical considerations related to the available rocks and to the tasks for which the implements were made. Did certain rocks demand certain knapping techniques and, if so, were rock types and the essential technology that went with each of them the major factors in determining the morphology of handaxes and cleavers, rather than 'cultural traditions' and 'time-trends'? How were the sizes, shapes, weights and thicknesses of bifaces related to the tasks the implements had to perform? Does rock type control the 'typology' as well as the 'technology' of bifaces to an important extent? How does choice of rock type relate to the intended function of an implement at Olduvai?

If these interrelated questions were some of Peter Jones' starting points, his study subsequently broadened out to include quite different aspects of the Olduvai stone artefact collections, to the extent that he was able to offer explanations of the role of such pieces as pitted anvils, punches, *outils écaillés* and certain classes of debitage. Thus the addition of Callow's chapter 9 and Jones' chapter 10 to Roe's chapter 8 shows some of the changing interests amongst students of lithic artefacts over two decades, and should help to make the present publication worth waiting for.

Metrical and statistical studies

On the basis of his metrical studies on the bifaces, Roe concludes that the differences between Developed Oldowan and Acheulean in Bed IV are substantial and genuine; also, that the 'Developed Oldowan C' of Bed IV is indeed close to the Developed Oldowan B of Bed II. He suggests that HEB East should be regarded as a Developed Oldowan rather than as an Acheulean site; that the Acheulean of Bed IV shows highly variable biface morphology from site to site, with only a few exceptions such as HEB Levels 2a and 2b; and that, overall, weak time-trends can perhaps be recognised in the Acheulean from Bed II to the Masek and later occurrences.

Callow's task was to add to the morphological study a consideration of how the different types of rock might be contributing to the pattern of morphological variation recognised by Roe. He used scar-counts and records of surviving cortex (or unflaked primary surface) patches, supplied by Dr Leakey, and he also worked out the surface area of each biface, devising new quantitative indices for each implement, based on all this information. To these new data were added rock type identifications for each biface, provided by Dr Leakey, a number of qualitative variables, and Roe's measurements for each piece. Complex multivariate analyses were performed, and factors identified which accounted for the variance observed, as described by Callow in chapter 9 (including statistical tables). Callow concludes that there are indeed consistent differences between the sets of Developed Oldowan and Acheulean bifaces. These do not merely reflect size, and the two technologies remain distinct even when raw material choices have been allowed for, the frequent Acheulean habit of using flake blanks for biface manufacture clearly being one of the most important differences. Callow sees little likelihood that Developed Oldowan and Acheulean are parts of a continuous spectrum of variation. His study, like Roe's, also confirms the general similarity of Developed Oldowan B from Bed II and Developed Oldowan C from Bed IV, and he too picks out HEB East as a Developed Oldowan rather than an Acheulean assemblage. Dr Leakey was happy to accept this redesignation.

Experimental studies relating to the stone industries

I think it likely that Peter Jones' report in chapter 10 of this volume will attract at least as much attention as any other contribution. It is the most recent of them to be produced, and may prove to be closest to current interests in the discipline of lithic studies. It is a splendidly practical piece of work, far removed from remorseless classification, laborious piece-by-piece measuring, or computer manipulation of vast quantities of that mysterious commodity, 'input'. It incorporates many interesting ideas and a lot of clear information, and offers conclusions which will certainly stimulate discussion. Quite apart from all that, few people these days ever get to cut up elephants with stone tools.

In this section, Jones reports seven years' experimental work, designed to explore aspects of the Olduvai stone artefact assemblages – mainly replication of tool types and of characteristic debitage, and testing of the efficiency of such implements as handaxes, cleavers and unretouched flakes (made from different rocks) in the kinds of tasks, notably butchery, which they were probably designed to perform. As he observes, the kind of studies presented in chapters 8 and 9 are essentially descriptive: direct experiments, on the other hand, can help break through the barrier

SUMMARY AND OVERVIEW

between description and interpretation. He first lists all the main raw materials actually used in the Olduvai assemblages, describes their qualities and records their sources. It is important to note that his experimental work was done with exactly (not 'approximately') the same rocks. They are: Precambrian quartzite from Naibor Soit and Naisiusiu Hill; phonolite from the Engelosin volcanic neck; basalt and trachyandesite from Lemagrut, obtainable in stream channels running down from the volcano itself; nephelinite from Sadiman volcano, also obtainable from stream channels; and trachyte, from boulders scattered over the area east of what is now the confluence of the Main and Side Gorges. Chert, important in some of the Bed II assemblages, does not feature in any of those of Bed III or later date. Jones' experiments concern the manufacture and use of bifaces (handaxes and cleavers); the group of polyhedrons, spheroids and subspheroids; and *outils écaillés*, punches and pitted anvils, which can also be taken together. He discusses his actual working procedures in sufficient detail to assist others carrying out similar projects.

As regards bifaces, Jones' work shows how each of the main rock types requires a particular approach to yield appropriate blanks for the making of handaxes and cleavers. The Acheulean workers made large flake blanks in most cases, even when suitable natural slabs or cobbles seem to have been available. The weight and raw material of the hammerstones is important in the consistent production of such flakes. Biface morphology and efficiency is profoundly affected by the nature of the blanks for the implements; so is the time required to produce the finished tool. If large bifaces with long working edges, shallow edge angles and a fair degree of weight are consistently desired, as seems to have been the case in most Acheulean industries, then large flake blanks are virtually essential. Jones goes on to discuss handaxe shapes more generally in these terms, concluding that the classic Acheulean handaxe of tear-drop shape, found widely over the Old World, is simply an optimum solution to functional and structural problems: weight plus long edges are needed, but tool mass increases at about six times the rate of edge increase. The flat, narrow, tear-drop shape was simply the most economic form. Such handaxes were useful for many tasks, and re-sharpenable; they can also be seen as transportable units of stone of manageable weight, from which flakes could be removed or quite different implements made.

Turning to the question of Acheulean versus Developed Oldowan industries, particularly the biface component, Jones confirms the profound differences which others have noted in the sizes, shapes and technology of the two sets of handaxes, though he comments that both sets do use the full range of available rocks. He emphasises the contrast between the two series in weights, edge angles and edge lengths. Many of the morphological differences noted relate ultimately to blank type. Assuming that the two sets of bifaces were designed to perform the same functions, there are various possible explanations of the morphological differences: Jones discusses these, and adds a suggestion of his own, namely that many Developed Oldowan handaxes may simply be reworked and indeed now flaked-out Acheulean bifaces. If so, they could be implements at the end of a long cycle of use by the same human group, rather than inferior versions made by a different population. Such intense reworking of Acheulean bifaces would presumably imply some temporary or local scarcity of raw materials, which did not apply at other times or in other areas.

In his work on polyhedrons, spheroids and subspheroids, Jones concludes that the first are a separate tool class, commonly made of lava, while the last two, mostly made from quartzite, are related: both show battering, which is probably from use not manufacture, and increased battering turns an angular subspheroid into a more rounded spheroid. Though these pieces have been seen by previous workers as bolas stones or pounders for vegetable processing, Jones regards the majority of them as pieces of quartzite debitage which were selected for their chunky shapes, flaked to a subspherical form, and then used as light-duty hammerstones, for which purpose he found them very effective. The longer the use, the more spherical the hammerstone tended to become. Tiny examples flaked to a subspherical shape also exist, however, and their use remains unknown: they were too small to be hammerstones, and lack the typical damage.

The *outils écaillés*, punches and pitted anvils were considered together. They all have characteristic damage patterns, which seem to be related, and all of them were consistently produced experimentally when small flakes of quartzite were battered between an anvil and a hammerstone. The punches and *outils écaillés* are characteristically broken quartzite flakes (damage products rather than actual tools), and the pitted pieces are anvils and hammerstones used and worn in the process. It has not so far proved possible to determine the nature of the human activity that demanded such bipolar battering of quartzite flakes, but experiments of other kinds failed to produce these very typical artefacts.

Chapter 10 concludes with Jones' account of his experimental use of replicated tools made from the various rock types, especially in animal butchery. He was fortunate enough to be able to work on elephant carcasses after an official cull of elephants in southern Tanzania. He gives tables showing the times taken to cut through elephant skin with unretouched flakes of different raw materials, and draws conclusions about their effectiveness and the desirable edge characteristics when such flakes are selected. He also reports meat-cutting experiments, using various large cutting tool types, and again comments on the nature of the most effective working edges. Though a little other work of this kind has been done by previous researchers, these data are unique in their quantity and quality, and will repay careful study. A small amount of work was also done on peeling and cutting up tubers and on working wood.

Appendices

There are two appendices to this volume. The first records some interesting modified bones from Beds III and IV. No systematic microscopic search for cut-marks has yet been attempted, and these are artefacts showing flaking or edge damage, not unlike certain material from the lower beds at Olduvai. Two elephant pelves, apparently used as mortars, are included. The second appendix records the field methods devised by Celia Nyamweru in surveying the JK West site, where the pits were, in 1973-4.

Conclusion

So, the archaeology of Olduvai Gorge Beds III, IV and Masek is published at last, to add to the volume on Beds I and II. This has been very much an 'internal' publication, in the sense that the contributors present the material and consider its significance, but they do not attempt to make detailed comparisons with, or indeed more than passing reference to, contemporary sites in East Africa, let alone regions further afield. On the basis of this volume, others can now begin that process.

These upper beds at Olduvai contain stone tool assemblages of which the majority can be described as later Acheulean, in a local sense at least, bearing in mind that, in Bed II, Olduvai has produced some of the oldest Acheulean industries known anywhere. Given that the Brunhes/Matuyama polarity change, adopted as the baseline for the Middle Pleistocene, is located well into Bed IV, then the Acheulean of Bed III and the lower parts of Bed IV is hardly 'Later Acheulean' in a world sense. The dating of Acheulean sites in Africa has rarely proved easy and, given the nature of the stream channel occurrences in the upper beds at Olduvai, it would be hard to say just how the various Acheulean assemblages described in this volume relate chronologically to such sites as Olorgesailie, Isimila or Kalambo Falls, for all of which chronometric dates have been obtained. In any case, the latter are all multiple occurrences, with a certain time-depth of their own to be considered. Maybe only FLK Masek, HK and TK Fish Gully should really be seen as part of the Late Acheulean of sub-Saharan Africa.

At least, formal comparisons can now begin to

SUMMARY AND OVERVIEW

be made between the Olduvai Acheulean industries and other African examples, but what should we expect to find? One of the main lessons of Olduvai is the strong effect of raw materials on artefact technology and morphology. Is there any reason why even demonstrably contemporary Acheulean industries in different parts of Africa should closely resemble each other? We must expect local responses to local situations, using local resources, though it is perhaps predictable that, at the general level, the life-styles may well be much the same. As ever, the need is for good, clear, well-dated sites, where the more fragile evidence is well preserved. Will it be the differences, or the similarities, that are most striking, when comparisons come to be made?

But the interest is far more than local. In these higher beds at Olduvai, we enter a period during which sub-Saharan Africa becomes an ever smaller part of the whole human distribution. The manufacture of handaxes may indeed have originated in East or central southern Africa, but by Olduvai Bed IV and Masek times that tradition had spread (by whatever means) far and wide over the Old World, from India to Britain. There are more and more local situations to be examined, absorbed and compared. In several of them, we encounter again a phenomenon with which this volume has been much concerned: a duality in the nature of contemporary Lower Palaeolithic stone tool assemblages, which is easy to demonstrate but hard to explain. Here, we have been considering Acheulean and Developed Oldowan C in Olduvai Gorge Beds III and IV: are these the products of the same, or different, human groups? If the same, in what way should the differences in typology and technology be explained?

Elsewhere, such duality may be on a somewhat broader geographical scale, while that at Olduvai could perhaps be given a purely local explanation. For example, in the Indian subcontinent, there are Acheulean and Soan; in northwest Europe, Acheulean and Clactonian. Uses of different raw materials can hardly account for these, in the way that, for instance, the nature of the rocks in east and southeast Asia has been held by some workers as sufficient to account for the absence of typical Acheulean assemblages there. It is also interesting to note that, on present evidence, the first humans to enter Europe may not have been handaxe makers, even though the date of entry may be later than one million years ago, contemporary perhaps with Bed III or Lower Bed IV at Olduvai. The site of Isernia in west central Italy (Cremaschi and Peretto 1988) will serve to illustrate both the chronology and the type of industry I have in mind. Nor is the situation a simple one in the Middle East: for example, amongst more than sixty separate archaeological levels at Ubeidiya in Israel, there are certainly some typical Acheulean artefact sets, but there are others with – dare I say it? – a very 'Developed Oldowan' look. The question of contemporary alternatives to the typical Acheulean in many parts of the latter's distribution seems unlikely to go away. The answers, at least, are far more likely to involve human behaviour or human populations than stone artefact typology, so they are worth pursuing.

I will not protract these concluding remarks by adding to, or even by more fully documenting, the deliberately broad comments just made. My intention is simply to suggest that this volume about Olduvai Gorge Beds III, IV and Masek is of varied interest, much of it local to East Africa, but some of it much wider. The next thing is for other workers to start using the information here presented and, in whatever way, thereby to broaden our understanding of the whole Lower Palaeolithic period.

APPENDIX A

MODIFIED BONES FROM BEDS III AND IV

M. D. LEAKEY

Most sites excavated in Bed IV and site JK in Bed III, have yielded bones with evidence of modification by flaking and edge damage that appears to have been caused by use. In relation to the total faunal remains from any particular site, however, such bones are less numerous than in Middle and Upper Bed II (Leakey 1971).

Facilities for scanning electron microscopy have not been available to the writer during the preparation of this report and the following notes are merely brief descriptions of bones with clear evidence of modification. A detailed study, such as that carried out by Dr P. Shipman on bones from earlier levels at Olduvai (Shipman 1981) would undoubtedly be rewarding.

The modified bones from Beds III and IV fall into the following broad categories:

(1) elephant pelves in which the acetabula have been extensively battered on the interior
(2) pieces of elephant limb bones flaked to pointed extremities
(3) distal parts of large humeri, probably of hippopotamus, with the condyles almost entirely removed by battering
(4) heads of large femora, broken off at the necks, with battering on the condylar surfaces
(5) fragments of large limb bones with one end flaked to a point.

There are also the tip of a hippopotamus incisor or canine with the end chipped off, an elephant calcaneum pitted on the convex face, a utilised costal scute of a tortoise and a pestle-shaped bone fragment in which the outer surface has been stripped off with the exception of one slightly convex extremity.

The elephant pelves (Pls 22 and 23)

These two bones were found at JK and at HEB West. Both are left sides of the pelvis and both show extensive wear on the interior of the acetabulum. The pubis is missing in both specimens, while the border of the ilium has been removed and the ischium reduced to a stump.

The specimen from JK is the most damaged (Pl. 22). It now measures 60 cm from the preserved part of the ischium to the opposite broken edge of the ilium. The width across the preserved part of the ilium is 71 cm. There is now a worn hollow where the pubis was originally and the edge of the acetabulum in this area has also been broken off. The interior has been so heavily battered that the acetabular fossa has been partly obliterated and now measures only 20 mm in depth instead of 40 mm in the specimen from HEB West, which is comparable in size but less damaged in this area. A deep groove has also been worn across the ischium near its base. Within the acetabulum the surface has been worn away in several places and the pitting has penetrated into the cancellous structure. The wear is more marked on the ischial and pubic sides of the acetabulum than on the opposite side. Owing to damage of the rim accurate measurements cannot be taken, but it is estimated that the diameter from the rim at the ischial side to the opposite border is 21 cm. The transverse diameter is approximately 17 cm. The depth is estimated at 7 cm.

The specimen from HEB West (Pl. 23) measures 83 cm from the tip of the preserved part of the ischium to the opposite broken border of the ilium. Although the pubis has been removed as in the specimen from JK a stump still remains and there is no groove across the ischium. In general, this specimen has not been so heavily damaged as the first. Wear on the interior of the acetabulum is confined to relatively shallow pitting and the outer surface has only been penetrated in a few areas. The acetabular groove remains almost intact although the edges have been slightly chipped. Pitting is more pronounced on the pubic side of the acetabulum than elsewhere. Measurements of the acetabulum are: diameter from the ischial to the opposite edge 23 cm, transverse diameter 21 cm, depth 5.5 cm.

The modification and wear on these two bones is very similar, in particular the removal of the pubis. This is a substantial bone in an elephant pelvis and its removal must have involved considerable effort if it was removed intentionally. It is possible, however, that it was merely reduced further and further by repeated battering when the bones were in use. If the acetabula were used as mortars, as seems probable, the retention of the pubis but removal of the inferior part of the ilium would be advantageous in keeping the acetabulum in an upright position while in use, but the removal of the pubis takes away support from one side so that the acetabulum lies sideways whichever way the bone rests on the ground.

Pieces of elephant bones with pointed ends
(Pl. 24)

There are three fragments of elephant bones in which one extremity has been flaked to a point. All three are from Acheulean sites in Bed IV, namely HEB, WK and LK (where a small trial trench was excavated in 1970).

The specimen from HEB appears to have been made on a large flake struck from a thick limb bone. It is 258 mm long, 128 mm wide and 49 mm thick. At one end three flakes have been removed from the outer surface of the bone, forming a symmetrical point. Some chipping is present along the edges of the flaked area, apparently caused by use. A broad, flat flake extending across the width of the specimen and 115 mm long has been removed from the opposite end. The edge at this end also shows some chipping. In many respects this artefact suggests a handaxe made from bone in place of stone.

The bone artefact from LK is large and flat, possibly part of a pelvis or scapula. It has been flaked at one extremity only and is 280 mm long, 164 mm wide and 20 mm thick. Flakes have been removed from both the inner and outer surfaces in order to form the point which is burin-like in that it lies at right angles to the outer and inner surfaces of the bone. Longitudinal cracks radiating from the flaked extremity are visible throughout the length of the bone, indicating that it was used with force.

The third specimen, from the Upper Channel at WK, consists of a fragment of particularly massive bone split lengthwise and with a longitudinal ridge in the centre. It measures 315 mm in length, 118 mm in width and 63 mm in thickness. It is similar to the previous specimen in that the point is at right angles to the surfaces of the bone. It has been formed by detaching a single flake from either side. These measure 142 mm and 40 mm in length. The point itself has been chipped and blunted by use. As in the specimen from LK, there are longitudinal cracks throughout the length of the artefact, originating from the pointed end. The opposite extremity exhibits battering and evidence of use but not of preparatory flaking.

Humeri (Pl. 25)

These four specimens were recovered from PDK Trenches I–III, HEB West and WK East C (two examples).

The specimen from PDK is the largest in the series. It consists of the distal part of a right humerus, possibly of a large hippopotamus, 235 mm long, 122 mm wide, measured across the remains of the articular end, and 35 mm thick, measured at the most central part of the shaft. Both ends of this specimen have been battered but not flaked. The coronoid fossa is also pitted and enlarged by some form of utilisation.

One of the two bones from WK East is very similar, although it is shorter and less massive. It consists of the distal part of a left humerus 176 mm long, 95 mm wide at the damaged articular end and 76 mm thick, measured on the shaft. The broken edges of the shaft are blunted and slightly chipped but the principal utilisation has been at the articular end where some flakes have been removed and the whole battered down to a stump. The second example from this site consists of quite a small fragment of left humerus, lacking nearly all the articular end and the shaft as well. It is of about the same thickness as a hippopotamus humerus but too fragmentary to identify. It measures 118 mm in length, 107 mm in width and 78 mm in thickness. Both ends exhibit flake scars, but the ridges are sharp and not abraded as in the previous example. This specimen seems to represent the final stage of use for a humerus when it became too reduced to serve any further useful purpose and was discarded.

The fragment of left humerus from HEB West was found adjacent to a costal scute of a tortoise. These two specimens were close together but isolated from other remains and their association may be of some significance. The humerus is probably of a hippopotamus and exhibits flaking and battering on the fractured shaft as well as at the articular end. In common with two of the pointed fragments of elephant bones described above, this specimen bears many longitudinal cracks. Length 147 mm, width 97 mm, thickness 72 mm.

Heads of femora (Pl. 26)

There are three examples, from HEB, HEB West and WK East C. The two specimens from HEB and HEB West appear to be hippopotamus femora. Both are broken obliquely across the neck. They are 136 mm and 103 mm long respectively. The heads measure 99 mm and 95 mm in diameter where the original surface still remains, but the second specimen has been so damaged that the diameter has been reduced to 91 mm in most areas. In one example the area most severely damaged is at the top of the femur head where the entire surface has been removed and the underlying cancellous tissue flattened. In the other specimen the damage is not so extensive and occurs only in an area 30 mm × 26 mm where the surface of the bone has been removed. The third example is an unfused epiphysis, larger than the first two, with a diameter of 105 mm. It may be of an immature elephant. The bruising of the surface is in two areas, measuring 25 × 26 mm and 38 × 24 mm. The circumference is chipped on the underside, indicating that the head was used after it became detached from the neck of the femur.

Limb bone shaft fragments

There are ten shaft fragments of large limb bones in which one end has been flaked and chipped. No anatomical features remain, but on size and thickness it seems likely that they are of hippopotamus or rhinoceros. In four specimens the tips are pointed and in six they are transverse, either straight or slightly convex. These specimens range in length from 158 to 53 mm.

There are also two fragments of elephant limb bone shafts. One shows chipping and flaking along one lateral edge which is thin and must originally have been sharp. The opposite lateral edge is thick and flat so that the specimen resembles a knife. The tip has also been utilised. Length 183 mm, width 70 mm, thickness 48 mm. The second fragment of elephant limb bone measures 226 mm long, 134 mm wide and 40 mm thick. One end is broken obliquely and shows no trace of wear but the opposite end is thin and rounded. It shows chipping and blunting of the edge for a length of 166 mm, measured on the curve.

Among other specimens showing evidence of modification there are the following:

(a) Part of a large radius from JK, possibly of hippopotamus, in which the broken shaft has been flaked. These flake scars do not follow any pattern and seem likely to have resulted from heavy blows during use. The articular surface at the head of the bone has also been damaged on one side. Length 152 mm, width 106 mm, thickness 93 mm.

(b) There are two bone flakes, one of which is from the Upper Channel at WK. It shows a striking platform, bulb of percussion and hinge fracture at the distal end. Length 85 mm, width 89 mm. It may be noted that a number of the negative flake scars on the bones described above also terminate in a hinge fracture. A second, smaller example is an end flake and measures 64 × 22 mm.

(c) The costal scute of a tortoise referred to earlier in connection with the utilised humerus from HEB West. This specimen is weathered and in such fragile condition that it has not been possible to clean it thoroughly. Evidence of utilisation is, therefore, uncertain, but it appears that the sutures on the lateral edges have been worn down to some extent. Length measured on the curve 172 mm, maximum width at the distal end 43 mm. It should be noted that no other parts of a tortoise occurred in the vicinity of this specimen, which was found close to the humerus.

(d) The tip of a hippopotamus incisor or canine from HEB West has been chipped and battered. Length 91 mm, width 36 mm, thickness 32 mm. The type of wear resembles that on a hippopotamus canine from Bed II at FLK figured in Volume 3 of the Olduvai monographs (p. 247).

(e) The 'pestle'-shaped bone fragment referred to earlier. It consists almost entirely of cancellous substance with the surface of the bone removed except for one extremity and a small area on one side. The extremity retaining the outer surface is almost exactly circular.

(f) A complete elephant astragalus from JK exhibits a number of pits or depressions on the convex surface. There are nine pits, eight of which are rough on the interior and similar to the pits on the pitted stone anvils and hammerstones, but in one example the surface has been crushed inwards. This depression measures 26 × 18 mm and is 3 mm deep. The eight other pits range in diameter from 21 × 20 mm to 7 × 6 mm and in depth from 2 to 1.5 mm.

It is probable that many more fragments of bone than those noted here were utilised and flaked, but only those with the most obvious modification have been described, in particular specimens from which large flakes have been detached to a discernible pattern. The nature of the archaeological occurrences in Beds III and IV, either in sandstones or conglomerates, could have been responsible for abrasion and chipping on the edges of bone fragments. For this reason specimens with merely chipped or abraded edges have not been included in this preliminary description. But as Shipman has shown (1981) scanning electron microscopy of edge damage on bones can often determine whether it is natural or artificial and sometimes indicates the agency responsible.

In most cases it seems that, as in Bed II, massive bones were flaked in the same fashion as stone. The two halves of elephant pelves are an exception and are unique. A single specimen could possibly have been fortuitous, but the occurrence of two from different sites and the fact that both have been similarly broken so that the acetabula are level when the bones rest on the ground gives reason to believe that the bowl-shaped acetabula, so heavily damaged on the interiors, were probably used as mortars.

APPENDIX B

MAPPING OF AN ARCHAEOLOGICAL SITE AT OLDUVAI GORGE*

CELIA K. NYAMWERU

An archaeological site at Olduvai Gorge, Tanzania, consists of a roughly quadrilateral surface of consolidated silt, about 25 ft by 47 ft. It shows a complex microtopography of pits and grooves. In order to record all these features a plan was made with a horizontal scale of 1:4 and contour interval of 0.1 ft (1.2 in). The contours were interpolated between a large number of spot-heights. The positions of the spot-heights in the horizontal plane were fixed by distance measurement along the lines in a rectangular grid laid out from points on the margins of the surveyed area. The elevations of the spot-heights were determined by levelling with an automatic level. When the spot-heights had been plotted the map was mounted and taken into the field. Constant comparison was necessary to ensure that the interpolated contours provided an accurate representation of the complex topography.

Field work was carried out during May–June 1973 and 1974. The greater part of the work in the first season was carried out by the author; during the second season she was greatly assisted by Mr David Kamau. Tracing of the completed plans was done by Mr Michael Kivuva of Kenyatta University College.

The author is grateful to the Department of Surveying and Photogrammetry, University of Nairobi, for the loan of survey equipment. Dr Leakey's work at Olduvai Gorge, including this particular project, is supported by the National Geographic Society.

Maps were needed to provide a detailed record of a unique set of features whose nature and origin are not yet fully understood. This record could then serve as the basis of analysis by Dr Leakey and other archaeologists. It was necessary to make this map as quickly as possible since the silt of the surfaces of the pits had shown itself to be vulnerable to cracking and crumbling when it became wet, and the available methods for excluding water were not totally adequate. Maps were also needed to provide a basis for the design

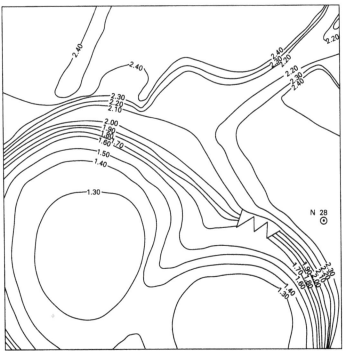

Fig. A.1 Detail of map of the main Pits surface. A large double pit (Pits 9 and 10) occupies the lower half of the plan, with a shallow groove running towards the upper right-hand corner of the plan

---

* The original version of this article appeared in the *Journal of Field Archaeology* 4 (1977), under the author's previous married name, Celia K. Kamau.

of a shelter to give more adequate protection to the main Pits surface.

Two plans were produced, one at a scale of 1:120 of the surface with pits and surrounding area (see Fig. 2.2) and the other at a scale of 1:4 of the details exposed on the main Pits surface, Trench VI and Trench VII (Figs A.1 and 2.3).

Choice of method

It should be stressed that the planning and design of the surveying operation was done in the field, as work began. Certain improvements in technique were incorporated while the work was in progress. Other possibilities could not be tried, chiefly because the necessary equipment could not be transported to Olduvai within the time available. It is felt that the method finally evolved, while still capable of improvement, was an efficient one and that an account of it would be of value to archaeologists who might encounter comparable survey and mapping problems. In the planning stages a number of methods were considered.

The method eventually used involved the use of a plane table and open sight rule (alidade) for making the map at a scale of 1:120. A telescopic alidade was not available at Olduvai, or it might well have been used. A surveyor's level and a linen tape were used for making the plan at scale 1:4. These methods are discussed in greater detail in the following sections.

The mapping was done in Imperial (British) units, since these were the units with which the surveyors and subordinate staff were most familiar, and in which the levelling staff was calibrated.

The plan at scale 1:120 (see Fig. 2.2)

The plan was intended to show the surfaces with pits and their immediate surroundings, including the following features.

(1) the outline of the main Pits surface and Trenches VI and VII;

(2) a slope to the north of the main Pits surface in which the outlines of more pits could be seen in cross section;
(3) two small concrete markers (H 29 and H 34) marking the sites at which pieces of hominid bone had been found;
(4) a survey beacon, marked JK 1961, that had been established in this area during an earlier period of excavation;
(5) the surrounding slopes. (These are rather steep and made of loose, easily eroded silts. There is very little vegetation cover and when heavy rains fall, usually several times a year, intensive erosion takes place. As a result, the slopes change rapidly from year to year. In 1973, parts of the area next to the main Pits surface were also being lowered by excavation. No attempt, therefore, was made to draw a contour map of these slopes.)

The method used to make the plan was as follows.

A network of points for triangulation covering the whole area was established. Each point was located by a concrete marker set flush with ground level. An iron nail was placed in the centre of the marker, its head flush with the concrete surface. The distances between the different points on this network were measured with a Rabone Chesterman 100-foot linen tape. Each line was measured three times (except for the inner quadrilateral CDEF, which was measured four times) and the averages were taken. Where the line was almost horizontal the levelling tripod was centred over the nail at one end of the line and a plumb bob and ranging rod were used to centre the end of the tape over the nail at the other end of the line. Where the ends of a line were at very different altitudes (the maximum difference was about 7 ft) the following method was used: at the upper end of the line, the end of the tape was held on the nail itself; at the lower end of the line, a ranging rod and plumb bob were used to position the tape vertically over the nail.

When reading, the tape was stretched by hand to the maximum possible tautness and the reading was taken to the nearest eighth of an inch. Only

APPENDIX B. MAPPING OF A SITE

very small differences among the three readings for each line were found. Sources of error probably included variations in the tension applied to the tape, because of the effects of the wind. Errors might also have arisen from failure to make the tape perfectly horizontal or to centre the ends of the tape vertically above the nails. The errors, however, were not of an order to show up when plotted at a scale of 1:120.

When the average lengths of the lines had been obtained, the network was plotted at the scale of 1:120 using a beam compass. This network was then used as the basis for the actual plan, which was surveyed with a plane table and an alidade (open sight rule).

The plan at scale 1:4 (Figs A.1 and 2.3)

The choice of scale for this plan depended largely on the dimensions of the features to be shown. The smallest features of interest were grooves that might be as small as $1\frac{1}{2}$ in. wide and about the same depth. At a scale of 1:4, these would be $\frac{3}{8}$ in. wide. They would show on a map with contour interval of 0.1 ft (1.2 in.). These horizontal and vertical scales also permitted the depiction of the larger features (pits some 2 ft deep and 6 ft in diameter) without overcrowding the map. A scale of 1:4 was also better suited to the British system of units than, for example, 1:5. The shortest distance measured in the field ($\frac{1}{8}$ in.) was represented at a scale of 1:4 by $\frac{1}{32}$ in., which could easily be drawn from scales available in the laboratory at Olduvai. Some difficulties were experienced because the horizontal scale was so large, in terms of handling very large sheets of paper and the necessity of joining sheets of paper to obtain long enough strips. Another problem was lack of a sufficiently large beam compass.

The establishment of horizontal control

The main Pits surface was subdivided into strips as shown in Fig. A.2; Trenches VI and VII were each surveyed separately as single strips. Horizontal control was provided by a network of points marked by short nails which were driven in flush with the surface, holding down paper tags on which the identification numbers of the points were written. The nails did not cause noticeable damage to the surface and when the paper tags were removed were very inconspicuous. Some difficulty was experienced in establishing markers at the upper and lower edges of the main Pits surface, which at the time ended abruptly in vertical sections of rather crumbly material. Here, 6 in. nails were driven in horizontally; in most cases they provided fairly firm markers. Each strip was about 3 ft wide and 25 ft long. The available drawing paper was in sheets measuring 20 in. by 27 in. At a scale of 1:4, each strip required three sheets of paper, which were carefully joined end to end.

For each strip, and for Trenches VI and VII, the mapping procedure was as follows.

(1) Control points were established, forming an approximate rectangle (Fig. A.3). Where possible the points were placed on a straight line, but because of the various hollows on the surfaces it was impossible always to do this or to make the figures into exact rectangles.

(2) Distances between the points were mea-

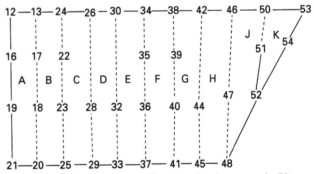

Fig. A.2 Arrangement of surveyed strips on main Pits surface

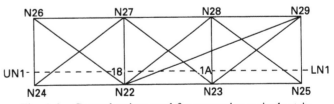

Fig. A.3 Control points used for measuring a single strip

sured, using the linen tape to measure the distances to the nearest $\frac{1}{8}$ in. Each length was measured twice, once in each direction. In some cases, where the line was not horizontal, a plumb bob and a short stick were used to position the end of the tape vertically above the lower end of the line. Where possible, the points were positioned so that there was no intermediate rise between the two ends of a line.

(3) Plotting usually started at the 'upper' end of the rectangle (i.e. with N26–N24 on Fig. A.3) and proceeded by triangulation to the other end. One problem that arose involved the lengths of the lines to be plotted. They ranged from 3 ft to a maximum of over 12 ft, and the largest available beam compass could only draw arcs up to 1 ft long (i.e. representing 4 ft on the given scale). A substitute 'beam compass' was made; a straight edge of stiff paper on which were marked, at the scale of 1:4, distances from 3 ft to 12 ft, subdivided to $\frac{1}{32}$ in. (i.e. $\frac{1}{8}$ in. on the ground). The zero point of this scale was pricked with a pin and held onto one point (with the left hand); the scale could then be rotated freely and an arc of the desired length could be drawn with a pencil held against the scale with the right hand. Repeated tests showed that the accuracy of this home-made beam compass was quite high. In most cases, the results of the measurements and plotting were very accurate, and the arcs intersected in a single point or in a very small triangle. In some cases, however, a larger triangle was obtained even after checking the lengths of the lines. If this occurred, an attempt was made to find a fourth line that could be measured to the point in question; if three of the four lines then intersected, the divergent one of the four was rejected. Where a fourth line could not be measured, the point was positioned within the triangle on the assumption that the error in measurements of the lines was proportionate to their lengths.

(4) When the control network had been plotted it was necessary to cover the surfaces with a network of spot-heights, with the aid of which contours could be plotted. The next step, therefore, was to position the spot-heights in both a horizontal and vertical plane.

For the horizontal positioning, a number of methods were tried.

One possibility was to position the spot-heights on a rectangular grid, e.g. every 6 in. on both axes. This method would have made distance measurement much quicker and easier, but it was rejected because much of the important detail, such as edges of pits and grooves, would not have shown up unless its position coincided with a point on the grid.

For mapping of Trench VI (the first strip to be done) the levelling staff was placed at the points considered to be significant (mostly the upper edges and bases of the various hollows). A reading on the staff was taken (see the following section on height determination) and the position of the staff was fixed by measurement with the tape to three of the control points of that strip. This provided an accurate way of locating each point, as three arcs were drawn, but it soon became clear that it was too time consuming.

The next modification was to fix the position of the staff by measurements to only two control points; this gave only two arcs to be drawn for each point. All the same, the measurement and plotting for each point took quite a long time.

The final method decided on was much quicker, and was used for the whole of the main Pits surface. A certain amount of accuracy was sacrificed, but it was felt that the level of accuracy obtained was still adequate. The method involved the stretching of strings along the pits surface, from nails put in at either end. Parallel lines about 9 in. apart were thus marked out. The levelling staff was then moved along each string in turn, being positioned at the points where the string crossed important features. At each point, three determinations were recorded: distance along string from the control point; reading on levelling staff; what the point was, i.e. 'edge of groove', 'centre of pit', etc.

At intervals a second distance reading was also taken for a particular point, to one of the nearby control points for that strip. This measurement provided a check on the accuracy of the distance measurements along the string. Where a particularly important point (for example the centre

Table A.1 *Booking of the control points for strip C of the main Pits surface: first measurement, 18 June 1974; line 1, 19 June 1974*

N24 to N26	3 ft $10\frac{3}{4}$ in.	UN1 to N24	$8\frac{1}{4}$ in.
N24 to N27	10 ft $3\frac{7}{8}$ in.	UN1 to N26	3 ft $2\frac{1}{2}$ in.
		UN1 to N27	10 ft $1\frac{1}{2}$ in.
N22 to N27	2 ft $10\frac{1}{2}$ in.	UN1 to N22	9 ft 2 in.
N22 to N26	9 ft $9\frac{5}{8}$ in.		
N22 to N28	7 ft $2\frac{1}{8}$ in.	1B to N26	9 ft $10\frac{3}{8}$ in.
		1B to N27	1 ft $11\frac{3}{4}$ in.
N23 to N27	6 ft $10\frac{1}{4}$ in.	1B to N22	$11\frac{1}{8}$ in.
N23 to N28	3 ft $2\frac{3}{4}$ in.		
N23 to N29	9 ft 5 in.	1A to N27	6 ft $9\frac{1}{4}$ in.
		1A to N28	2 ft $5\frac{1}{2}$ in.
N22 to N29	16 ft $2\frac{1}{2}$ in.	1A to N23	11 in.
N26 to N27	9 ft $9\frac{7}{8}$ in.	LN1 to N29	2 ft $3\frac{3}{8}$ in.
N27 to N28	5 ft $4\frac{1}{4}$ in.	LN1 to N28	9 ft $8\frac{1}{4}$ in.
N28 to N29	9 ft $5\frac{1}{2}$ in.	LN1 to N23	8 ft 10 in.
N29 to N25	3 ft $1\frac{7}{8}$ in.		
N28 to N25	10 ft $\frac{3}{8}$ in.		

of a pit) occurred between the parallel strings, a reading at right angles to the strings could also be taken, and this was indicated in the booking. Such points were thus plotted with reference to a distance along the string and a distance at right angles from the string.

The nails that marked the ends of the strings were located by measurement to three or four of the nearby control points. Two intermediate points on the string were marked with tightly knotted thread and were used as points from which distances could be measured. These points were so spaced that it was never necessary to measure a line more than about 4 ft 6 in. long. This was important as it meant that only one man need be used for measuring with the tape, rather than having a man at each end of the tape. Examples of the booking of the data are shown in Tables A.1 and A.2.

The establishment of vertical control

Vertical control was maintained by reference to two points on the original triangulation network, namely C and D. For the purposes of this mapping, the elevation of point C was taken as 0.000 ft. Repeated levelling showed that D was 3.235 ft higher. Before and after levelling along each line across the pits surface, the staff was placed on either C or D and a reading taken. Intermediate readings were also taken near the middles of the lines, especially during windy days or when there was reason to suspect that the line of sight might have shifted. The equipment used was a surveyor's level and a 14 ft staff. In the first season the level used was a Wild NK-10 tilting level. In the second season we were fortunate to obtain a Japanese Sokkisha automatic level, which greatly increased the speed of the work.

Table A.2 *Booking of the levelling*

Line	Ht.	Dist.	19 June 1974 Check N23	RL	Notes
D	4.981	3 ft $9\frac{7}{8}$ in.		2.448	edge transv. groove
	4.195	CiA			
	4.986	3 ft $11\frac{3}{4}$ in.		2.443	edge corner of hollow
	4.912	3 ft $4\frac{1}{4}$ in.	$3.9\frac{1}{4}$	2.517	side edge round hole
	5.097	3 ft $4\frac{1}{4}$ in.	$0.3\frac{1}{8}$	2.332	b C1 B1 bottom g hole
	4.949	2 ft $9\frac{7}{8}$ in.		2.480	edge of hole
	4.933	1 ft $8\frac{5}{8}$ in.		2.496	edge of transv. groove
	5.172	1 ft $6\frac{1}{8}$ in.		2.257	in groove
	4.993	1 ft $4\frac{1}{2}$ in.		2.436	edge of groove
	4.934	0		2.495	crest
	4.950	$3\frac{1}{8}$ in.		2.479	edge of 3-headed groove
	5.039	$4\frac{5}{8}$ in.		2.390	centre of 3-headed groove
	4.941	7 in.		2.488	ridge
	4.974	$10\frac{1}{4}$ in.	$1.0\frac{1}{2}$	2.455	corner above groove and pit

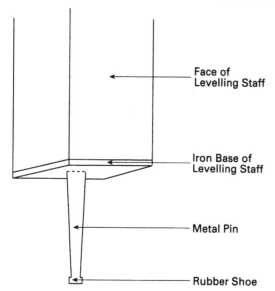

Fig. A.4 Modification to levelling staff in order to make measurements in narrow grooves

Some modification to the levelling staff was necessary, as the dimensions of its base (3 in. × 2 in.) made it difficult for the staff to be positioned precisely on the string and impossible for it to fit into most of the grooves. A metal pin 6 in. long was welded onto the metal shoe that forms the base of the staff. This pin (see Fig. A.4) had a point less than $\frac{1}{2}$ in. in diameter that was shod with strong rubber to prevent it from damaging the pits-surface. The staff could thus be positioned precisely on the surface and could be stood in the narrow grooves. Since the same staff was used throughout, no adjustment in levels because of the extra length of the staff was necessary. Readings on the levelling staff were taken to the nearest 0.001 ft. The final figure was estimated since the maximum line of sight was only about 45 ft.

The plotting of the spot-heights

The levels were reduced using a Facit manual calculator; at this stage each height was correct to three decimal places. The spot-heights were then plotted along the parallel lines, scaling off the distances from the ends of the lines or the intermediate points. At this stage the heights were plotted correct to two decimal places, e.g. 2.01 ft, 0.98 ft.

The drawing of the contours

When all the spot-heights for a given strip had been drawn in, the paper was fixed on a board and was taken out onto the pits surface for the contours to be sketched in. This was a time-consuming job that required a good deal of concentration and judgement. The task was made easier than it might have been because the strings which had marked the lines of levelling were left in place, so that visual comparison between the map and the ground was facilitated. Also, each spot-height had a brief description booked, so that one could be reminded what type of feature had been measured in each case.

The drawing of the final map

The map was traced in pencil and the contours numbered at intervals of 0.10 ft. Some of the pits and grooves had overhanging sides, and it was necessary to develop a special symbol to depict this (see Fig. A.1). In two cases there were even small underground 'tunnels' below the pits surface, which also had to have a special symbol. A tracing, in red, of the outlines of the most important pits and grooves was also made. All the above work was done in the drawing office at the Olduvai Gorge camp. Later, in the drawing office of the Department of Geography, Kenyatta University College, a set of ink tracings was made from which copies were run on a Rowe 111 printing machine. These copies were then sent to the National Geographic Society for redrawing, reduction and final printing.

REFERENCES

Blackith, R. E. and Reyment, R. A. 1971. *Multivariate Morphometrics*. London: Academic Press.

Brock, A., Hay, R. L. and Brown, F. H. 1972. Magnetic stratigraphy of Olduvai Gorge and Ngorongoro, Tanzania (abstract). *Geological Society of America Abstracts* 4: 457.

Cahen, D. and Martin, P. 1972. Classification formelle automatique et industries lithiques: interprétation des hacheraux de la Kamoa. *Annales du Musée Royal de l'Afrique Centrale, Série IN 80, Sciences Humaines* 76. Tervuren.

Cerling, T. E., Hay, R. L. and O'Neil, J. 1977. Isotopic evidence for dramatic climate changes in East Africa during the Pleistocene. *Nature* 267: 137–8.

Clark, J. D. 1970. *The Prehistory of Africa*. London: Thames and Hudson.

1980. Raw material and African lithic technology. *Man and Environment* 4: 44–55.

Cranshaw, S. 1983. *Handaxes and Cleavers: Selected English Acheulean Industries*. British Archaeological Reports (British Series), 113. Oxford.

Cremaschi, M. and Peretto, C. 1988. Les sols d'habitat du site paléolithique d'Isernia in Pineta (Molise, Italie centrale). *L'Anthropologie* 92: 1017–40.

Davis, D. D. 1980. Further consideration of the Developed Oldowan at Olduvai Gorge. *Current Anthropology* 21: 840–3.

Day, M. H. 1971. Postcranial remains of *Homo erectus* from Bed IV, Olduvai Gorge, Tanzania. *Nature* 232: 383–4.

Doran, J. E. and Hodson, F. R. 1975. *Mathematics and Computers in Archaeology*. Edinburgh: Edinburgh University Press.

Gentry, A. W. and Gentry, A. 1978. Fossil Bovidae (Mammalia) of Olduvai Gorge, Tanzania, Parts 1 and 2. *Bulletin of the British Museum (Natural History) Geological Series* 29 and 30.

Harris, J. M. and White, T. D. 1979. Evolution of the Plio-Pleistocene African Suidae. *Transactions of the American Philosophical Society* 69: 1–118.

Hay, R. L. 1976. *Geology of the Olduvai Gorge: A Study of Sedimentation in a Semiarid Basin*. Berkeley: University of California Press.

Jaeger, J.-J. 1976. Les Rongeurs (Mammalia, Rodentia) du Pléistocène inférieur d'Olduvai Bed I, Tanzanie. Part I: General Introduction and Muridae. In *Fossil Vertebrates of Africa* Vol. 4, R. J. G. Savage and S. C. Coryndon (eds.), London: Academic Press.

Jones, P. 1979. Effects of raw materials on biface manufacture. *Science* 204: 835–6.

1980. Experimental butchery with modern stone tools and its relevance for archaeology. *World Archaeology* 12: 153–65.

1981. Experimental implement manufacture and use: a case study from Olduvai Gorge, Tanzania. In *The Emergence of Man*, J. Z. Young, E. M. Jope and K. P. Oakley (eds.), pp. 189–95. (*Philosophical Transactions of the Royal Society of London*, Series B, vol. 292, no. 1057).

Kleindienst, M. R. 1961. Variability within the Late Acheulean assemblages in Eastern Africa. *South African Archaeological Bulletin* 16: 35–52.

1962. Components of the East African Acheulean assemblage: an analytical approach. In *Actes du IVe Congrès Panafricain de Préhistoire et de l'Etude du Quaternaire*, G. Mortelmans and J. Nenquin (eds.), pp. 81–111. Tervuren.

1964. Summary report on excavations at site JK2, Olduvai Gorge, Tanganyika, 1961–1962. *Annual Report, Antiquities Division, Tanganyika, for the Year 1962*, pp. 4–6.

Leakey, L. S. B. 1931. *The Stone Age Cultures of Kenya Colony*. London: Frank Cass.

1948. The bolas in Africa. *Man* 48: 48.

1951. *Olduvai Gorge: A Report on the Evolution of the Hand-Axe Culture in Beds I–IV*. Cambridge: Cambridge University Press.

1965. *Olduvai Gorge 1951–1961. Volume 1: A Preliminary Report on the Geology and Fauna*. Cambridge: Cambridge University Press.

Leakey, M. D. 1971. *Olduvai Gorge: Excavations in Beds I and II, 1960–1963*. Cambridge: Cambridge University Press.

1975a. Cultural patterns in the Olduvai sequence. In *After the Australopithecines: Stratigraphy, Ecology and Culture Change in the Middle Pleistocene*, K. W. Butzer and G. Ll. Isaac (eds.), pp. 447–93. The Hague: Mouton.

1975b. *Olduvai Gorge: My Search for Early Man*. London: Collins.

1976. The early stone industries of Olduvai Gorge. In *Les plus anciennes industries en Afrique*, J. D. Clark and G. Ll. Isaac (eds.), pp. 24–41. Colloque V, UISPP Congrès, Nice, France, 13–18 Sept. 1976.

1978. Olduvai fossil hominids: their stratigraphic positions and associations. In *Early Hominids of Africa*, C. J. Jolly (ed.), 3–16. London.

1984. *Disclosing the Past: An Autobiography*. New York: Doubleday.

1989. Foreword. In *The Archaeology of Human Origins: Papers by Glynn Isaac*, B. Isaac (ed.) p. xv. Cambridge: Cambridge University Press.

Leakey, M. D., Clarke, R. J. and Leakey, L. S. B. 1971. New hominid skull from Bed I Olduvai Gorge, Tanzania. *Nature* 232: 308–12.

REFERENCES

Leakey, M. D. and Harris, J. M. (eds.) 1987. *Laetoli: A Pliocene Site in Northern Tanzania.* Oxford: Clarendon Press.

Macintyre, R. M., Mitchell, J. G. and Dawson, J. B. 1974. Age of fault movements in Tanzanian sector of East African rift valley system. *Nature* 247: 354–6.

Maglio, V. J. 1970. Early Elephantidae of Africa and a tentative correlation of African Plio-Pleistocene deposits. *Nature* 225: 328–32.

1973. Origin and evolution of the Elephantidae. *Transactions of the American Philosophical Society* 63 (3): 1–149.

Merrick, M. V. and Brown, F. H. 1984. Obsidian sources and patterns of source utilization in Kenya and northern Tanzania: some initial findings. *African Archaeological Review* 2: 129–52.

Nie, N. H., Hull, C. H., Jenkins, J. G., Steinbrenner, K. and Bent, D. H. 1975. *Statistical Package for the Social Sciences.* Second edn. New York: McGraw-Hill.

O'Brien, E. M. 1981. The origin and evolution of the human species: what was the Acheulean hand-axe? *Natural History* 93: 20.

Price Williams, D. and Lindsey, N. E. (eds.) 1978. Hlalakahle/Kufika: Stone Age sites in north western Swaziland. *Swaziland Archaeological Research Association Preliminary Series* 1.

Reck, H. 1951. A preliminary survey of the tectonics and stratigraphy of Olduvai. In Leakey 1951, pp. 5–19.

Roe, D. A. 1964. The British Lower and Middle Palaeolithic: some problems, methods of study and preliminary results. *Proceedings of the Prehistoric Society* 30: 245–67.

1968. British Lower and Middle Palaeolithic handaxe groups. *Proceedings of the Prehistoric Society* 34: 1–82.

n.d. Kalambo Falls large cutting tools: a metrical analysis. Unpublished manuscript (1973).

1981. *The Lower and Middle Palaeolithic Periods in Britain.* London: Routledge and Kegan Paul.

Sainty, J. E. 1927. An Acheulean Palaeolithic workshop site at Whitlingham, near Norwich. *Proceedings of the Prehistoric Society of East Anglia* 5: 176–213.

Shipman, P. 1981. Application of scanning electron microscopy to taphonomic problems. *Annals of the New York Academy of Sciences* 376: 357–86.

Stiles, D. N. 1977. Acheulean and Developed Oldowan: the meaning of variability in the Early Stone Age. *Mila* 6: 1–35.

1979. Early Acheulean and Developed Oldowan. *Current Anthropology* 20: 126–9.

White, T. D. and Harris, J. M. 1977. Suid evolution and correlation of African hominid localities. *Science* 198: 13–27.

1. The north side of the gorge showing Beds I, II and the red Bed III overlain by Bed IV. The volcanic neck of Engelosen, source of the green phonolite, may be seen in the background and, beyond, the Precambrian Ogol mountains. (Photo: Diana Saltoon)

2. Bed III, JK photographic mosaic of the pits and furrows. (Photos: R. I. M. Campbell)

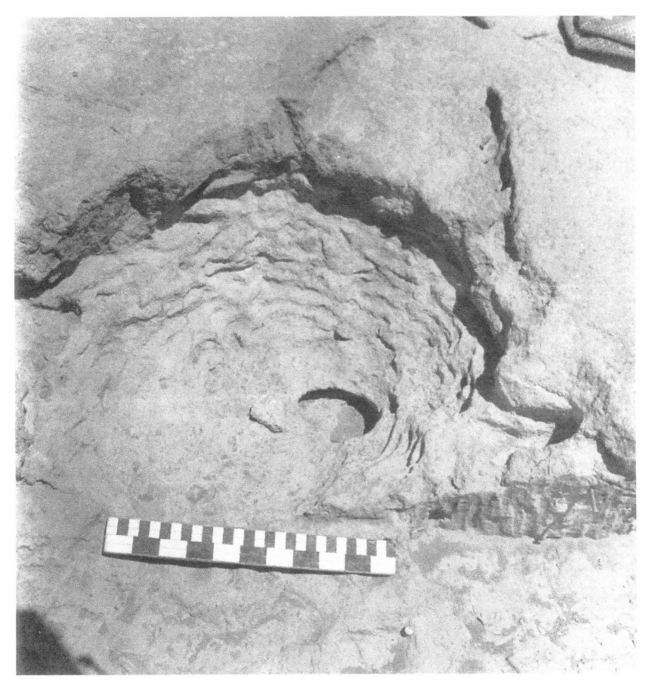

3. JK, Pit 2. Five sets of four parallel grooves with intervening ridges can be seen on the right and on the back wall of the pit. The quartzite flake was found beside the small hole in the bottom of the pit. Note the previous trench cut into the pit just above the scale. (Photo: M. D. Leakey)

4. JK: two pairs of convergent furrows. The pair parallel with the ranging pole has been cut through by the second pair leading into Pit 12, on the right. Pit 11 is to the left between the second pairs of furrows. The present erosion slope has truncated the surface with the pits at the top and right side of the photo. Unexcavated overburden is at the bottom of the photo. (Photo: R. I. M. Campbell)

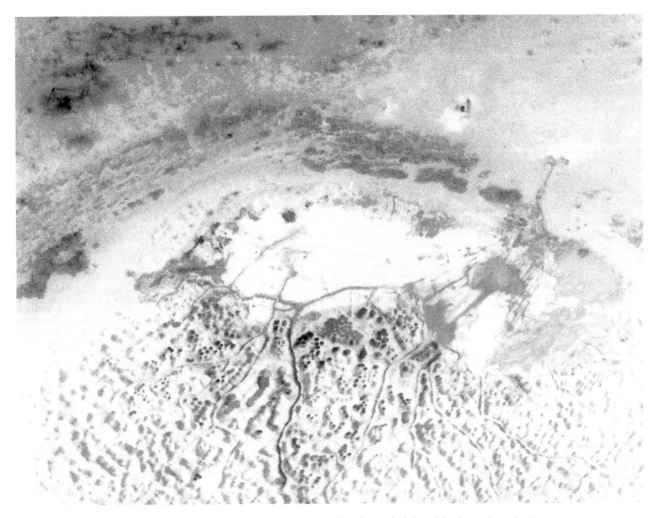

5. Aerial photograph of Magado Crater showing the partially dry soda lake with channels and salt evaporation pits in the foreground. (Photo: Philip Leakey)

6. Magado Crater: salt evaporation pits. Note gaps in the pit walls to permit water to filter in. (Photo: R. Lowis)

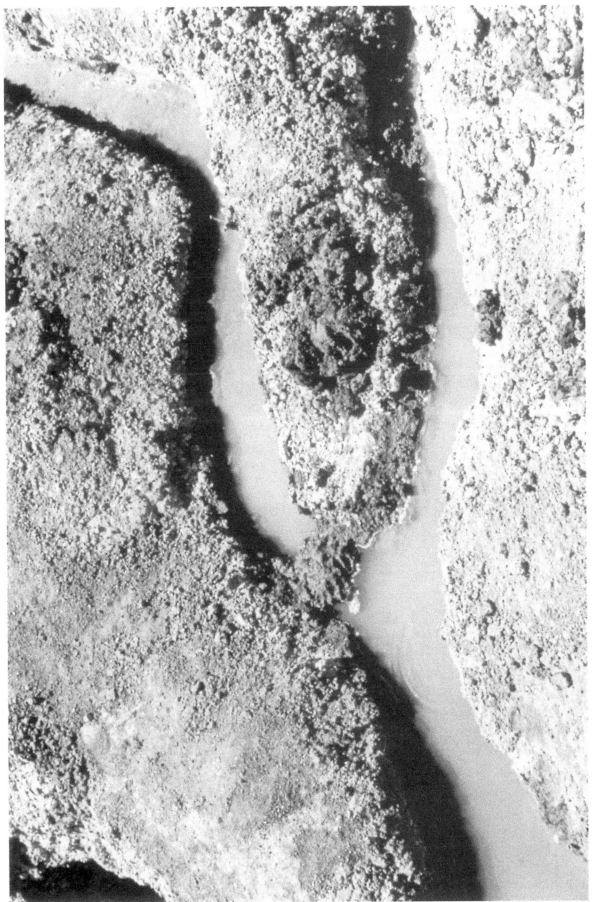

7. Magado Crater: irrigation channels to lead water into the pits from the soda lake. Note that one channel has been blocked to divert water elsewhere. (Photo: R. Lowis)

8. Bed IV, HEB Level 3: one cleaver and five handaxes made from the fine-grained green phonolite. The rough surface of some specimens, such as that in the centre at the top, is due to weathering. (Photo: John Reader)

9. Bed IV, HEB West, Level 2b: two cleavers and three handaxes made from basalt and trachyandesite. (Photo: John Reader)

10. Looking west down the gorge from WK East. Tuff IVB can be seen as a small cliff on the left, below the calcrete that is on the surface of the plains. WK Hippo site is a light patch to the left of Tuff IVB. At the extreme left two small overhanging cliffs mark Bed III. (Photo: M. D. Leakey)

11. Bed IV, site WK at an early stage in the excavation. The ranging pole, marked in feet, stands in the deepest part of the Intermediate Channel with the surface of the Upper Channel about 6 in. above its top. One of the grey siltstones used as a marker horizon between Upper and Lower Bed IV and which has reversed polarity can be seen between $1\frac{1}{2}$ and 3 ft from the top of the ranging pole. (Photo: M. D. Leakey)

12. Bed IV, WK Upper Channel. Pitted anvils, handaxes and other artefacts *in situ* in the conglomerate within the channel. (Photo: M. D. Leakey)

13. Bed IV, WK Upper Channel. Handaxes, cleavers and other artefacts with faunal remains *in situ*.

14. Handaxes from WK Upper Channel. The top left and centre bottom specimens are of quartzite, the remainder are of lava. (Photo: John Reader)

15. Cleavers from WK Upper Channel. The specimen at the top left is of quartzite and that at the bottom right of phonolite, the remainder are of basalt or trachyandesite. (Photo: John Reader)

16. WK Upper Channel. Four handaxes made on flakes and six small rolled bifaces of Developed Oldowan facies. (Photo: John Reader)

17. Three pitted hammerstones or anvils with single pits from the WK Upper Channel. (Photo: John Reader)

18. Pitted hammerstones or anvils from WK Upper Channel, both with twin pits. (Photo: John Reader)

19. A pitted anvil with a single pit on one face and twin pits on the left side, from WK Upper Channel. (Photo: John Reader)

20. Quartzite handaxes from FLK Masek Beds. (Photo: John Reader)

21. Two large quartzite handaxes from FLK Masek Beds. (Photo: John Reader)

22. Elephant acetabulum from JK showing extensive battering on the interior. Note the worn hollow in the area of the pubis. (Photo: John Reader)

23. Elephant acetabulum from HEB Level 3. Note the battering and pitting on the interior of the hollow. (Photo: John Reader)

24. Fragments of elephant limb bone shafts flaked to pointed ends. In the first specimen the butt is also flaked so that it resembles a handaxe. From HEB and LLK respectively. (Photo: John Reader)

25. Three distal ends of humeri, probably of hippopotamus, in which the condyles have been removed by battering. No 1 from PDK Trenches I–III. No 2 from WK East C. No 3 from HEB West. (Photo: John Reader)

26. Three proximal condyles of hippopotamus femora in which the surface is battered. In No 1 there is a single irregular area of battering. No 2 is the most extensively damaged and all the surface has been removed; the battering has also formed a shallow hollow in the cancellous tissue. No 3 has three smaller areas of battering. From HEB, HEB West and WK East C, respectively. (Photo: John Reader)

INDEX

Italic page numbers indicate figures, bold indicate tables.

Acheulean industry, 301, 306
 bifaces, 271–5, **272**, 295–8
 cleavers, 203–22
 handaxes, 203–22, 246–7, 252–3
Alcelaphini, **132–3**, 144–5
Amphibia **131**, **133**, **139**, 142
animal bone, 59, 90–1, 123, 311–14, pls. 22–6 *and see* fauna, fish bones
animals, butchering experiments, *292*, 292–4, *293*, **294**, 308
Antidorcas recki, 46, **133**, 138, 145
Antilopini, **132–3**, 145
anvils
 defined, 7
 FLK Masek Beds, 121
 HEB East, 51
 HEB West, 67
 HK, 125
 WK, 85
 and see pitted anvils
artiodactyl footprints, 26–7
Arvicanthus niloticus, **131**, 140
Aves **131**, **133**, **137**, **139–41**, 140, **143–4**
awls
 defined, 5
 HEB, 57
 HEB East, 50–1
 JK, 22–3
 PDK Trenches I–III, 110
 WK East, 96, 105

basalt for tools, 257–8, 262–4, **264**, **268**
Beatragus antiquus, 145
Bed II, pl. 1, *and see* BK, EF–HR, MLK, TK Upper Level, TK Lower Level
Bed III, 1–5, 302, pls. 1, 10
 dating, 2, 10, 301, 308
 distribution, 8–10
 drainage, 10–11, *11*, *16*
 environment, 10–11
 fauna, **131–4**, 133–4, 141–2, 144–5, 304
 geology, 8, *9*, 10–11
 palaeogeography, *11*
 sites excavated, *2*
 stratigraphy, *3*, 8–10, 301
 and see JK
Bed IV, 1–5, 302–4, pl. 1
 dating, 2, 10, 301, 308
 distribution, 8–10
 drainage, 10–11, *11*, *16*
 environment, 10–11
 fauna, **131–4**, 134, 136, 141–2, 144–5, 304
 geology, 8, *9*, 10–11

handaxes, 246–7, 252
palaeogeography, *11*
section, *76*
sites excavated, *2*
stratigraphy, *3*, 8–10, 301
and see HEB, HEB East, HEB West, PDK Trenches I–III, PDK Trench IV, WK, WK East, WK Hippo Cliff, WK Intermediate Channel, WK Lower Channel, WK Upper Channel
bifaces
 Acheulean industry, 271–5, **272**, 295–8
 analysis, **150**, *152*, 235–8, **238–9**, 240–1, **241–5**, *246–7*, 246–7, **248–52**, 252–3
 defined, 5, 240, 261
 experimental manufacture, 261–75, 307
 FLK Masek Beds, *118*, 119–20, 266, 297
 HEB, 55, 59, 62, 71, 265–6
 HEB East, 48, 265, 267
 HEB West, 64–5, 69, 72, 113, 267
 HK, 124, 267
 JK, 18, 21
 metrical analysis, 149–51, **158–201**
 PDK Trenches I–III, 109
 PDK Trench IV, 40–1, 267
 raw materials, **236**, **238**, 238, 240–1, 246, **252**
 shapes, 269–71, **270**
 technological considerations, **239**, **241**, 267–9
 TK Fish Gully, 128–9
 WK, 76
 WK East, 91–3, 104, 263, 265
 WK Hippo Cliff, 38–9
 WK Intermediate Channel, 72
 WK Lower Channel, 42–3
 WK Upper Channel, 263, 266
 and see cleavers, experimental work, handaxes
bird bones, *see* Aves
bivalves, 46, 137, 141
BK
 cleavers and handaxes, metrical analysis, 216, *217*, *231*
 outils écaillés, 283
 subspheroids, 276
blocks
 defined, 7
 HEB, 57, 60, 63–4
 HEB West, 67, 72
 JK, 23, 25
 PDK Trenches I–III, 111, 113
 TK Fish Gully, 129

WK, 85
WK East, 100, 106
WK Intermediate Channel, 74
WK Lower Channel, 43
Bovidae, 130, **132–44**, 137, 140, 144–5
 HEB East, 46
 JK, 134
burins
 defined, 5
 HEB, 57
 JK, 19
 PDK Trenches I–III, 110
 WK East, 96

Carnivora, **131**, **133–44**, 137
catfish, *see* Clariidae
Cephalophini, **132**, **139**, 145
Ceratotherium simum, **132–3**, 142
Cercopithecidae, **135–44**, 137
Chelonia, **131**, **133**, **133–44**
choppers
 defined, 5–6
 FLK Masek Beds, 120
 HEB, 55, 59, 62–3, 71
 HEB East, 48
 HEB West, 65–6, 69, 72, 113
 HK, 124
 JK, 18, 21, 25
 PDK Trenches I–III, 109
 PDK Trench IV, 42
 WK, 79
 WK East, 93, 104
 WK Hippo Cliff, 39
 WK Intermediate Channel, 72
 WK Lower Channel, 43
Clariidae, **131**, **133**, **135–44**, 136–7, 140–2
cleavers, 149, 151–3
 Acheulean industry, 203–22
 BK, 216, *231*
 EF–HR, 218–19, *233*
 FLK Masek Beds, metrical analysis, 207–8, *224*
 HEB, pl. 8
 HEB East, 214–15, *229*
 HEB West, pl. 9
 metrical analysis, 211–14, *226–8*
 HK, metrical analysis, 205–6, *223*
 JK, 18
 MLK, 219–20, *234*
 PDK Trenches I–III, metrical analysis, 208–9
 PDK Trench IV, metrical analysis, 215–16, *230*

INDEX

shape diagrams, *156–7, 223–34*
SHK, 217, *232*
TK Fish Gully, metrical analysis, 206–7
TK Lower Level, metrical analysis, 218
TK Upper Level, metrical analysis, 217
WK, 77–9
    metrical analysis, 210–11, *225*
WK East, metrical analysis, 209–10
WK Upper Channel, pl. 15
climate, 11, 13–14, *and see* environmental evidence
cobbles
    defined, 7
    FLK Masek Beds, 121
    HEB, 57, 60, 63–4
    HEB East, 51
    HEB West, 67, 70, 72, 114
    JK, 19, 23, 25
    PDK Trenches I–III, 111
    PDK Trench IV, 41
    WK, 85
    WK East, 100
    WK Intermediate Channel, 74
    WK Lower Channel, 43
concavo-convex edges, *287*, 287–8
Connochaetes, 136
core fragments
    defined, 7
    FLK Masek Beds, 122
    HEB West, 70–2
    HK, 126
    JK, 20, 24–5
    PDK Trenches I–III, 112
    TK Fish Gully, 129
    WK East, 102–3, 107
    WK Hippo Cliff, 39
    and see cores
cores
    defined, 7
    FLK Masek Beds, 122–3
    HEB, 60–1
    HEB East, 52
    HEB West, 67, 70–1
    HK, 126
    JK, 20–1, 24
    WK, 86–7
    WK East, 103, 107
    and see core fragments
crocodiles, **131**, 142
Crocodylidae, **135–44**, 137, 140

Damaliscus niro, **132–3**, 145
dating of Beds, 2, 10, 12–13, 301, 308
debitage
    defined, 7
    FLK Masek Beds, **121–2**, 122–3
    HEB, **57–8**, 57–8, **60–3**, 60–1, 64
    HEB East, **52–3**, 52
    HEB West, **67–8**, 67, **70–1**, 70–2, **113–14**, 114–15
    HK, **125–6**, 125–6
    JK, **20**, 20–1, 22, **24–5**, 24
    PDK Trenches I–III, **111–12**, 112–13
    PDK Trench IV, **41–2**, 41–2
    TK Fish Gully, **128**, 129
    WK, 86–7
    WK East, **101–2**, 101–3, **107–8**, 107

WK Hippo Cliff, **38**, 39
WK Intermediate Channel, **73–4**, 74
WK Lower Channel, **43–4**, 43
WK Upper Channel, **86–7**
and see core fragments, cores, flakes
definition of terms, 5–7
Developed Oldowan industry, 203, 220, 247, 252–3, 271–5, **272**, 295–8, 301, 306
    Developed Oldowan B, 216–17, 220, 306
    Developed Oldowan C, 208–10, 220, 301, 303, 306
Diceros bicornis, **132–3**, 142
discoids
    defined, 6
    HEB, 56, 59, 63
    HEB East, 49
    HEB West, 66, 69, 72, 113
    HK, 124–5
    JK, 18, 21, 25
    PDK Trenches I–III, 109
    PDK Trench IV, 42
    TK Fish Gully, 129
    WK, 79–80
    WK East, 94, 104
    WK Intermediate Channel, 72–3
DK, polyhedrons, 275
drainage of Beds, 10–11, *11, 13*, 13, 16

EF-HR, cleavers and handaxes, metrical analysis, 218–19, *221, 233*
elephant, 134, 144, 310–12, pls. 22–4
Elephantidae, **132–44**
Elephas recki, **132**, 144
environmental evidence, 10–11, 13–14, *and see* climate
Equidae, **132–44**, 136–7, 142
Equus, 142
experimental work, 254
    tool manufacture, 259–61, *287–90*, **291**, 305–8
    tuber preparation, 294
    wood working, 294–5

fauna
    Bed III, **131–4**, 133–4, 141–2, 144–5, 304
    Bed IV, **131–4**, 134, 136, 141–2, 144–5, 304
    FLK Masek Beds, 139, **144**
    HEB, 138, **139**, 142, 311–12
    HEB East, *50*, 137–8, **138**, 141–2
    HEB West, 138, **139**, 142
    HK, 140
    JK, 133–4, **135–6**, 311–14
    LK, 312
    Masek Beds, **131–4**, 139, 144–5, 304
    PDK Trenches I–III, 139, **143**, 312–13
    PDK Trench IV, 136–7
    WK, 142
    WK East, *94*, **105**, 138–9, **142–3**, 142, 313
    WK Hippo Cliff, 136
    WK Intermediate Channel, **140**
    WK Lower Channel, **137**, 137
    WK Upper Channel, **141**, 312, 314
    and see animal bone, name of animal

fish bones, 126–7, *and see* animal bone and Clariidae
flakes
    breakage, *288*, 288
    defined, 7
    FLK Masek Beds, 122
    HEB, 60–1, 64, 71
    HEB East, 52
    HEB West, 67, 70–2, 114–15
    HK, 125–6
    JK, 20, 24
    PDK Trenches I–III, 112–13
    PDK Trench IV, 41–2
    TK Fish Gully, 129
    WK, 86–7
    WK East, 101–2, 107
    WK Hippo cliff, 39
    WK Intermediate Channel, 74
    WK Lower Channel, 43
    and see laterally trimmed flakes, light-duty flakes
FLK Masek Beds, 4, 116–23, *117*
    cleavers, metrical analysis, 207–8, *224*
    debitage, **121–2**, 122–3
    fauna, 139, **144**
    handaxes, metrical analysis, *207*, 207–8
    tools, *118, 119*, 119–21, **121–2**, 266, 275, 278–9, 297, pls. 20–1
    utilised material, **121–2**, 121–2
footprints
    artiodactyl, 26–7
    hominid, 28

Gazellini, **132**, *133*, 145
geology and topography, 8, *9*, 10–11, 13–14
Giraffa jumae, 90, **132–3**, 144
Giraffa stillei, **132–3**, 144
Giraffidae, **132–44**, 144

hammerstones, 282
    defined, 7, 281–2
    FLK Masek Beds, 121
    HEB, 57, 60, 63
    HEB East, 51
    HEB West, 67, 72
    HK, 125
    JK, 23
    PDK Trenches I–III, 111
    size ranges, *281*
    WK, 85
    WK East, *98*, 100
    WK Intermediate Channel, 73–4
    WK Lower Channel, 43
    and see pitted hammerstones
handaxes
    Acheulean industry, 203–22, 246–7, 252–3
    Bed IV, 246–7, 252
    BK, 216, *217*
    Developed Oldowan industry, 247, 252
    EF-HR, 218–19, *221*
    FLK Masek Beds, pls. 20–1
        metrical analysis, *207*, 207–8
    HEB, pl. 8
    HEB East, metrical analysis, 214–15, *215*
    HEB West, pl. 9
        metrical analysis, 211–14, *212–14*

INDEX

HK, metrical analysis, *205*, 205–6
JK, 18
MLK, metrical analysis, 219–20, *222*
PDK Trenches I–III, metrical analysis, *208*, 208–9
PDK Trench IV, metrical analysis, 215–16, *216*
shape diagrams, *155*, 205–22
SHK, metrical analysis, 217, *220*
TK Fish Gully, metrical analysis, *206*, 206–7
TK Lower Level, metrical analysis, 218, *219*
TK Upper Level, metrical analysis, 217, *218*
weight:edge length ratios, **274**
WK, 76–9
  metrical analysis, 210–11, *211*
WK East, metrical analysis, *209–10*, 209–10
WK Upper Channel, pls. 14, 16
HEB, 4, 45, 52–5, 61–4, 71
  animal bone, pls. 24, 26
  debitage, **62–3**, 64
  fauna, 138, **139**, 142, 312–13
  tools, **62–3**, 62–3, 71
  utilised material, **62–3**, 63–4, 71
HEB, Level 3, 58–61
  animal bone, pl. 23
  debitage, **60–1**, 60–1
  tools, 59–60, **60–1**, 265–6
  utilised material, **60–1**, 60
HEB, Level 4, 55–8
  debitage, **57–8**, 57–8
  tools, 55–7, **57–8**
  utilised material, **57–8**, 57
HEB East, 2, 45–52, *46–50*, 53, 141
  bovidae, 46
  cleavers, metrical analysis, 214–15, *229*
  debitage, **52–3**, 52
  fauna, *50*, 137–8, **138**, 141–2
  handaxes, metrical analysis, 214–15, *215*
  tools, 48–51, **52–3**, 265, 267, 277–8
  utilised material, 51–2, **52–3**
HEB West, 4, 45, 52–5, *53*, 65
  animal bone, 310–13, pls. 25–6
  debitage, 72
  fauna, 138, **139**, 142
  tools, 71–2, 267
  utilised material, 72
HEB West, Level 1, 113–15
  debitage, **113–14**, 114–15
  tools, **113–14**, 113–14
  utilised material, **113–14**, 114
HEB West, Level 2a, 67–71
  cleavers, metrical analysis, 211–12, *226*
  debitage, **70–1**, 70–1
  handaxes, metrical analysis, 211–12, *212*
  tools, 69–70, **70–1**
  utilised material, **70–1**, 70–1
HEB West, Level 2b, 64–7
  cleavers, metrical analysis, 212–13, *227*
  debitage, **67–8**, 67
  handaxes, metrical analysis, 212–13, *213*
  tools, 64–6, **67–8**
  utilised material, **67–8**, 67
HEB West, Level 3

cleavers and handaxes, pl. 8
  metrical analysis, 213–14, *214*, *228*
Hipparion, 142
Hippopotamidae, **132–44**, 137, 140
hippopotamus, 36, 123, 133, 136, 140, 142, 311–13, pls. 25–6
Hippopotamus gorgops, 36, **132**, **133**, 136, 141–2
Hippotragus gigas, 145
HK, 4, 116, 123–6, *124*
  cleavers, metrical analysis, 205–6, *223*
  debitage, **125–6**, 125–6
  fauna, 140
  handaxes, metrical analysis, *205*, 205–6
  tools, 124–5, **125–6**, 267
  utilised material, **125–6**, 125
hominids
  footprints, 28
  O.H.28, 75, 116
  O.H.34, 15
horn cores, 90–1, 130
HWK East Level 3, spheroids, 275

Jaculus, **131**, **133**, 140
jerbils, 139
JK, 4, 15–35, 315–20
  animal bone, pl. 22
  bovidae, 134
  dating, 16–17
  debitage, **20**, 20–1, **22**, **24–5**, 24
  fauna, 133–4, **135–6**, 311–14
  pits and furrows, *26*, 26–34, *315*, 315–20, *317*, *319*, pls. 2–4
  soil samples, **33**
  stratigraphy, 17–28
  tools, 18–19, **20**, 21–3, **22**, **24–5**, 25
  utilised material, 19–20, **20**, **22**, 23–4, **24–5**, 25–6

Kerimasi, dating, 12–13
Kobus ellipsiprymnus, **132**, **133**, 136
Kobus kob, **132**, **133**, 136, 145

Lagomorpha, **131**, **133**, **135**, **141**
laterally trimmed flakes
  defined, 6
  FLK Masek Beds, 121
  HEB, 57
  HEB East, 51
  HEB West, 66, 72
  JK, 19, 23
  PDK Trenches I–III, 110
  TK Fish Gully, 129
  WK, 81
  WK East, 96
  and see flakes, light-duty flakes
Leporidae, **131**, **133**
light-duty flakes
  defined, 7
  FLK Masek Beds, 121–2
  HEB, 57, 60, 64
  HEB East, 51–2
  HEB West, 67, 70, 72, 114
  JK, 19–20, 23–6
  PDK Trenches I–III, 111–12
  PDK Trench IV, 41
  WK, 85–6

WK East, 100, 106–7
WK Hippo Cliff, 39
WK Intermediate Channel, 74
WK Lower Channel, 43
and see flakes, laterally trimmed flakes
lions, 140
LK, fauna, **137**, 312, pl. 24

Magado Crater salt workings, 34–5, pls. 5–7
Masek Beds, 1–5, 304
  dating, 2, 12–13, 301
  distribution, 11–12
  drainage, *13*, 13
  environment, 13–14
  fauna, **131–4**, 139, 144–5, 304
  geology, 13–14
  palaeogeography, *13*
  sites excavated, *2*
  stratigraphy, *3*, 11–12
  and see FLK: post-Masek beds *see* HK, TK Fish Gully
Mastomys, **131**, **133**, 140
Megalotragus kattwinkeli, 126, **132–3**, 137, 141
MLK, cleavers and handaxes, metrical analysis, 219–20, *222*, *234*
MNK, subspheroids, 276
Mollusca, **131**, **133**, **138**, **141**, 141, **144**

Naisiusiu Beds, stratigraphy, *3*
Ndutu Beds, stratigraphy, *3*
Neotragini, **132**, **133**, 139, 145
nephelinite, 258–9
Norkilili Member, 12–13

Ostrich, *see* Struthionidae
outils écaillés
  BK, 283
  defined, 6, 283, 288
  distribution, **283**
  experimental manufacture, 283–91, 308
  FLK Masek Beds, 121
  HEB, 60
  HEB East, 51
  HEB West, 66, 70, 114
  JK, 19, 23
  PDK Trenches I–III, 110
  size ranges, *284–6*
  WK, 81
  WK East, 96–7, 105
  WK Intermediate Channel, 73

palaeogeography, *11*, *13*
Panthera leo, **131**, **133**, 141
PDK Trenches I–III, 4, 87, 89, 107–13
  animal bone, pl. 25
  cleavers, metrical analysis, 208–9
  debitage, **111–12**, 112–13
  fauna, 139, **143**, 311–12
  handaxes, metrical analysis, *208*, 208–9
  tools, 109–10, **111–12**, 284
  utilised material, 110–12, **111–12**, 284
PDK Trench IV, 4, 39–42
  cleavers, metrical analysis, 215–16, *230*
  debitage, **41–2**, 41–2
  fauna, 136–7

INDEX

handaxes, metrical analysis, 215–16, *216*
tools, 40–2, **41–2**, 267
utilised material, **41–2**, *41*
Pelorovis antiquus, 90–1, **132**, **133**, 136, **142**, 144
Pelorovis oldowayensis, **132**, **133**, 134, 144–5
phonolite, 58, 257, 264–5, **265**
picks
  defined, 6
  WK, 79
pits and furrows (? salt extraction), JK, 26–34, *315*, 315–20, *317*, **319**, pls. 2–4
pitted anvils
  defined, 7, 283
  distribution, **283**
  experimental manufacture, 283–91, 308
  FLK Masek Beds, 121
  HEB, 60
  HEB East, 51
  HEB West, 67, 70, 114
  JK, 19, 23
  PDK Trenches I–III, 110–11, 284
  PDK Trench IV, 41
  WK, 81–5, *83*, 284
  WK East, 97–100, *98*, 105–6
  WK Intermediate Channel, 73
  WK Upper Channel, pls. 17–19
  and see anvils
pitted hammerstones
  defined, 7
  HEB, 60
  HEB East, 51
  HEB West, 67, 70, 114
  JK, 19, 23
  PDK Trenches I–III, 110–11
  PDK Trench IV, 41
  WK, 81–5, *83*, 284
  WK East, 97–100, 105–6
  WK Intermediate Channel, 73
  WK Upper Channel, pls. 17–18
  and see hammerstones
polyhedrons
  defined, 6, 275
  distribution, **275**, **278**
  DK, 275
  experimental manufacture, 275–83, 307
  FLK Masek Beds, 278–9
  HEB, 55–6, 63
  HEB East, 48–9, 277–8
  HEB West, 66, 69, 72, 113
  JK, 18
  PDK Trench IV, 41
  size ranges, **276–7**
  WK, 79
  WK East, 93–4
  WK Intermediate Channel, 72
Primates, **131**, **133–4**
punches
  defined, 6, 283, 288
  distribution, **283**
  experimental manufacture, 283–91, 308
  FLK Masek Beds, 121
  HEB West, 70
  JK, 23
  PDK Trenches I–III, 110, 284
  PDK Trench IV, 41

size ranges, *284–6*
WK, 81, 284
WK East, 96–7, *97*, 105
WK Intermediate Channel, 73

quartzite, 256–7, **266**, 266–7, **269**, *280*

raw materials, 254–9, *255 and see* basalt, nephelinite, phonolite, quartzite, trachyandesite, trachyte
Redunca sp., **133**, 145
Reptilia, **131**, **133**, **144**
Rhinocerotidae, **132–44**, 137, 140, 142, 312
Rodentia, **131**, **133–8**, **140–4**
rodents, 130, 138, 140

salt extraction, 4, 34–5, pls. 5–7, *and see* JK, pits and furrows
scrapers, defined, 6
scrapers, heavy-duty
  FLK Masek Beds, *119*, 120
  HEB, 56, 59 60, 63, 71
  HEB East, 49–50
  HEB West, 66, 69, 72, 114
  HK, 125
  JK, 19, 22
  PDK Trenches I–III, 109–10
  WK, 80
  WK East, 94
  WK Hippo Cliff, 39
  WK Intermediate Channel, 73
scrapers, light-duty
  FLK Masek Beds, 120–1
  HEB, 56–7, 60, 63
  HEB East, 50
  HEB West, 66, 69–70, 114
  HK, 125
  JK, 19, 22
  PDK Trenches I–III, 110
  PDK Trench IV, 41
  TK Fish Gully, 129
  WK, 80–1
  WK East, *95*, 95–6, 104
  WK Hippo Cliff, 39
  WK Intermediate Channel, 73
  WK Lower Channel, 43
SHK
  cleavers and handaxes, metrical analysis, 217, *220*, *232*
  subspheroids, 276
soil samples, JK, **33**
spheroids
  defined, 7, 275
  distribution, **275**, **278**
  experimental manufacture, 275–83, 307
  FLK Masek Beds, 120, 275
  HEB, 56, 59, 71
  HEB East, 49
  HEB West, 66, 69, 72, 113
  HK, 125
  HWK East Level 3, 275
  JK, 18–19, 21–2
  PDK Trenches I–III, 109
  PDK Trench IV, 41
  size ranges, **276–7**
  WK, 80
  WK East, 94, 104

WK Hippo Cliff, 39
WK Intermediate Channel, 73
and see subspheroids
stratigraphy, *3*, 8–10, 300–1
Struthionidae, **135**, **137–44**, 140
subspheroids
  BK, 276
  defined, 7, 275
  distribution, **275**, **278**
  experimental manufacture, 275–83, 307
  FLK Masek Beds, 120
  HEB, 56, 59
  HEB East, 49
  HEB West, 66, 69, 72, 113
  JK, 18–19, 21–2, 25
  MNK, 276
  PDK Trench IV, 41
  SHK, 276
  size ranges, 276–7, *278–9*, *281*
  TK Fish Gully, 129
  WK, 80
  WK East, 94, 104
  WK Hippo Cliff, 39
  WK Intermediate Channel, 73
  and see spheroids
Suidae, 130, **132–44**, 140

Tatera, **131**, **133**, 140
Thaleroceros radiciformis, **133**, 145
TK Fish Gully, 4, 116, 126–9, *127*
  cleavers, metrical analysis, 206–7
  debitage, **128**, 129
  fauna, 140–1
  handaxes, metrical analysis, *206*, 206–7
  tools, **128**, 128–9
  utilised material, **128**, 129
TK Lower Level, cleavers and handaxes, metrical analysis, 218, *219*
TK Upper Level, cleavers and handaxes, metrical analysis, 217, *218*
tools
  defined, 5–7
  experimental manufacture, 259–61, *287–90*, **291**, 305–8
  experimental use, 291–5
  FLK Masek Beds, 119–21, **121–2**
  HEB, 55–7, **57–8**, 59–60, **60–3**, 62–3, 71
  HEB East, 48–51, **52–3**
  HEB West, 64–6, **67–8**, 69–72, **70–1**, **113–14**, 113–14
  HK, 124–5, **125–6**
  JK, 18–19, **20**, 21–3, **22**, **24–5**
  manufacture, 259–61, 306–8
  metrical analysis, 146–59, 202–22, 305–6
  PDK Trenches I–III, 109–10, **111–12**
  PDK Trench IV, 40–1, **41–2**
  summary of reports, 304–8
  TK Fish Gully, **128**, 128–9
  WK, 76–85
  WK East, 91–7, **101–2**, 104–5, **107–8**
  WK Hippo Cliff, **38**, 38–9
  WK Intermediate Channel, 72–3, **73–4**
  WK Lower Channel, 42–3, **43–4**
  WK Upper Channel, **86–7**
  and see awls, bifaces, burins, choppers, discoids, laterally trimmed flakes,

326

INDEX

outils écaillés, picks, polyhedrons, punches, scrapers, spheroids, subspheroids
tortoise, 313–14
trachyandesite, 257–8, 262–4, **264**
trachyte, 259, 267
Tragelaphus strepsiceros grandis, **132**, **133**, 145
tuber preparation experiments, 294
Tuff IVA, 8–9, 40
Tuff IVB, 9–10, pl. 10

utilised material
  defined, 7
  FLK Masek Beds, **121–2**, 121–2
  HEB, **57–8**, 57, **60–3**, 60, 63–4
  HEB East, 51–2, **52–3**
  HEB West, **67–8**, 67, **70–1**, 70, 72, **113–14**, 114
  HK, **125–6**, 125
  JK, 19–20, **20**, **22**, 23–4, **24–5**
  PDK Trenches I–III, 110–12, **111–12**
  PDK Trench IV, **41–2**, 41
  TK Fish Gully, **128**, 129
  WK, 85–6
  WK East, 97–101, **101–2**, 105–7, **107–8**
  WK Hippo Clif, **38**, 39

WK Intermediate Channel, **73–4**, 73–4
WK Lower Channel, **43–4**, 43
WK Upper Channel, **86–7**
and see anvils, blocks, cobbles, hammerstones, light-duty flakes, pitted anvils, pitted hammerstones

vegetation, 11, *and see* environmental evidence

WK, 75–87, pl. 11
  animal bone, 311
  cleavers, metrical analysis, 210–11, *225*
  debitage, 86–7
  fauna, 142
  handaxes, metrical analysis, 210–11, *211*
  stratigraphy, *77*
  tools, 76–85, 284
WK East A, 87–107, *88–94*
  cleavers, metrical analysis, 209–10
  debitage, **101–2**, 101–3
  fauna, *94*, 138–9, **142**
  handaxes, metrical analysis, *209*, 209–10
  tools, 91–7, *97*, **101–2**, 263, 265
  utilised material, 97–101, *98*, **101–2**
WK East C, 87–107, *102*, *105*
  animal bone, pls. 25–6

  cleavers, metrical analysis, 210
  debitage, **107–8**, 107
  fauna, *105*, 138–9, 142, **143**, 313
  handaxes, metrical analysis, *210*, 210
  tools, 104–5, **107–8**
  utilised material, 105–7, **107–8**
WK Hippo Cliff, 36–9, *37*, pl. 10
  debitage, **38**, 39
  tools, **38**, 38–9
  utilised material, **38**, 39
WK Intermediate Channel, 72–4, pl. 11
  debitage, **73–4**, 74
  fauna, **140**
  tools, 72–3, **73–4**
  utilised material **73–4**, 73–4
WK Lower Channel, 42–4
  debitage, **43–4**, 43
  fauna, **137**, 137
  tools, 42–3, **43–4**
  utilised material, **43–4**, 43
WK Upper Channel, 303, pls. 11–13
  debitage, **86–7**
  fauna, **141**, 312, 314
  tools, **86–7**, pls. 14–16
  utilised material, **86–7**, pls. 17–19
wood working experiments, 294–5

For EU product safety concerns, contact us at Calle de José Abascal, 56–1°, 28003 Madrid, Spain or eugpsr@cambridge.org.

www.ingramcontent.com/pod-product-compliance
Ingram Content Group UK Ltd.
Pitfield, Milton Keynes, MK11 3LW, UK
UKHW030904150625
459647UK00025B/2887